THE GEOMETRY OF
HESSIAN STRUCTURES

T0331567

HIROHIKO SHIMA

Yamaguchi University, Japan

THE GEOMETRY OF
HESSIAN STRUCTURES

World Scientific

NEW JERSEY • LONDON • SINGAPORE • BEIJING • SHANGHAI • HONG KONG • TAIPEI • CHENNAI

Published by

World Scientific Publishing Co. Pte. Ltd.

5 Toh Tuck Link, Singapore 596224

USA office: 27 Warren Street, Suite 401-402, Hackensack, NJ 07601

UK office: 57 Shelton Street, Covent Garden, London WC2H 9HE

British Library Cataloguing-in-Publication Data
A catalogue record for this book is available from the British Library.

ISBN-13 978-981-270-031-5
ISBN-10 981-270-031-5

Printed in Singapore.

Dedicated to
Professor Jean Louis Koszul
I am grateful for his interest in my studies and constant encouragement.
The contents of the present book finds their origin in his studies.

Preface

This book is intended to provide a systematic introduction to the theory of Hessian structures. Let us first briefly outline Hessian structures and describe some of the areas in which they find applications. A manifold is said to be flat if it admits local coordinate systems whose coordinate changes are affine transformations. For flat manifolds, it is natural to pose the following fundamental problem:

> Among the many Riemannian metrics that may exist on a flat manifold, which metrics are most compatible with the flat structure ?

In this book we shall explain that it is the Hessian metrics that offer the best compatibility. A Riemannian metric on a flat manifold is called a *Hessian metric* if it is locally expressed by the Hessian of functions with respect to the affine coordinate systems. A pair of a flat structure and a Hessian metric is called a *Hessian structure*, and a manifold equipped with a Hessian structure is said to be a *Hessian manifold*. Typical examples of these manifolds include regular convex cones, and the space of all positive definite real symmetric matrices.

We recall here the notion of Kählerian manifolds, which are formally similar to Hessian manifolds. A complex manifold is said to be a Kählerian manifold if it admits a Riemannian metric such that the metric is locally expressed by the complex Hessian of functions with respect to the holomorphic coordinate systems. It is well-known that Kählerian metrics are those most compatible with the complex structure.

Thus both Hessian metrics and Kählerian metrics are similarly expressed by Hessian forms, which differ only in their being *real* or *complex* respectively. For this reason S.Y. Cheng and S.T. Yau called Hessian metrics affine Kähler metrics. These two types of metrics are not only formally similar, but also intimately related. For example, the tangent bundle of a

Hessian manifold is a Kählerian manifold.

Hessian geometry (the geometry of Hessian manifolds) is thus a very close relative of Kählerian geometry, and may be placed among, and finds connection with important pure mathematical fields such as affine differential geometry, homogeneous spaces, cohomology and others. Moreover, Hessian geometry, as well as being connected with these pure mathematical areas, also, perhaps surprisingly, finds deep connections with information geometry. The notion of flat dual connections, which plays an important role in information geometry, appears in precisely the same way for our Hessian structures. Thus Hessian geometry offers both an interesting and fruitful area of research.

However, in spite of its importance, Hessian geometry and related topics are not as yet so well-known, and there is no reference book covering this field. This was the motivation for publishing the present book.

I would like to express my gratitude to the late Professor S. Murakami who, introduced me to this subject, and suggested that I should publish the Japanese version of this book.

My thanks also go to Professor J.L. Koszul who has shown interest in my studies, and whose constant encouragement is greatly appreciated. The contents of the present book finds their origin in his studies.

Finally, I should like to thank Professor S. Kobayashi, who recommended that I should publish the present English version of this book.

Introduction

It is well-known that for a bounded domain in a complex Euclidean space $\mathbf{C}^n$ there exists the Bergman kernel function $K(z, w)$, and that the corresponding complex Hessian form

$$\sum_{i,j} \frac{\partial^2 \log K(z, \bar{z})}{\partial z^i \partial \bar{z}^j} dz^i d\bar{z}^j,$$

is positive definite and invariant under holomorphic automorphisms. This is the so-called Bergman metric on a bounded domain. E. Cartan classified all bounded symmetric domains with respect to the Bergman metrics. He found all homogeneous bounded domains of dimension 2 and 3, which are consequently all symmetric. He subsequently proposed the following problem [Cartan (1935)].

> *Among homogeneous bounded domains of dimension greater than 3, are there any non-symmetric domains ?*

A. Borel and J.L. Koszul proved independently by quite different methods that homogeneous bounded domains admitting transitive semisimple Lie groups are symmetric [Borel (1954)][Koszul (1955)]. On the other hand I.I. Pyatetskii-Shapiro gave an example of a non-symmetric homogeneous bounded domain of dimension 4 by constructing a Siegel domain [Pyatetskii-Shapiro (1959)]. Furthermore, E.B. Vinberg, S.G. Gindikin and I.I. Pyatetskii-Shapiro proved the fundamental theorem that any homogeneous bounded domain is holomorphically equivalent to an affine homogeneous Siegel domain [Vinberg, Gindikin and Pyatetskii-Shapiro (1965)].

A Siegel domain is defined by using a regular convex cone in a real Euclidean space $\mathbf{R}^n$. The domain is holomorphically equivalent to a bounded domain. It is known that a regular convex cone admits the characteristic function $\psi(x)$ such that the Hessian form given by

$$\sum_{i,j} \frac{\partial^2 \log \psi(x)}{\partial x^i \partial x^j} dx^i dx^j$$

is positive definite and invariant under affine automorphisms. Thus the Hessian form defines a canonical invariant Riemannian metric on the regular convex cone.

These facts suggest that there is an analogy between Siegel domains and regular convex cones as follows:

Siegel domain	$\longleftrightarrow$	*Regular convex cone*
Holomorphic coordinate system $\{z^1, \cdots, z^n\}$	$\longleftrightarrow$	*Affine coordinate system* $\{x^1, \cdots, x^n\}$
Bergman kernel function $K(z,w)$	$\longleftrightarrow$	*Characteristic function* $\psi(x)$
Bergman metric $\sum_{i,j} \frac{\partial^2 \log K(z,\bar{z})}{\partial z^i \partial \bar{z}^j} dz^i d\bar{z}^j$	$\longleftrightarrow$	*Canonical metric* $\sum_{i,j} \frac{\partial^2 \log \psi}{\partial x^i \partial x^j} dx^i dx^j$

A Riemannian metric g on a complex manifold is said to be Kählerian if it is locally expressed by a complex Hessian form

$$g = \sum_{i,j} \frac{\partial^2 \phi}{\partial z^i \partial \bar{z}^j} dz^i d\bar{z}^j.$$

Hence Bergman metrics on bounded domains are Kählerian metrics. For this reason it is natural to ask the following fundamental open question.

Which Riemannian metrics on flat manifolds are an extension of canonical Riemannian metrics on regular convex cones, and analogous to Kählerian metrics ?

In this book we shall explain that Hessian metrics fulfil these requirements. A Riemannian metric g on a flat manifold is said to be a *Hessian metric* if g can be locally expressed in the Hessian form

$$g = \sum_{i,j} \frac{\partial^2 \varphi}{\partial x^i \partial x^j} dx^i dx^j,$$

with respect to an affine coordinate system. Using the flat connection D, this condition is equivalent to

$$g = Dd\varphi.$$

A pair (D, g) of a flat connection D and a Hessian metric g is called a *Hessian structure*.

J.L. Koszul studied a flat manifold endowed with a closed 1-form α such that $D\alpha$ is positive definite, whereupon $D\alpha$ is a Hessian metric. This is the ultimate origin of the notion of Hessian structures [Koszul (1961)]. However, not all Hessian metrics are globally of the form $g = D\alpha$. The more general definition of Hessian metric given above is due to [Cheng and Yau (1982)] and [Shima (1976)]. In [Cheng and Yau (1982)], Hessian metrics are called *affine Kähler metrics*.

A pair (D, g) of a flat connection D and a Riemannian metric g is a Hessian structure if and only if it satisfies the *Codazzi equation*,

$$(D_X g)(Y, Z) = (D_Y g)(X, Z).$$

The notion of Hessian structure is therefore easily generalized as follows. A pair (D, g) of a torsion-free connection D and a Riemannian metric g is said to be a *Codazzi structure* if it satisfies the Codazzi equation. A Hessian structure is a Codazzi structure (D, g) whose connection D is flat. We note that a pair (∇, g) of a Riemannian metric g and the Levi-Civita connection ∇ of g is of course a Codazzi structure, and so the geometry of Codazzi structures is, in a sense, an extension of Riemannian geometry.

For a Codazzi structure (D, g) we can define a new torsion-free connection D' by

$$Xg(Y, Z) = g(D_X Y, Z) + g(Y, D'_X Z).$$

Denoting by ∇ the Levi-Civita connection of g, we obtain

$$D' = 2\nabla - D,$$

and the pair (D', g) is also a Codazzi structure. The connection D' and the pair (D', g) are called the *dual connection* of D with respect to g, and the *dual Codazzi structure* of (D, g), respectively.

For a Hessian structure $(D, g = Dd\varphi)$, the dual Codazzi structure (D', g) is also a Hessian structure, and $g = D'd\varphi'$, where φ' is the Legendre transform of φ,

$$\varphi' = \sum_i x^i \frac{\partial \varphi}{\partial x^i} - \varphi.$$

Historically, the notion of dual connections was obtained by quite distinct approaches. In affine differential geometry the notion of dual connections was naturally obtained by considering a pair of a non-degenerate affine hypersurface immersion and its conormal immersion [Nomizu and

Sasaki (1994)]. In contrast, S. Amari and H. Nagaoka found that smooth families of probability distributions admit dual connections as their natural geometric structures. Information geometry aims to study information theory from the viewpoint of the dual connections. It is known that many important smooth families of probability distributions, for example normal distributions and multinomial distributions, admit flat dual connections which are the same as Hessian structures [Amari and Nagaoka (2000)].

Contents

Chapter 1

Affine spaces and connections

Although most readers will have a good knowledge of manifolds, we will begin this chapter with a summary of the basic results required for an understanding of the material in this book. In section **1.1** we summarize affine spaces, affine coordinate systems and affine transformations in affine geometry. Following Koszul, we define affine representations of Lie groups and Lie algebras which will be seen to play an important role in the following chapters. In sections **1.2** and **1.3**, we outline some important fundamental results from differential geometry, including connections, Riemannian metrics and vector bundles, and assemble necessary formulae.

1.1 Affine spaces

In this section we give a brief outline of the concepts of affine spaces, affine transformations and affine representations which are necessary for an understanding of the contents of subsequent chapters of this book.

Definition 1.1. Let V be an n-dimensional vector space and Ω a non-empty set endowed with a mapping,

$$(p, q) \in \Omega \times \Omega \longrightarrow \overrightarrow{pq} \in V,$$

satisfying the following conditions.

(1) For any $p, q, r \in \Omega$ we have $\overrightarrow{pr} = \overrightarrow{pq} + \overrightarrow{qr}$.
(2) For any $p \in \Omega$ and any $v \in V$ there exists a unique $q \in \Omega$ such that $v = \overrightarrow{pq}$.

Then Ω is said to be an n-dimensional **affine space** associated with V.

Example 1.1. Let V be an n-dimensional vector space. We define a mapping

$$(p, q) \in V \times V \longrightarrow \overrightarrow{pq} = q - p \in V.$$

Then the set V is an n-dimensional affine space associated with the vector space V.

Example 1.2. An affine space associated with the standard vector space $\mathbf{R}^n = \{p = (p^1, \cdots, p^n) \mid p^i \in \mathbf{R}\}$ is said to be the **standard affine space**.

A pair $\{o; e_1, \cdots, e_n\}$ of a point $o \in \Omega$ and a basis $\{e_1, \cdots, e_n\}$ of V is said to be an **affine frame** of Ω with origin o. An affine frame $\{o; e_1, \cdots, e_n\}$ defines an n-tuple of functions $\{x^1, \cdots, x^n\}$ on Ω by

$$\overrightarrow{op} = \sum_i x^i(p) e_i, \quad p \in \Omega,$$

which is called an **affine coordinate system** on Ω with respect to the affine frame.

Let $\{\bar{x}^1, \cdots, \bar{x}^n\}$ be another affine coordinate system with respect to an affine frame $\{\bar{o}; \bar{e}_1, \cdots, \bar{e}_n\}$. If $e_j = \sum_i a^i_j \bar{e}_i$, $\overrightarrow{oo} = \sum_i a^i \bar{e}_i$, then

$$\bar{x}^i = \sum_j a^i_j x^j + a^i.$$

Representing the column vectors $[x^i]$, $[\bar{x}^i]$ and $[a^i]$ by $x = [x^i]$, $\bar{x} = [\bar{x}^i]$ and $a = [a^i]$ respectively, and the matrix $[a^i_j]$ by $A = [a^i_j]$, we have

$$\bar{x} = Ax + a,$$

or

$$\begin{bmatrix} \bar{x} \\ 1 \end{bmatrix} = \begin{bmatrix} A & a \\ 0 & 1 \end{bmatrix} \begin{bmatrix} x \\ 1 \end{bmatrix}.$$

Let e_i be a vector in the **standard vector space** $\mathbf{R}^n = \{p = (p^1, \cdots, p^n) \mid p^i \in \mathbf{R}\}$ whose j-th component is the Kronecker's δ_{ij}, then $\{e_1, \cdots, e_n\}$ is called the **standard basis** of $\mathbf{R}^n$. An affine coordinate system with respect to the affine frame $\{0; e_1, \cdots, e_n\}$, with origin the zero vector 0, is called the **standard affine coordinate system** on $\mathbf{R}^n$.

Let $\mathbf{R}^*_n$ be the dual vector space of $\mathbf{R}^n$, and let $\{e^{*1}, \cdots, e^{*n}\}$ be the dual basis of the standard basis $\{e_1, \cdots, e_n\}$ of $\mathbf{R}^n$. The affine coordinate system $\{x^*_1, \cdots, x^*_n\}$ on $\mathbf{R}^n$ with respect to the affine frame $\{0^*; e^{*1}, \cdots, e^{*n}\}$, with origin the zero vector 0^*, is said to be the **dual affine coordinate system** of $\{x^1, \cdots, x^n\}$.

Let Ω and $\tilde{\Omega}$ be affine spaces associated to vector spaces V and $\tilde{V}$ respectively. A mapping $\varphi : \Omega \longrightarrow \tilde{\Omega}$ is said to be an **affine mapping**, if there exists a linear mapping $\varphi' : V \longrightarrow \tilde{V}$ satisfying

$$\varphi'(\overrightarrow{pq}) = \overrightarrow{\varphi(p)\varphi(q)} \quad \text{for } p, q \in \Omega.$$

The mapping φ' is called a **linear mapping associated with** φ.

Let us consider vector spaces V and $\tilde{V}$ to be affine spaces as in Example 1.1. Let $\varphi : V \longrightarrow \tilde{V}$ be an affine mapping and let φ' be its associated linear mapping. Since $\varphi'(v) = \varphi'(\overrightarrow{0v}) = \overrightarrow{\varphi(0)\varphi(v)} = \varphi(v) - \varphi(0)$, we have

$$\varphi(v) = \varphi'(v) + \varphi(0).$$

Conversely for a linear mapping φ' from V to $\tilde{V}$ and $v_0 \in V$, we define a mapping $\varphi : V \longrightarrow \tilde{V}$ by

$$\varphi(v) = \varphi'(v) + v_0.$$

Then φ is an affine mapping with associated linear mapping φ' and $\varphi(0) = v_0$.

For an affine mapping $\varphi : V \longrightarrow \tilde{V}$, the associated linear mapping φ' and the vector $\varphi(0)$ are called the **linear part** and the **translation part** of φ respectively. A bijective affine mapping from Ω into itself is said to be an **affine transformation** of Ω. A mapping $\varphi : \Omega \longrightarrow \Omega$ is an affine transformation if and only if there exists a regular matrix $[a^i_j]$ and a vector $[a^i]$ such that

$$x^i \circ \varphi = \sum_j a^i_j x^j + a^i.$$

Let $A(V)$ be the set of all affine transformations of a real vector space V. Then $A(V)$ is a Lie group, and is called the **affine transformation group** of V. The set $GL(V)$ of all regular linear transformations of V is a subgroup of $A(V)$.

Definition 1.2. Let G be a group. A pair $(\boldsymbol{f}, \boldsymbol{q})$ of a homomorphism $\boldsymbol{f} : G \longrightarrow GL(V)$ and a mapping $\boldsymbol{q} : G \longrightarrow V$ is said to be an **affine representation** of G on V if it satisfies

$$\boldsymbol{q}(st) = \boldsymbol{f}(s)\boldsymbol{q}(t) + \boldsymbol{q}(s) \quad \text{for } s, t \in G. \tag{1.1}$$

For each $s \in G$ we define an affine transformation $\boldsymbol{a}(s)$ of V by

$$\boldsymbol{a}(s) : v \longrightarrow \boldsymbol{f}(s)v + \boldsymbol{q}(s).$$

Then the above condition (1.1) is equivalent to requiring the mapping

$$\boldsymbol{a} : s \in G \longrightarrow \boldsymbol{a}(s) \in A(V)$$

to be a homomorphism.

Let us denote by $\mathfrak{gl}(V)$ the set of all linear endomorphisims of V. Then $\mathfrak{gl}(V)$ is the Lie algebra of $GL(V)$. Let G be a Lie group, and let $\mathfrak{g}$ be its Lie algebra. For an affine representation $(\boldsymbol{f}, \boldsymbol{q})$ of G on V, we denote by f and q the differentials of $\boldsymbol{f}$ and $\boldsymbol{q}$ respectively. Then f is a linear representation of $\mathfrak{g}$ on V, that is, $f : \mathfrak{g} \longrightarrow \mathfrak{gl}(V)$ is a Lie algebra homomorphism, and q is a linear mapping from $\mathfrak{g}$ to V. Since

$$q(\mathrm{Ad}(s)Y) = \frac{d}{dt}\bigg|_{t=0} \boldsymbol{q}(s(\exp tY)s^{-1}) = \boldsymbol{f}(s)f(Y)\boldsymbol{q}(s^{-1}) + \boldsymbol{f}(s)q(Y),$$

it follows that

$$\begin{aligned}
q([X,Y]) &= \frac{d}{dt}\bigg|_{t=0} q(\mathrm{Ad}(\exp tX)Y) \\
&= f(X)q(Y)\boldsymbol{q}(e) + \boldsymbol{f}(e)f(Y)(-q(X)) + f(X)q(Y),
\end{aligned}$$

where e is the unit element in G. Since $\boldsymbol{f}(e)$ is the identity mapping and $\boldsymbol{q}(e) = 0$, we have

$$q([X,Y]) = f(X)q(Y) - f(Y)q(X). \tag{1.2}$$

A pair (f, q) of a linear representation f of a Lie algebra $\mathfrak{g}$ on V and a linear mapping q from $\mathfrak{g}$ to V is said to be an **affine representation** of $\mathfrak{g}$ on V if it satisfies the above condition (1.2).

1.2 Connections

In this section we summarize fundamental results concerning connections and Riemannian metrics. Let M be a smooth manifold. We denote by $\mathfrak{F}(M)$ the set of all smooth functions, and by $\mathfrak{X}(M)$ the set of all smooth vector fields on M. In this book the geometric objects we consider, for example, manifolds, functions, vector fields and so on, will always be smooth.

Definition 1.3. A **connection** on a manifold M is a mapping

$$D : (X,Y) \in \mathfrak{X}(M) \times \mathfrak{X}(M) \longrightarrow D_X Y \in \mathfrak{X}(M)$$

satisfying the following conditions,

(1) $D_{X_1+X_2}Y = D_{X_1}Y + D_{X_2}Y,$
(2) $D_{\varphi X}Y = \varphi D_X Y,$
(3) $D_X(Y_1 + Y_2) = D_X Y_1 + D_X Y_2,$
(4) $D_X(\varphi Y) = (X\varphi)Y + \varphi D_X Y,$

where $\varphi \in \mathfrak{F}(M)$. The term $D_X Y$ is called the **covariant derivative** of Y in the direction X.

Henceforth, we always assume that a manifold M is endowed with a connection D. A tensor field F of type $(0, p)$ is identified with a $\mathfrak{F}(M)$-valued p-multilinear function on $\mathfrak{F}(M)$-module $\mathfrak{X}(M)$;

$$F : \overbrace{\mathfrak{X}(M) \times \cdots \times \mathfrak{X}(M)}^{p \ terms} \longrightarrow \mathfrak{F}(M).$$

In the same way a tensor field of type $(1, p)$ is identified with a $\mathfrak{X}(M)$-valued p-multilinear mapping on $\mathfrak{F}(M)$-module $\mathfrak{X}(M)$.

Definition 1.4. For a tensor field F of type $(0, p)$ or $(1, p)$, we define a tensor field $D_X F$ by

$$(D_X F)(Y_1, \cdots, Y_p)$$

$$= D_X(F(Y_1, \cdots, Y_p)) - \sum_{i=1}^{p} F(Y_1, \cdots, D_X Y_i, \cdots, Y_p).$$

The tensor field $D_X F$ is called the **covariant derivative** of F in the direction X. A tensor field DF defined by

$$(DF)(Y_1, \cdots, Y_p, Y_{p+1}) = (D_{Y_{p+1}} F)(Y_1, \cdots, Y_p),$$

is said to be a **covariant differential** of F with respect to D.

Let $\{x^1, \cdots, x^n\}$ be a local coordinate system on M. The **components** or the **Christoffel's symbols** $\Gamma^k{}_{ij}$ of the connection D are defined by

$$D_{\partial/\partial x^i} \partial/\partial x^j = \sum_{k=1}^{n} \Gamma^k{}_{ij} \frac{\partial}{\partial x^k}.$$

The **torsion tensor** T of D is by definition

$$T(X, Y) = D_X Y - D_Y X - [X, Y].$$

The component $T^k{}_{ij}$ of the torsion tensor T given by

$$T\left(\frac{\partial}{\partial x^i}, \frac{\partial}{\partial x^j}\right) = \sum_{k} T^k{}_{ij} \frac{\partial}{\partial x^k}$$

satisfies

$$T^k{}_{ij} = \Gamma^k{}_{ij} - \Gamma^k{}_{ji}.$$

The connection D is said to be **torsion-free** if the torsion tensor T vanishes identically.

The **curvature tensor** R of D is defined by

$$R(X,Y)Z = D_X D_Y Z - D_Y D_X Z - D_{[X,Y]}Z.$$

The component $R^i{}_{jkl}$ of R given by

$$R\left(\frac{\partial}{\partial x^k}, \frac{\partial}{\partial x^l}\right)\frac{\partial}{\partial x^j} = \sum_i R^i{}_{jkl}\frac{\partial}{\partial x^i},$$

is expressed in the form

$$R^i{}_{jkl} = \frac{\partial \Gamma^i{}_{lj}}{\partial x^k} - \frac{\partial \Gamma^i{}_{kj}}{\partial x^l} + \sum_m (\Gamma^m{}_{lj}\Gamma^i{}_{km} - \Gamma^m{}_{kj}\Gamma^i{}_{lm}). \qquad (1.3)$$

The **Ricci tensor** Ric of D is by definition

$$\mathrm{Ric}(Y, Z) = \mathrm{Tr}\{X \longrightarrow R(X,Y)Z\}.$$

The component R_{jk} of Ric given by

$$R_{jk} = \mathrm{Ric}\left(\frac{\partial}{\partial x^j}, \frac{\partial}{\partial x^k}\right)$$

satisfies

$$R_{jk} = \sum_i R^i{}_{kij}. \qquad (1.4)$$

Definition 1.5. A curve $\sigma = x(t)$ in M is called a **geodesic** if it satisfies:

$$D_{\dot{x}(t)}\dot{x}(t) = 0,$$

where $\dot{x}(t)$ is the tangent vector of the curve σ at $x(t)$.

Using a local coordinate system $\{x^1, \cdots, x^n\}$, the equation of the geodesic is expressed by

$$\frac{d^2 x^i(t)}{dt^2} + \sum_{j,k}^n \Gamma^i{}_{jk}(x^1(t), \cdots, x^n(t))\frac{dx^j(t)}{dt}\frac{dx^k(t)}{dt} = 0,$$

where $x^i(t) = x^i(x(t))$D

Theorem 1.1. *For any point $p \in M$ and for any tangent vector X_p at p, there exists locally a unique geodesic $x(t)$ $(-\delta < t < \delta)$ satisfying the initial conditions (p, X_p), that is,*

$$x(0) = p, \qquad \dot{x}(0) = X_p.$$

A geodesic satisfying the initial conditions (p, X_p) is denoted by $\exp tX_p$.

If a geodesic $x(t)$ is defined for $-\infty < t < \infty$, then we say that the geodesic is **complete**. A connection D is said to be **complete** if every geodesic is complete.

Theorem 1.2. *For a tangent space T_pM at any point $p \in M$ there exists a neighbourhood, N_p, of the zero vector in T_pM such that: For any $X_p \in N_p$, $\exp tX_p$ is defined on an open interval containing $[-1, 1]$.*

A mapping on N_p given by
$$X_p \in N_p \longrightarrow \exp X_p \in M$$
is said to be the **exponential mapping** at p.

Definition 1.6. A connection D is said to be **flat** if the tosion tensor T and the curvature tensor R vanish identically. A manifold M endowed with a flat connection D is called a **flat manifold**.

The following results for flat manifolds are well known. For the proof see section **8.1**.

Proposition 1.1.

(1) *Suppose that M admits a flat connection D. Then there exist local coordinate systems on M such that $D_{\partial/\partial x^i}\partial/\partial x^j = 0$. The changes between such coordinate systems are affine transformations.*

(2) *Conversely, if M admits local coordinate systems such that the changes of the local coordinate systems are affine transformations, then there exists a flat connection D satisfying $D_{\partial/\partial x^i}\partial/\partial x^j = 0$ for all such local coordinate systems.*

For a flat connection D, a local coordinate system $\{x^1, \cdots, x^n\}$ satisfying $D_{\partial/\partial x^i}\partial/\partial x^j = 0$ is called an **affine coordinate system** with respect to D.

A flat connection D on $\mathbf{R}^n$ defined by
$$D_{\partial/\partial x^i}\partial/\partial x^j = 0,$$
where $\{x^1, \cdots, x^n\}$ is the standard affine coordinate system on $\mathbf{R}^n$, is called the **standard flat connection** on $\mathbf{R}^n$.

Definition 1.7. Two torsion-free connections D and $\bar{D}$ with symmetric Ricci tensors are said to be **projectively equivalent** if there exists a closed 1-form ρ such that
$$\bar{D}_X Y = D_X Y + \rho(X)Y + \rho(Y)X.$$

Definition 1.8. A torsion-free connection D with symmetric Ricci tensor is said to be **projectively flat** if D is projectively equivalent to a flat connection around each point of M.

Theorem 1.3. *A torsion-free connection D with symmetric Ricci tensor is projectively flat if and only if the following conditions hold* (cf. [Nomizu and Sasaki (1994)]).

(1) $R(X,Y)Z = \dfrac{1}{n-1}\{\mathrm{Ric}(Y,Z)X - \mathrm{Ric}(X,Z)Y\}$, *where* $n = \dim M$,

(2) $(D_X\mathrm{Ric})(Y,Z) = (D_Y\mathrm{Ric})(X,Z)$.

A non-degenerate symmetric tensor g of type $(0,2)$ is said to be an **indefinite Riemannian metric**. If g is positive definite, it is called a **Riemannian metric**.

Theorem 1.4. *Let g be an indefinite Riemannian metric. Then there exists a unique torsion-free connection ∇ such that*

$$\nabla g = 0.$$

Proof.   Suppose that there exists a torsion-free connection ∇ satisfying $\nabla g = 0$. Since

$$0 = \nabla_X Z - \nabla_Z X - [X, Z],$$
$$0 = (\nabla_X g)(Y, Z) = Xg(Y, Z) - g(\nabla_X Y, Z) - g(Y, \nabla_X Z),$$

and we have

$$Xg(Y, Z) = g(\nabla_X Y, Z) + g(\nabla_Z X, Y) + g([X, Z], Y).$$

Cycling X, Y, Z in the above formula, we obtain

$$Yg(Z, X) = g(\nabla_Y Z, X) + g(\nabla_X Y, Z) + g([Y, X], Z)$$
$$Zg(X, Y) = g(\nabla_Z X, Y) + g(\nabla_Y Z, X) + g([Z, Y], X).$$

Eliminating $\nabla_Y Z$ and $\nabla_Z X$ from the above relations, we have

$$2g(\nabla_X Y, Z) = Xg(Y, Z) + Yg(X, Z) - Zg(X, Y) \qquad (1.5)$$
$$+ g([X, Y], Z) + g([Z, X], Y) - g([Y, Z], X).$$

Given that g is non-degenerate and the right-hand side of equation (1.5) depends only on g, the connection ∇ is uniquely determined by g. For a given indefinite Riemannian metric g we define $\nabla_X Y$ by equation (1.5). It is then easy to see that ∇ is a torsion-free connection satisfying $\nabla g = 0$.

$\square$

The connection ∇ given in Theorem 1.4 is called the **Riemannian connection** or the **Levi-Civita connection** for g. We denote by g_{ij} the components of an indefinite Riemannian metric g with respect to a local coordinate system $\{x^1, \cdots, x^n\}$;

$$g_{ij} = g\left(\frac{\partial}{\partial x^i}, \frac{\partial}{\partial x^j}\right).$$

Let $\Gamma^k{}_{ij}$ be the Christoffel's symbols of ∇. Upon substituting for X, Y and Z in equation (1.5) using $X = \partial/\partial x^i C Y = \partial/\partial x^j$ and $Z = \partial/\partial x^k$, we obtain

$$2\sum_l \Gamma^l{}_{ij} g_{lk} = \frac{\partial g_{jk}}{\partial x^i} + \frac{\partial g_{ik}}{\partial x^j} - \frac{\partial g_{ij}}{\partial x^k},$$

and hence

$$\Gamma^k{}_{ij} = \frac{1}{2}\sum_l g^{kl}\left(\frac{\partial g_{jl}}{\partial x^i} + \frac{\partial g_{il}}{\partial x^j} - \frac{\partial g_{ij}}{\partial x^l}\right). \tag{1.6}$$

For a Riemannian metric g the **sectional curvature** K for a plane spanned by tangent vectors X, Y is given by

$$K = \frac{g(R(X,Y)Y,X)}{g(X,X)g(Y,Y) - g(X,Y)^2}. \tag{1.7}$$

A Riemannian metric g is said to be of **constant curvature** c if the sectional curvature is a constant c for any plane. This condition is equivalent to

$$R(X,Y)Z = c\{g(Z,Y)X - g(Z,X)Y\}. \tag{1.8}$$

1.3   Vector bundles

In this section we generalize the notion of connections defined in section **1.2** to that on vector bundles.

Definition 1.9. A manifold E is said to be a **vector bundle** over M, if there exists a surjective mapping $\pi : E \longrightarrow M$, and a finite-dimensional real vector space F satisfying the following conditions.

(1) For each point in M there exists a neighbourhood U and a diffeomorphism

$$\hat{\phi}_U : u \in \pi^{-1}(U) \longrightarrow (\pi(u), \phi_U(u)) \in U \times F.$$

(2) Given two neighbourhoods U and V satisfying (1) above, if $U \cap V$ is non-empty, then there is a mapping

$$\psi_{UV} : U \cap V \longrightarrow GL(F)$$

such that

$$\phi_V(u) = \psi_{VU}(\pi(u))\phi_U(u), \quad for \; all \; u \in \pi^{-1}(U \cap V).$$

π is called the **projection** and F is called the **standard fiber**.

A mapping s from an open set $U \subset M$ into E is said to be a **section** of E on U if $\pi \circ s$ is the identity mapping on U. The set $\mathfrak{S}(U)$ consisting of all sections on U is a real vector space and an $\mathfrak{F}(U)$-module.

Example 1.3. Let M be a manifold and let T_pM be the tangent space at $p \in M$. We set $TM = \bigcup_{p \in M} T_pM$, and define a mapping $\pi : TM \longrightarrow M$ by $\pi(X) = p$ for $X \in T_pM$. Let $\{x^1, \cdots, x^n\}$ be a local coordinate system on U. A mapping given by

$$X \in \pi^{-1}(U) \longrightarrow ((x^1 \circ \pi)(X), \cdots, (x^n \circ \pi)(X), dx^1(X), \cdots, dx^n(X)) \in \mathbf{R}^{2n}$$

is injective. The $2n$-tuple $\{x^1 \circ \pi, \cdots, x^n \circ \pi, dx^1, \cdots, dx^n\}$ then defines a local coordinate system on $\pi^{-1}(U)$, and TM is a manifold. Upon setting

$$\hat{\phi}_U : X \in \pi^{-1}(U) \longrightarrow (\pi(X), dx^1(X), \cdots, dx^n(X)) \in U \times \mathbf{R}^n,$$

we have that TM is a vector bundle over M with the standard fiber $\mathbf{R}^n$, and is said to be the **tangent bundle** over M. A section of TM on M is a vector field on M.

Example 1.4. Let T_p^*M be the dual space of the tangent space T_pM at $p \in M$. We set $T^*M = \bigcup_{p \in M} T_p^*M$, and define a mapping $\pi : T^*M \longrightarrow M$ by $\pi(\omega) = p$ for $\omega \in T_p^*M$. Let $\{x^1, \cdots, x^n\}$ be a local coordinate system on U. A mapping given by

$$\omega \in \pi^{-1}(U) \longrightarrow ((x^1 \circ \pi)(\omega), \cdots, (x^n \circ \pi)(\omega), i_{\partial/\partial x^1}(w), \cdots, i_{\partial/\partial x^n}(w)) \in \mathbf{R}^{2n}$$

is injective, where $i_{\partial/\partial x^i}(\omega) = \omega(\partial/\partial x^i)$. Then $\{x^1 \circ \pi, \cdots, x^n \circ \pi, i_{\partial/\partial x^1}, \cdots, i_{\partial/\partial x^n}\}$ defines a local coordinate system on $\pi^{-1}(U)$, and T^*M is a manifold. Upon setting

$$\hat{\phi}_U : \omega \in \pi^{-1}(U) \longrightarrow (\pi(\omega), i_{\partial/\partial x^1}(w), \cdots, i_{\partial/\partial x^n}(w)) \in U \times \mathbf{R}_n^*,$$

we have that T^*M is a vector bundle over M with the standard fiber $\mathbf{R}_n^*$, and is said to be the **cotangent bundle** over M. A section of T^*M on M is a 1-form on M.

Definition 1.10. A **connection** D on a vector bundle E over M is a mapping

$$D : (X, s) \in \mathfrak{X}(M) \times \mathfrak{S}(M) \longrightarrow D_X s \in \mathfrak{S}(M),$$

satisfying the following conditions,

(1) $D_{X+Y} s = D_X s + D_Y s,$
(2) $D_{\varphi X} s = \varphi D_X s,$
(3) $D_X(s + t) = D_X s + D_X t,$
(4) $D_X(\varphi s) = (X\varphi)s + \varphi D_X s,$

where X, $Y \in \mathfrak{X}(M)$, s, $t \in \mathfrak{S}(M)$ and $\varphi \in \mathfrak{F}(M)$.

Example 1.5. A connection on the tangent bundle TM over M is a connection on M in the sense of Definition 1.3.

Example 1.6. Let D be a connection on the tangent bundle TM over M. We denote by $\mathfrak{S}^*(M)$ the set of all sections of the cotangent bundle T^*M over M, and define a mapping

$$D^* : (X, \omega) \in \mathfrak{X}(M) \times \mathfrak{S}^*(M) \longrightarrow D_X \omega \in \mathfrak{S}^*(M)$$

by $(D_X^* \omega)(Y) = X(\omega(Y)) - \omega(D_X Y)$. Then D^* is a connection on T^*M.

Chapter 2

Hessian structures

A Riemannian metric g on a flat manifold is said to be a Hessian metric if it can be expressed by the Hessian form with respect to the flat connection D. The pair (D, g) is called a Hessian structure. Of all the Riemannian metrics that can exist on a flat manifold, Hessian metrics appear to be the most compatible metrics with the flat connection D. In this chapter we will study the fundamental properties of Hessian structures. In section **2.1** we derive basic identities for a Hessian structure. In section **2.2** we proceed to show that the tangent bundle over a Hessian manifold (a manifold with a Hessian structure) is a Kählerian manifold and investigate the relation between a Hessian structure and a Kählerian structure. In section **2.3** we define the gradient mapping, which is an affine immersion, and show the duality of Hessian structures. In section **2.4** we define the divergence of a Hessian structure, which is particularly useful for applications in statistics. By extending the notion of Hessian structures, we define in section **2.5** Codazzi structures.

2.1 Hessian structures

We denote by (M, D) a flat manifold M with a flat connection D. In this section we consider a class of Riemannian metrics compatible with the flat connection D. A Riemannian metric g on M is said to be a Hessian metric if g is locally expressed by the Hessian with respect to D, and the pair (D, g) is called a Hessian structure. A pair (D, g) of a flat connection D and a Riemannian metric g is a Hessian structure if and only if it satisfies the Codazzi equation. The difference tensor γ between the Levi-Civita connection ∇ of a Hessian metric g and a flat connection D defined by $\gamma = \nabla - D$ plays various important roles in the study of Hessian structures.

13

Definition 2.1. A Riemannian metric g on a flat manifold (M, D) is called a **Hessian metric** if g can be locally expressed by

$$g = Dd\varphi,$$

that is,

$$g_{ij} = \frac{\partial^2 \varphi}{\partial x^i \partial x^j},$$

where $\{x^1, \cdots, x^n\}$ is an affine coordinate system with respect to D. Then the pair (D, g) is called a **Hessian structure** on M, and φ is said to be a **potential** of (D, g). A manifold M with a Hessian structure (D, g) is called a **Hessian manifold**, and is denoted by (M, D, g).

Definition 2.2. A Hessian structure (D, g) is said to be of **Koszul type**, if there exists a closed 1-form ω such that $g = D\omega$.

Let (M, D) be a flat manifold, g a Riemannian metric on M, and ∇ the Levi-Civita connection of g. We denote by γ the **difference tensor** of ∇ and D ;

$$\gamma_X Y = \nabla_X Y - D_X Y.$$

Since ∇ and D are torsion-free it follows that

$$\gamma_X Y = \gamma_Y X. \tag{2.1}$$

It should be remarked that the components $\gamma^i{}_{jk}$ of γ with respect to affine coordinate systems coincide with the Christoffel symbols $\Gamma^i{}_{jk}$ of ∇.

Proposition 2.1. *Let (M, D) be a flat manifold and g a Riemannian metric on M. Then the following conditions are equivalent.*

(1) *g is a Hessian metric.*
(2) *$(D_X g)(Y, Z) = (D_Y g)(X, Z)$* D
(3) *$\dfrac{\partial g_{ij}}{\partial x^k} = \dfrac{\partial g_{kj}}{\partial x^i}$* D
(4) *$g(\gamma_X Y, Z) = g(Y, \gamma_X Z)$* D
(5) *$\gamma_{ijk} = \gamma_{jik}$* D

Proof.   By the definition of Hessian metrics (1) implies (3). The conditions (3) and (5) are the local expressions of (2) and (4) respectively. From (1.6) the Christoffel symbols of g are given by

$$\gamma^i{}_{jk} = \frac{1}{2} g^{is} \left(\frac{\partial g_{sj}}{\partial x^k} + \frac{\partial g_{sk}}{\partial x^j} - \frac{\partial g_{jk}}{\partial x^s} \right),$$

$$\gamma_{ijk} = \frac{1}{2}\left(\frac{\partial g_{ij}}{\partial x^k} + \frac{\partial g_{ik}}{\partial x^j} - \frac{\partial g_{jk}}{\partial x^i}\right).^1$$

This demonstrates that conditions (3) and (5) are equivalent. Finally, we will show that condition (3) implies (1). Upon setting $h_j = \sum_i g_{ij}dx^i$, we have $dh_j = \sum_i dg_{ij} \wedge dx^i = \sum_{k<i}\left(\frac{\partial g_{ij}}{\partial x^k} - \frac{\partial g_{kj}}{\partial x^i}\right)dx^k \wedge dx^i = 0$. Hence, by Poincarè's lemma, there exists φ_j such that $h_j = d\varphi_j$. If we put $h = \sum_j \varphi_j dx^j$, then $dh = \sum d\varphi_j \wedge dx^j = 0$. Upon applying Poincarè's lemma again, there exists φ such that $h = d\varphi$. Therefore we have $\frac{\partial \varphi}{\partial x^j} = \varphi_j$ and $\frac{\partial^2 \varphi}{\partial x^i \partial x^j} = \frac{\partial \varphi_j}{\partial x^i} = g_{ij}$. $\square$

The equation (2) of Proposition 2.1 is said to be the **Codazzi equation** of g with respect to D. In the course of the proof of Proposition 2.1 we have proved the following proposition.

Proposition 2.2. *Let (D, g) be a Hessian structure. Then we have*

$$g(\gamma_X Y, Z) = \frac{1}{2}(D_X g)(Y, Z),$$
$$\gamma^i{}_{jk} = \frac{1}{2}g^{ir}\frac{\partial g_{rj}}{\partial x^k}, \qquad \gamma^{ij}{}_k = -\frac{1}{2}\frac{\partial g^{ij}}{\partial x^k}, \qquad \gamma_{ijk} = \frac{1}{2}\frac{\partial g_{ij}}{\partial x^k}.$$

Proposition 2.3. *Let R be the curvature tensor of a Hessian metric g. Then we have*

(1) $R(X, Y) = -[\gamma_X, \gamma_Y], \qquad R^i{}_{jkl} = \gamma^i{}_{lr}\gamma^r{}_{jk} - \gamma^i{}_{kr}\gamma^r{}_{jl}.$

(2) *The sectional curvature K for a plane spanned by X and Y is given by*
$$K = \frac{g(\gamma_X Y, \gamma_X Y) - g(\gamma_X X, \gamma_Y Y)}{g(X, X)g(Y, Y) - g(X, Y)^2}.$$

Proof. By equation (1.3) and Proposition 2.2 we have

$$\begin{aligned}
R^i{}_{jkl} &= \frac{\partial \gamma^i{}_{jl}}{\partial x^k} - \frac{\partial \gamma^i{}_{jk}}{\partial x^l} - \gamma^i{}_{lr}\gamma^r{}_{jk} + \gamma^i{}_{kr}\gamma^r{}_{jl} \\
&= \frac{1}{2}\left(\frac{\partial g^{ir}}{\partial x^k}\frac{\partial g_{rj}}{\partial x^l} + g^{ir}\frac{\partial^2 g_{rj}}{\partial x^l \partial x^k}\right) - \frac{1}{2}\left(\frac{\partial g^{ir}}{\partial x^l}\frac{\partial g_{rj}}{\partial x^k} + g^{ir}\frac{\partial^2 g_{rj}}{\partial x^l \partial x^k}\right) \\
&\quad (-\gamma^i{}_{lr}\gamma^r{}_{jk} + \gamma^i{}_{kr}\gamma^r{}_{jl}) \\
&= 2(-\gamma^{ir}{}_k\gamma_{rjl} + \gamma^{ir}{}_l\gamma_{rjk}) + (-\gamma^i{}_{lr}\gamma^r{}_{jk} + \gamma^i{}_{kr}\gamma^r{}_{jl}) \\
&= \gamma^i{}_{lr}\gamma^r{}_{jk} - \gamma^i{}_{kr}\gamma^r{}_{jl}.
\end{aligned}$$

---
[1] We use Einstein's summation convention.

This proves (1). From equation (1.7) we have

$$K = \frac{g(R(X,Y)Y,X)}{g(X,X)g(Y,Y) - g(X,Y)^2},$$

while from Proposition 2.1 and (1) above we have

$$g(R(X,Y)Y,X) = g(-[\gamma_X, \gamma_Y]Y, X) = g(-\gamma_X\gamma_Y Y + \gamma_Y\gamma_X Y, X)$$
$$= -g(\gamma_Y Y, \gamma_X X) + g(\gamma_X Y, \gamma_X Y).$$

Upon substituting into the expression for K above for $g(R(X,Y)Y,X)$ we derive (2). □

Lemma 2.1. *A vector field X is a Killing vector field with respect to a Hessian metric g if and only if*

$$2g(\gamma_X Y, Z) = g(A_X Y, Z) + g(Y, A_X Z), \quad \text{for all } Y, Z \in \mathfrak{X}(M),$$

where $A_X = \mathcal{L}_X - D_X$ and $\mathcal{L}_X$ is the Lie derivative with respect to X.

Proof. By Proposition 2.2 we have

$$0 = (\mathcal{L}_X g)(Y, Z)$$
$$= X(g(Y,Z)) - g(\mathcal{L}_X Y, Z) - g(Y, \mathcal{L}_X Z)$$
$$= (D_X g)(Y,Z) + g(D_X Y, Z) + g(Y, D_X Z) - g(\mathcal{L}_X Y, Z) - g(Y, \mathcal{L}_X Z)$$
$$= 2g(\gamma_X Y, Z) - g(A_X Y, Z) - g(Y, A_X Z). \qquad \square$$

Lemma 2.2. *Let (D, g) be a Hessian structure. Then we have*

(1) *The difference tensor γ is ∇-parallel if and only if*

$$[\nabla_X, \gamma_Y] = \gamma_{\nabla_X Y}, \quad \text{for all } X, Y \in \mathcal{X}(M).$$

(2) *The curvature tensor R for g is ∇-parallel if and only if*

$$[\nabla_X, [\gamma_X, \gamma_Z]] = [\gamma_{\nabla_X Y}, \gamma_Z] + [\gamma_Y, \gamma_{\nabla_X Z}], \quad \text{for all } X,Y,Z \in \mathcal{X}(M).$$

Proof. (1) follows from

$$(\nabla_X \gamma)(Y, Z) = \nabla_X(\gamma_Y Z) - \gamma_{\nabla_X Y} Z - \gamma_Y \nabla_X Z$$
$$= ([\nabla_X, \gamma_Y] - \gamma_{\nabla_X Y})Z.$$

Applying Proposition 2.3 (1), we have

$$(\nabla_X R)(Y,Z)W$$
$$= \nabla_X(R(Y,Z)W) - R(\nabla_X Y, Z)W - R(Y, \nabla_X Z)W - R(Y,Z)\nabla_X W$$
$$= -(\nabla_X[\gamma_Y, \gamma_Z] - [\gamma_{\nabla_X Y}, \gamma_Z] - [\gamma_Y, \gamma_{\nabla_X Z}] - [\gamma_Y, \gamma_Z]\nabla_X)W$$
$$= -([\nabla_X, [\gamma_Y, \gamma_Z]] - [\gamma_{\nabla_X Y}, \gamma_Z] - [\gamma_Y, \gamma_{\nabla_X Z}])W.$$

This implies (2). □

Example 2.1. Let g be a Riemannian metric and ∇ the Levi-Civita connection for g. If ∇ is flat, then the pair (∇, g) is a Hessian structure.

Example 2.2. Let $\mathbf{R}^n$ be the standard affine space with the standard flat connection D and the standard affine coordinate system $\{x^1, \cdots, x^n\}$. Let Ω be a domain in $\mathbf{R}^n$ equipped with a convex function φ, that is, the Hessian $g = Dd\varphi$ is positive definite on Ω. Then the pair $(D, g = Dd\varphi)$ is a Hessian structure on Ω. Important examples of these structures include:

(1) Let $\Omega = \mathbf{R}^n$ and $\varphi = \dfrac{1}{2} \sum_{i=1}^{n} (x^i)^2$, then $g_{ij} = \delta_{ij}$ (Kronecker's delta) and g is a Euclidean metric.

(2) Let $\Omega = \{x \in \mathbf{R}^n \mid x^1 > 0, \cdots, x^n > 0\}$ and $\varphi = \sum_{i=1}^{n} (x^i \log x^i - x^i)$, then $g_{ij} = \delta_{ij} \dfrac{1}{x^i}$.

(3) Let $\Omega = \left\{ x \in \mathbf{R}^n \mid x^n > \dfrac{1}{2} \sum_{i}^{n-1} (x^i)^2 \right\}$ and $\varphi = - \log \left(x^n - \dfrac{1}{2} \sum_{i=1}^{n-1} (x^i)^2 \right)$.

Then $[g_{ij}] = \dfrac{1}{f^2} \begin{bmatrix} \delta_{ij} f + x^i x^j & -x^i \\ -x^j & 1 \end{bmatrix}$, where $f = x^n - \dfrac{1}{2} \sum_{i=1}^{n-1} (x^i)^2$.

(4) Let $\Omega = \mathbf{R}^n$ and $\varphi = \log \left(1 + \sum_{i=1}^{n} e^{x^i} \right)$. Then $g_{ij} = \dfrac{1}{f} \delta_{ij} e^{x^j} - \dfrac{1}{f^2} e^{x^i + x^j}$, where $f = 1 + \sum_{i=1}^{n} e^{x^i}$.

(5) Let $\Omega = \left\{ x \in \mathbf{R}^n \mid x^n > \left(\sum_{i=1}^{n-1} (x^i)^2 \right)^{1/2} \right\}$ and $\varphi = - \log \left((x^n)^2 - \sum_{i=1}^{n-1} (x^i)^2 \right)$, then $g_{ij} = \dfrac{2}{f} \epsilon_i \delta_{ij} + \dfrac{4}{f^2} \epsilon_i \epsilon_j x^i x^j$, where $f = (x^n)^2 - \sum_{i=1}^{n-1} (x^i)^2$, $\epsilon_i = 1$ for $1 \leq i \leq n-1$, and $\epsilon_n = -1$.

(6) Let $\Omega = \left\{ x \in \mathbf{R}^n \mid 1 > \sum_{i=1}^{n} (x^i)^2 \right\}$ and $\varphi = - \log \left(1 - \sum_{i=1}^{n} (x^i)^2 \right)$. Then $g_{ij} = \dfrac{2}{f} \left(\delta^{ij} + \dfrac{2}{f} x^i x^j \right)$, where $f = 1 - \sum_{i=1}^{n} (x^i)^2$.

2.2   Hessian structures and Kählerian structures

As stated in section **2.1**, a Riemannian metric on a flat manifold is a Hessian metric if it can be locally expressed by the Hessian with respect to an affine coordinate system. On the other hand, a Riemannian metric on a complex manifold is said to be a Kählerian metric if it can be locally given by the complex Hessian with respect to a holomorphic coordinate system. This suggests that the following set of analogies exists between Hessian structures and Kählerian structures:

>*Flat manifolds* $\longleftrightarrow$ *Complex manifolds*
>*Affine coordinate systems* $\longleftrightarrow$ *Holomorphic coordinate systems*
>*Hessian metrics* $\longleftrightarrow$ *Kählerian metrics*

In this section we show that the tangent bundle over a Hessian manifold admits a Kählerian metric induced by the Hessian metric. We first give a brief summary of Kählerian manifolds, for more details the interested reader may refer to [Kobayashi (1997, 1998)][Weil (1958)].

Definition 2.3. A Hausdorff space M is said to be an n-dimensional **complex manifold** if it admits an open covering $\{U_\lambda\}_{\lambda \in \Lambda}$ and mappings $f_\lambda : U_\lambda \longrightarrow \mathbf{C}^n$ satisfying the following conditions.

(1) Each $f_\lambda(U_\lambda)$ is an open set in $\mathbf{C}^n$, and $f_\lambda : U_\lambda \longrightarrow f_\lambda(U_\lambda)$ is a homeomorphism.
(2) If $U_\lambda \cap U_\mu \neq \phi$, then

$$f_\mu \circ f_\lambda^{-1} : f_\lambda(U_\lambda \cap U_\mu) \longrightarrow f_\mu(U_\lambda \cap U_\mu)$$

is a holomorphic mapping.

An n-tuple of functions $\{z_\lambda^1, \cdots, z_\lambda^n\}$ on U_λ defined by $f_\lambda(p) = (z_\lambda^1(p), \cdots, z_\lambda^n(p))$ for $p \in U_\lambda$ is called a **holomorphic coordinate system** on U_λ.

Let M be a complex manifold and let $\{z^1, \cdots, z^n\}$ be a holomorphic coordinate system in M. Denoting $z^k = x^k + \sqrt{-1}y^k$, we have

$$\frac{\partial}{\partial z^k} = \frac{1}{2}\left(\frac{\partial}{\partial x^k} - \sqrt{-1}\frac{\partial}{\partial y^k}\right), \quad \frac{\partial}{\partial \bar{z}^k} = \frac{1}{2}\left(\frac{\partial}{\partial x^k} + \sqrt{-1}\frac{\partial}{\partial y^k}\right)$$

and

$$dz^k = dx^k + \sqrt{-1}dy^k, \quad d\bar{z}^k = dx^k - \sqrt{-1}dy^k.$$

Define a tensor J of type $(1,1)$ by

$$J\left(\frac{\partial}{\partial x^k}\right) = \frac{\partial}{\partial y^k}, \quad J\left(\frac{\partial}{\partial y^k}\right) = -\frac{\partial}{\partial x^k}.$$

Note that, with this definition, J is independent of holomorphic coordinate systems that are selected. We have

$$J^2(X) = -X, \quad X \in \mathfrak{X}(M),$$

and J is said to be the **complex structure tensor** on M. A complex manifold M with a complex structure tensor J is denoted by (M, J).

Let g be a Riemannian metric on a complex manifold M. We denote by $T_p^c M = T_p M \otimes \mathbf{C}$ the complexification of the tangent space $T_p M$ at $p \in M$, and extend g to

$$g : T_p^c M \times T_p^c M \longrightarrow \mathbf{C},$$

so that $g(U, V)$ is complex linear and complex conjugate linear with respect to U and V respectively. We set

$$g_{ij} = g\left(\frac{\partial}{\partial z^i}, \frac{\partial}{\partial z^j}\right), \quad g_{i\bar{j}}\left(\frac{\partial}{\partial z^i}, \frac{\partial}{\partial \bar{z}^j}\right),$$

$$g_{\bar{i}j} = g\left(\frac{\partial}{\partial \bar{z}^i}, \frac{\partial}{\partial \bar{z}^j}\right), \quad g_{\bar{i}\bar{j}} = g\left(\frac{\partial}{\partial \bar{z}^i}, \frac{\partial}{\partial z^j}\right).$$

Definition 2.4. A Riemannian metric g on a complex manifold is said to be a **Hermitian metric** if

$$g_{ij} = g_{\bar{i}\bar{j}} = 0.$$

We denote the Hermitian metric by

$$g = \sum_{i\bar{j}} g_{i\bar{j}} dz^i d\bar{z}^j.$$

Proposition 2.4. *A Riemannian metric on a complex manifold (M, J) is a Hermitian metric if and only if*

$$g(JX, JY) = g(X, Y), \quad \text{for all } X, Y \in \mathfrak{X}(M).$$

The following fact is well known.

Theorem 2.1. *A complex manifold admits a Hermitian metric.*

The proof follows by applying a standard argument using a partition of unity.

Definition 2.5. A Hermitian metric g on a complex manifold (M, J) is said to be a **Kählerian metric** if g can be locally expressed by the complex Hessian of a function φ,

$$g_{i\bar{j}} = \frac{\partial^2 \varphi}{\partial z^i \partial \bar{z}^j},$$

where $\{z^1, \cdots, z^n\}$ is a holomorphic coordinate system. The pair (J, g) is called a **Kählerian structure** on M. A complex manifold M with a Kählerian structure (J, g) is said to be a **Kählerian manifold** and is denoted by (M, J, g).

For a Hermitian metric g we set

$$\rho(X, Y) = g(JX, Y).$$

Then the skew symmetric bilinear form ρ is called a **Kählerian form** for (J, g), and, using a holomorhic coordinate system, we have

$$\rho = \sqrt{-1} \sum_{i,j} g_{i\bar{j}} dz^i \wedge d\bar{z}^j.$$

Proposition 2.5. *Let g be a Hermitian metric on a complex manifold M. Then the following conditions are equivalent.*

(1) *g is a Kählerian metric.*
(2) *The Kählerian form ρ is closed; $d\rho = 0$.*

Let (M, D) be a flat manifold and let TM be the tangent bundle over M with projection $\pi : TM \longrightarrow M$. For an affine coordinate system $\{x^1, \cdots, x^n\}$ on M, we set

$$z^j = \xi^j + \sqrt{-1}\xi^{n+j}, \tag{2.2}$$

where $\xi^i = x^i \circ \pi$ and $\xi^{n+i} = dx^i$. Then n-tuples of functions given by $\{z^1, \cdots, z^n\}$ yield holomorphic coordinate systems on TM. We denote by J_D the complex structure tensor of the complex manifold TM. For a Riemannian metric g on M we put

$$g^T = \sum_{i,j=1}^{n} (g_{ij} \circ \pi) dz^i d\bar{z}^j. \tag{2.3}$$

Then g^T is a Hermitian metric on the complex manifold (TM, J_D).

Proposition 2.6. *Let (M, D) be a flat manifold and g a Riemannian metric on M. Then the following conditions are equivalent.*

(1) g is a Hessian metric on (M, D).

(2) g^T is a Kählerian metric on (TM, J_D).

Proof. Denoting by ρ^T the Kählerian form of the Hermitian metric g^T, we have

$$\rho^T = \sqrt{-1} \sum_{i,j=1}^{n} (g_{ij} \circ \pi) dz^i \wedge d\bar{z}^j = 2 \sum_{i,j=1}^{n} (g_{ij} \circ \pi) d\xi^i \wedge d\xi^{n+j}.$$

Differentiating both sides we have

$$d\rho^T = 2 \sum_{i,j=1}^{n} d(g_{ij} \circ \pi) \wedge d\xi^i \wedge d\xi^{n+j}$$

$$= 2 \sum_{i,j=1}^{n} \sum_{k=1}^{n} \frac{\partial(g_{ij} \circ \pi)}{\partial \xi^k} d\xi^k \wedge d\xi^i \wedge d\xi^{n+j}$$

$$= 2 \sum_{i,j,k=1}^{n} \left(\frac{\partial(g_{ij} \circ \pi)}{\partial \xi^k} - \frac{\partial(g_{kj} \circ \pi)}{\partial \xi^i} \right) d\xi^k \wedge d\xi^i \wedge d\xi^{n+j}$$

$$= 2 \sum_{i,j,k=1}^{n} \left(\left(\frac{\partial g_{ij}}{\partial x^k} - \frac{\partial g_{kj}}{\partial x^i} \right) \circ \pi \right) d\xi^k \wedge d\xi^i \wedge d\xi^{n+j}.$$

Hence the equations $d\rho^T = 0$ and $\dfrac{\partial g_{ij}}{\partial x^k} = \dfrac{\partial g_{kj}}{\partial x^i}$ are equivalent. Conditions (1) and (2) are therefore equivalent by Proposition 2.1. $\square$

Example 2.3.

(1) Let $\Omega = \mathbf{R}^n$ and $\varphi = \dfrac{1}{2} \sum_{i=1}^{n} (x^i)^2$. Then $g = \sum_{i=1}^{n} (dx^i)^2$ is a Euclidean metric on $\mathbf{R}^n$. The tangent bundle $T\mathbf{R}^n$ is identified with $\mathbf{C}^n$ by the complex coordinate system $\{z^1, \cdots, z^n\}$ given in (2.2). Since $g^T = \sum_i dz^i d\bar{z}^i$, $(T\mathbf{R}^n, J_D, g^T)$ is a complex Euclidean space.

(2) Let $\Omega = \mathbf{R}^+ = \{x \in \mathbf{R} \mid x > 0\}$ and $\varphi = \log x^{-1}$. We then have $g = \dfrac{1}{x^2} dx^2$. Let $\{\xi^1, \xi^2\}$ be the coordinate system on the tangent bundle $T\mathbf{R}^+$ as defined by equation (2.2). Then $T\mathbf{R}^+$ is identified with a half plane $\{(\xi^1, \xi^2) \mid \xi^1 > 0\}$, and the Kählerian metric g^T on $T\mathbf{R}^+$ induced by g is expressed by

$$g^T = \frac{(d\xi^1)^2 + (d\xi^2)^2}{(\xi^1)^2}.$$

Thus g^T is the Poincaré metric on the half plane.

Example 2.3 (2) is extended to regular convex cones as follows.

Example 2.4. Let Ω be a regular convex cone in $\mathbf{R}^n$, and let ψ be the characteristic function. Then $(D, g = Dd\log\psi)$ is a Hessian structure on Ω (cf. section **4.1**). The tangent bundle $T\Omega$ over Ω is identified with the tube domain $T_\Omega = \Omega + \sqrt{-1}\mathbf{R}^n$ over Ω in $\mathbf{C}^n = \mathbf{R}^n + \sqrt{-1}\mathbf{R}^n$. T_Ω is holomorphically equivalent to a bounded domain in $\mathbf{C}^n$, while g^T is isometric to the Bergman metric on the bounded domain (cf. Theorem 8.4).

Example 2.5. Let $(\Omega, D, g = Dd\varphi)$ be the Hessian domain of Example 2.2 (3), where $\Omega = \left\{ x \in \mathbf{R}^n \mid x^n > \dfrac{1}{2}\sum_{i=1}^{n-1}(x^i)^2 \right\}$ and $\varphi = -\log\Big(x^n - \dfrac{1}{2}\sum_{i=1}^{n-1}(x^i)^2\Big)$. Let $\{z^1, \cdots, z^n\}$ be a holomorphic coordinate system on $T_\Omega = \Omega + \sqrt{-1}\mathbf{R}^n$ as defined in equation (2.2). Consider the following holomorphic transformation defined by

$$w^j = z^j\Big(z^n - \frac{1}{4}\sum_{k=1}^{n-1}(z^k)^2 + 1\Big)^{-1}, \quad 1 \le j \le n-1,$$

$$w^n = \Big(z^n - \frac{1}{4}\sum_{k=1}^{n-1}(z^k)^2 - 1\Big)\Big(z^n - \frac{1}{4}\sum_{k=1}^{n-1}(z^k)^2 + 1\Big)^{-1}.$$

Then T_Ω is holomorphically equivalent to the bounded domain

$$\Big\{(w^1, \cdots, w^n) \in \mathbf{C}^n \mid \sum_{k=1}^{n}|w^k|^2 < 1\Big\}.$$

2.3 Dual Hessian structures

In this section we will establish the duality that exists for Hessian structures. Let $\mathbf{R}_n^*$ be the dual vector space of $\mathbf{R}^n$. We denote by D^* the standard flat connection on $\mathbf{R}_n^*$, and by $\{x_1^*, \cdots, x_n^*\}$ the dual affine coordinate system on $\mathbf{R}_n^*$ with respect to the standard affine coordinate system $\{x^1, \cdots, x^n\}$ on $\mathbf{R}^n$(cf.section 1.1) Let Ω be a domain in $\mathbf{R}^n$ with a Hessian structure $(D, g = Dd\varphi)$. We call this domain a **Hessian domain**, and denote it by $(\Omega, D, g = Dd\varphi)$. Let us define a mapping ι from Ω into $\mathbf{R}_n^*$ by

$$\iota = -d\varphi,$$

that is,

$$x_i^* \circ \iota = -\frac{\partial \varphi}{\partial x^i}. \tag{2.4}$$

Then since the Jacobian matrix $\left[\dfrac{\partial^2 \varphi}{\partial x^i \partial x^j} \right]$ of ι is regular, we know that ι is an immersion from Ω into $\mathbf{R}_n^*$. The mapping ι is called the **gradient mapping** for the Hessian domain $(\Omega, D, g = Dd\varphi)$.

Theorem 2.2. *Let $(\Omega, D, g = Dd\varphi)$ be a Hessian domain in $\mathbf{R}^n$ and ι the gradient mapping. We define locally a flat affine connection D' on Ω by*

$$\iota_*(D_X' Y) = D_{\iota_* X}^* \iota_*(Y).$$

Then

(1) $D' = 2\nabla - D$, *where ∇ is the* Levi-Civita *connection for g.*
    Hence D' is a globally defined flat connection on Ω.
(2) *Let $x_i' = \partial \varphi / \partial x^i$. Then $\{x_1', \cdots, x_n'\}$ is an affine coordinate system with respect to D' and*

$$g\left(\frac{\partial}{\partial x_i'}, \frac{\partial}{\partial x^j} \right) = \delta_j^i, \quad g\left(\frac{\partial}{\partial x_i'}, \frac{\partial}{\partial x_j'} \right) = g^{ij},$$

   where δ_j^i is the Kronecker's *delta and $[g^{ij}] = [g_{ij}]^{-1}$.*
(3) $Xg(Y, Z) = g(D_X Y, Z) + g(Y, D_X' Z)$.
(4) *The pair (D', g) is a Hessian structure.*

Proof. Denoting by $\Gamma^i{}_{jk}$ the Christoffel symbols of the Levi-Civita connection ∇ for g, it follows from Proposition 2.2 that

$$\Gamma^i{}_{jk} = \gamma^i{}_{jk} = \frac{1}{2} \sum_r g^{ir} \frac{\partial g_{rj}}{\partial x^k}.$$

Since ι is locally bijective and

$$\iota_* \left(\frac{\partial}{\partial x^i} \right) = -\sum_j (g_{ij} \circ \iota^{-1}) \frac{\partial}{\partial x_j^*}, \quad \iota_*^{-1} \left(\frac{\partial}{\partial x_i^*} \right) = -\sum_j g^{ij} \frac{\partial}{\partial x^j},$$

we have

$$D'_{\partial/\partial x^i}\frac{\partial}{\partial x^j} = \iota_*^{-1}\left(D^*_{\iota_*(\partial/\partial x^i)}\iota_*\left(\frac{\partial}{\partial x^j}\right)\right)$$

$$= \iota_*^{-1}\left(D^*_{\sum_k(g_{ik}\circ\iota^{-1})\partial/\partial x_k^*}\sum_l(g_{jl}\circ\iota^{-1})\frac{\partial}{\partial x_l^*}\right)$$

$$= \iota_*^{-1}\left(\sum_{k,l}(g_{ik}\circ\iota^{-1})\frac{\partial(g_{jl}\circ\iota^{-1})}{\partial x_k^*}\frac{\partial}{\partial x_l^*}\right)$$

$$= \sum_{k,l,r,s}g_{ik}\frac{\partial g_{jl}}{\partial x^r}g^{rk}g^{sl}\frac{\partial}{\partial x^s} = 2\sum_s\Gamma_{ij}^s\frac{\partial}{\partial x^s}$$

$$= (2\nabla - D)_{\partial/\partial x^i}\frac{\partial}{\partial x^j}.$$

We have thus proven statement (1). It follows from the definition of D', and the relation $\iota_*\left(\dfrac{\partial}{\partial x_i'}\right) = -\dfrac{\partial}{\partial x_i^*}$, that $\{x_1',\cdots,x_n'\}$ is an affine coordinate system with respect to D', and

$$g\left(\frac{\partial}{\partial x_i'},\frac{\partial}{\partial x^j}\right) = g\left(\sum_p\frac{\partial x^p}{\partial x_i'}\frac{\partial}{\partial x^p},\frac{\partial}{\partial x^j}\right) = \sum_p g^{pi}g_{pj} = \delta_j^i,$$

$$g\left(\frac{\partial}{\partial x_i'},\frac{\partial}{\partial x_j'}\right) = g\left(\sum_p\frac{\partial x^p}{\partial x_i'}\frac{\partial}{\partial x^p},\sum_p\frac{\partial x^q}{\partial x_j'}\frac{\partial}{\partial x^q}\right) = \sum_{pq}g^{pi}g_{pq}g^{qj} = g^{ij}.$$

This completes the proof of assertion (2). In seeking to prove assertion (3), it is sufficient to consider the case when $X = \dfrac{\partial}{\partial x^i}$, $Y = \dfrac{\partial}{\partial x^j}$, and $Z = \dfrac{\partial}{\partial x^k}$. By assertion (1) and Proposition 2.2 we have

$$\frac{\partial}{\partial x^i}g\left(\frac{\partial}{\partial x^j},\frac{\partial}{\partial x^k}\right) = 2\gamma_{jik} = 2\sum_r g_{jr}\gamma_{ik}^r = g\left(\frac{\partial}{\partial x^j},2\nabla_{\partial/\partial x^i}\frac{\partial}{\partial x^k}\right)$$

$$= g\left(D_{\partial/\partial x^i}\frac{\partial}{\partial x^j},\frac{\partial}{\partial x^k}\right) + g\left(\frac{\partial}{\partial x^j},D'_{\partial/\partial x^i}\frac{\partial}{\partial x^k}\right).$$

Thus assertion (3) is also proved. Since

$$d\left(\sum_i x^i\,dx_i'\right) = \sum_i dx^i\wedge dx_i' = \sum_{i,j}\frac{\partial x_i'}{\partial x^j}dx^i\wedge dx^j = \sum_{i,j}g_{ij}dx^i\wedge dx^j = 0,$$

by Poincaré's Lemma there exists a local function ψ' such that

$$\sum_i x^i\,dx_i' = d\psi'.$$

Therefore

$$x^i = \frac{\partial\psi'}{\partial x_i'}, \qquad g^{ij} = \frac{\partial x^i}{\partial x_j'} = \frac{\partial^2\psi'}{\partial x_i'\partial x_j'}. \tag{2.5}$$

This shows that g is a Hessian metric with respect to D'. $\square$

Corollary 2.1. *Let (M, D, g) be a Hessian manifold and let ∇ be the Levi-Civita connection for g. We define a connection D' on M by*

$$D' = 2\nabla - D.$$

Then

(1) *D' is a flat connection.*
(2) *$Xg(Y, Z) = g(D_X Y, Z) + g(Y, D'_X Z)$.*
(3) *(D', g) is a Hessian structure.*

Proof. By Theorem 2.2, we know that (1)-(3) hold on any small local coordinate neighbourhood. Hence they hold on M. □

Definition 2.6. The flat connection D' given in Corollary 2.1 is said to be the **dual connection** of D with respect to g, and the pair (D', g) is called the **dual Hessian structure** of (D, g).

Let us study the relation between the potentials of a Hessian structure (D, g) and its dual Hessian structures (D', g). Let φ be a potential of (D, g). Using the same notation as in the proof of the above theorem, we have

$$
\begin{aligned}
\frac{\partial^2 \psi'}{\partial x'_i \partial x'_j} &= g^{ij} = \sum_{k,l} g_{kl} \frac{\partial x^k}{\partial x'_i} \frac{\partial x^l}{\partial x'_j} = \sum_{k,l} \frac{\partial^2 \varphi}{\partial x^k \partial x^l} \frac{\partial x^k}{\partial x'_i} \frac{\partial x^l}{\partial x'_j} \\
&= \sum_l \frac{\partial}{\partial x'_i}\left(\frac{\partial \varphi}{\partial x^l}\right) \frac{\partial x^l}{\partial x'_j} = \frac{\partial}{\partial x'_i}\left(\sum_l \frac{\partial \varphi}{\partial x^l} \frac{\partial x^l}{\partial x'_j}\right) - \sum_l \frac{\partial \varphi}{\partial x^l} \frac{\partial^2 x^l}{\partial x'_i \partial x'_j} \\
&= \frac{\partial^2 \varphi}{\partial x'_i \partial x'_j} - \sum_l x'_l \frac{\partial^2 x^l}{\partial x'_i \partial x'_j},
\end{aligned}
$$

$$
\frac{\partial^2}{\partial x'_i \partial x'_j}\left(\sum_l x'_l x^l\right) = \frac{\partial}{\partial x'_i}\left(x^j + \sum_l x'_l \frac{\partial x^l}{\partial x'_j}\right) = 2\frac{\partial^2 \psi'}{\partial x'_i \partial x'_j} + \sum_l x'_l \frac{\partial^2 x^l}{\partial x'_i \partial x'_j}.
$$

Hence

$$
\frac{\partial^2 \psi'}{\partial x'_i \partial x'_j} = \frac{\partial^2}{\partial x'_i \partial x'_j}\left(\sum_l x'_l x^l - \varphi\right).
$$

Thus

$$
\psi' = \sum_i x'_i x^i - \varphi + \sum_i a^i x'_i + a,
$$

where a_i and a are constants. Differentiating both sides by x'_i, we obtain $a^i = 0$. Hence

$$
\psi' = \sum_i x'_i x^i - \varphi + a.
$$

Therefore, a function φ' defined by

$$\varphi' = \sum_i x^i \frac{\partial \varphi}{\partial x^i} - \varphi, \tag{2.6}$$

is a potential of the dual Hessian structure (D', g). The function φ' given by (2.6) is called the **Legendre transform** of φ.

Definition 2.7. A Hessian domain $(\bar{\Omega}, D, \bar{g} = Dd\bar{\varphi})$ in $\mathbf{R}^n$ is said to be a **deformation** of a Hessian domain $(\Omega, D, g = Dd\varphi)$ in $\mathbf{R}^n$, if there exists a diffeomorphism $f : \bar{\Omega} \longrightarrow \Omega$ satisfying $\bar{\varphi} = \varphi \circ f$.

Example 2.6. Let $\tilde{D}$ be the standard flat connection on $\mathbf{R}^{n+1}$ and let $\tilde{\psi} = \frac{1}{2} \sum_i^{n+1} (x^i)^2$. We set $(\mathbf{R}^+)^{n+1} = \{(x^1, \cdots, x^{n+1}) \mid x^i > 0, \text{ for all } i\}$ and restrict the Hessian structure $(\tilde{D}, \tilde{D}d\tilde{\psi})$ to $(\mathbf{R}^+)^{n+1}$. We define a diffeomorphism $f : \mathbf{R}^{n+1} \longrightarrow (\mathbf{R}^+)^{n+1}$ by

$$f(x^1, \cdots, x^{n+1}) = \sqrt{2}(e^{-x^1/2}, \cdots, e^{-x^{n+1}/2}),$$

and set

$$\tilde{\varphi} = \tilde{\psi} \circ f = \sum_{i=1}^{n+1} e^{-x^i}.$$

Then the Hessian domain $(\mathbf{R}^{n+1}, \tilde{D}, \tilde{D}d\tilde{\varphi})$ is a deformation of the Hessian domain $((\mathbf{R}^+)^{n+1}, \tilde{D}, \tilde{D}d\tilde{\psi})$.

Let $(\Omega, D, g = Dd\varphi)$ be a Hessian domain. Assume that the gradient mapping ι for $(\Omega, D, g = Dd\varphi)$ is injective, that is, we suppose $\iota : \Omega \longrightarrow \mathbf{R}_n^*$ to be an imbedding. We will also term a function φ^* on $\Omega^* = \iota(\Omega)$ defined by

$$\varphi^* = \varphi' \circ \iota^{-1} = -\sum_i (x^i \circ \iota^{-1}) x_i^* - \varphi \circ \iota^{-1} \tag{2.7}$$

the Legendre transform of φ.

From equations (2.4), (2.5) and (2.7) it follows that

$$\frac{\partial \varphi^*}{\partial x_i^*} = \sum_p \left(\frac{\partial \varphi'}{\partial x_p'} \circ \iota^{-1} \right) \frac{\partial (x_p' \circ \iota^{-1})}{\partial x_i^*} = -x^i \circ \iota^{-1}.$$

Upon introducing $g^* = D^* d\varphi^*$, we have

$$g^*\left(\frac{\partial}{\partial x_i^*}, \frac{\partial}{\partial x_j^*} \right) = \frac{\partial^2 \varphi^*}{\partial x_i^* \partial x_j^*} = \frac{\partial}{\partial x_i^*}(-x^j \circ \iota^{-1})$$

$$= -\sum_p \left(\frac{\partial x^j}{\partial x_p'} \circ \iota^{-1} \right) \frac{\partial (x_p' \circ \iota^{-1})}{\partial x_i^*} = \frac{\partial x^j}{\partial x_i'} \circ \iota^{-1}$$

$$= g^{ij} \circ \iota^{-1}.$$

Hence $(D^*, g^* = D^* d\varphi^*)$ is a Hessian structure on Ω^*. Using this result, together with the proof of Theorem 2.2, we obtain

$$g^*\left(\iota_*\left(\frac{\partial}{\partial x^i}\right), \iota_*\left(\frac{\partial}{\partial x^j}\right)\right) = g^*\left(\sum_p (g_{ip} \circ \iota^{-1})\frac{\partial}{\partial x_p^*}, \sum_q (g_{jq} \circ \iota^{-1})\frac{\partial}{\partial x_q^*}\right)$$

$$= \sum_{p,q} (g_{ip} g_{jq} g^{pq}) \circ \iota^{-1} = g\left(\frac{\partial}{\partial x^i}, \frac{\partial}{\partial x^j}\right) \circ \iota^{-1}.$$

This implies that $\iota : (\Omega, g) \longrightarrow (\Omega^*, g^*)$ is an isometry. Identifying the dual space of $\mathbf{R}_n^*$ with $\mathbf{R}^n$, the gradient mapping ι^* for $(\Omega^*, D^*, g^* = D^* d\varphi^*)$ satisfies

$$x^i \circ \iota^* = -\frac{\partial \varphi^*}{\partial x_i^*} = x^i \circ \iota^{-1},$$

so

$$\iota^* = \iota^{-1}.$$

Denoting by $(\varphi^*)^*$ the Legendre transform of φ^* we have

$$(\varphi^*)^* = -\sum_i x^i (x_i^* \circ \iota^{*-1}) - \varphi^* \circ \iota^{*-1}$$

$$= -\sum_i x^i (x_i^* \circ \iota) - \left\{-\sum_i (x_i^* \circ \iota)x^i - \varphi\right\}$$

$$= \varphi.$$

The following proposition summarizes these results.

Proposition 2.7. *Suppose that the gradient mapping ι for a Hessian domain $(\Omega, D, g = Dd\varphi)$ is an imbedding. Then $(\Omega^*, D^*, g^* = D^* d\varphi^*)$ is a Hessian domain. We denote by ι^* the gradient mapping for $(\Omega^*, D^*, g^* = D^* d\varphi^*)$ and identify the dual space of $\mathbf{R}_n^*$ with $\mathbf{R}^n$. We then have*

(1) $\iota^* = \iota^{-1} : \Omega^* \longrightarrow \Omega$.
(2) $\iota : (\Omega, g) \longrightarrow (\Omega^*, g^*)$ *is an isometry.*
(3) *The Legendre transform $(\varphi^*)^*$ of φ^* coincides with φ ;* $(\varphi^*)^* = \varphi D$

The Hessian domain (Ω^*, D^*, g^*) is said to be the **dual Hessian domain** of (Ω, D, g).

Example 2.7. Let $(\mathbf{R}^{n+1}, \tilde{D}, \tilde{g} = \tilde{D}d\tilde{\varphi})$ be the Hessian domain given in Example 2.6. We denote by $\tilde{D}^*$ the standard flat connection on $\mathbf{R}_{n+1}^*$ and set

$$(\mathbf{R}^{*+})_{n+1} = \{(x_1^*, \cdots, x_{n+1}^*) \in \mathbf{R}_{n+1}^* \mid x_i^* > 0, \ \text{for all } i \}.$$

Since the gradient mapping ι satisfies $x_i^* \circ \iota = e^{-x^i}$, we have

$$\iota(\mathbf{R}^{n+1}) = (\mathbf{R}^{*+})_{n+1}.$$

The Legendre transform $\tilde{\varphi}'$ of $\tilde{\varphi}$ is reduced to

$$\tilde{\varphi}' = -\sum(x^i + 1)e^{-x^i},$$

and so

$$\tilde{\varphi}^* = \tilde{\varphi}' \circ \iota^{-1} = \sum_{i=1}^{n+1}(x_i^* \log x_i^* - x_i^*).$$

Thus the dual Hessian domain of $(\mathbf{R}^{n+1}, \tilde{D}, \tilde{g} = \tilde{D}d\tilde{\varphi})$ is given by

$$((\mathbf{R}^{*+})_{n+1}, \tilde{D}^*, \tilde{g}^* = \tilde{D}^*d\tilde{\varphi}^*).$$

Example 2.8. We use the same notation as in Example 2.7. Upon introducing

$$\Delta_n^* = \left\{(x_1^*, \cdots, x_{n+1}^*) \in (\mathbf{R}^{*+})_{n+1} \mid \sum_{i=1}^{n+1} x_i^* = 1\right\},$$

we have that Δ_n^* is a flat manifold with an affine coordinate system $\{x_1^*, \cdots, x_n^*\}$. We denote by D^* the flat connection on Δ_n^*, and by $\varphi^* = \sum_{i=1}^{n+1} x_i^* \log x_i^* - 1$ the restriction of $\tilde{\varphi}^*$ to Δ_n^*. We then have

$$\frac{\partial \varphi^*}{\partial x_i^*} = \log \frac{x_i^*}{x_{n+1}^*}, \qquad \frac{\partial^2 \varphi^*}{\partial x_i^* \partial x_j^*} = \delta^{ij}\frac{1}{x_i^*} + \frac{1}{x_{n+1}^*}.$$

Hence $\left[\dfrac{\partial^2 \varphi^*}{\partial x_i^* \partial x_j^*}\right]$ is positive definite on Δ_n^*, and so $(D^*, g^* = D^*d\varphi^*)$ is a Hessian structure on Δ_n^*. The Legendre transform of φ^* is given by

$$\varphi = \sum_{i=1}^{n} x_i^* \frac{\partial \varphi^*}{\partial x_i^*} - \varphi^* = -\log x_{n+1}^* + 1 = \log\left(\sum_{i=1}^{n} e^{x^i} + 1\right) + 1, \text{ and so}$$

$(\mathbf{R}^n, D, g = Dd\varphi)$ is the dual Hessian domain of $(\Delta_n^*, D^*, g^* = D^*d\varphi^*)$ (cf. Example 2.2 (4) and Proposition 3.9). The Hessian structure $(D, g = Dd\varphi)$ coincides with the Hessian structure on the multinomial distributions (cf. Example 6.2).

Example 2.9. Let $\Omega = \{x \in \mathbf{R}^n \mid f(x) > 0\}$ and $\varphi(x) = -\log f(x)$, where $f(x) = x^n - \dfrac{1}{2}\sum_{i=1}^{n-1}(x^i)^2$. Then $(\Omega, D, g = Dd\varphi)$ is a Hessian domain (see

Example 2.2 (3)). Since $\dfrac{\partial \varphi}{\partial x^i} = \dfrac{x^i}{f}$ $(i \le n-1)$ and $\dfrac{\partial \varphi}{\partial x^n} = -\dfrac{1}{f}$, the gradient mapping ι satisfies $-\infty < x_i^* \circ \iota < \infty$ $(i \le n-1)$ and $x_n^* \circ \iota > 0$.

Hence $x^i \circ \iota^{-1} = -\dfrac{x_i^*}{x_n^*}$ $(i \le n-1)$ and $x^n \circ \iota^{-1} = \dfrac{1}{x_n^*} + \dfrac{1}{2}\displaystyle\sum_{i=1}^{n-1}\left(\dfrac{x_i^*}{x_n^*}\right)^2$ D

The Legendre transform φ' of φ is given by $\varphi' = \dfrac{x^n}{f} + \log f - 2$D Thus

$\varphi' \circ \iota^{-1} = \dfrac{x^n}{f} \circ \iota^{-1} + \log(f \circ \iota^{-1}) - 2 = \dfrac{1}{2}\displaystyle\sum_{i=1}^{n-1}\dfrac{(x_i^*)^2}{x_n^*} - \log x_n^* - 1$. The

dual Hessian domain is therefore given by $(\Omega^*, D^*, g^* = D^* d\varphi^*)$, where $\Omega^* = \{(x_1^*, \cdots, x_n^*) \in \mathbf{R}_n^* \mid -\infty < x_i^* < \infty$ for $i \le n-1$, $x_n^* > 0\}$ and

$\varphi^* = \dfrac{1}{2}\displaystyle\sum_{i=1}^{n-1}\dfrac{(x_i^*)^2}{x_n^*} - \log x_n^*$.

Example 2.10. Let $\Omega = \{x \in \mathbf{R}^n \mid f(x) > 0,\ x^n > 0\}$ and $\varphi(x) = -\log f(x)$, where $f(x) = (x^n)^2 - \displaystyle\sum_{i=1}^{n-1}(x^i)^2$. Then $(\Omega, D, g = Dd\varphi)$ is a Hessian domain (see Example 2.2 (5)). Since $\dfrac{\partial \varphi}{\partial x^i} = \dfrac{2x^i}{f}$ $(i \le n-1)$ and $\dfrac{\partial \varphi}{\partial x^n} = -\dfrac{2x^n}{f}$, we have $f^* \circ \iota = \dfrac{4}{f}$, $x^i \circ \iota^{-1} = -\dfrac{2x_i^*}{f^*}$ for $i \le n-1$ and $x^n \circ \iota^{-1} = \dfrac{2x_n^*}{f^*}$, where $f^*(x^*) = (x_n^*)^2 - \displaystyle\sum_{i=1}^{n-1}(x_i^*)^2$. The Legendre transform φ' of φ is given by $\varphi' = -\varphi - 2$. Hence $\varphi' \circ \iota^{-1} = \log(f \circ \iota^{-1}) - 2 = -\log f^* + \log 4 - 2$D This implies the dual Hessian domain is $(\Omega^*, D^*, g^* = D^* d\varphi^*)$, where $\Omega^* = \{x^* \in \mathbf{R}_n^* \mid f^*(x^*) > 0,\ x_n^* > 0\}$ and $\varphi^*(x^*) = -\log f^*(x^*)$.

2.4 Divergences for Hessian structures

In this section we define the divergence for a Hessian structure which plays an important role in statistics.

Let $(\Omega, D, g = Dd\varphi)$ be a Hessian domain in $\mathbf{R}^n$. For $p \in \Omega$ we define a function φ_p on Ω by

$$\varphi_p(x) = \sum_i (x^i - x^i(p))\dfrac{\partial \varphi}{\partial x^i}(x) - (\varphi(x) - \varphi(p)).$$

Then

$$\varphi_p(p) = 0, \qquad \frac{\partial \varphi_p}{\partial x^i}(p) = 0, \qquad \left[\frac{\partial^2 \varphi_p}{\partial x^i \partial x^j}(p)\right] = [g_{ij}(p)] > 0.$$

This implies that $\varphi_p(x)$ attains a unique relative minimum $\varphi_p(p) = 0$ at p. Therefore, defining a mapping $\mathcal{D}$ by

$$\mathcal{D} : (p, q) \in \Omega \times \Omega \longrightarrow \varphi_p(q) \in \mathbf{R},$$

we have

$$\mathcal{D}(p, q) \geq 0,$$
$$\mathcal{D}(p, q) = 0 \iff p = q,$$

where q is a point in a small neighbourhood of p. Using the Legendre transform φ' of φ, we obtain

$$\mathcal{D}(p, q) = \varphi(p) + \varphi'(q) - \sum_i x^i(p) x_i'(q),$$

where $x_i' = \dfrac{\partial \varphi}{\partial x^i}$. The mapping $\mathcal{D}$ is called the **divergence** for the Hessian structure $(D, g = Dd\varphi)$ [Amari and Nagaoka (2000)].

By making use of the Taylor expansion of $\varphi_p(x)$ at p, we have

$$\varphi_p(x) = \frac{1}{2} \sum_{ij} g_{ij}(p)(x^i - x^i(p))(x^j - x^j(p))$$

$$+ \frac{2}{3!} \sum_{ijk} \frac{\partial g_{ij}}{\partial x^k}(p)(x^i - x^i(p))(x^j - x^j(p))(x^k - x^k(p))$$

$$+ \frac{3}{4!} \sum_{ijkl} \frac{\partial^2 g_{ij}}{\partial x^k \partial x^l}(p)(x^i - x^i(p))(x^j - x^j(p))(x^k - x^k(p))(x^l - x^l(p))$$

$$+ \cdots .$$

Since $\dfrac{\partial g_{ij}}{\partial x^k}(p) = (Dg)_{ijk}(p)$, $\dfrac{\partial^2 g_{ij}}{\partial x^k \partial x^l}(p) = (D^2 g)_{ijkl}(p), \cdots$, we know that the definition of the divergence $\mathcal{D}$ is independent of the choice of an affine coordinate system $\{x^1, \cdots, x^n\}$ and a potential φ, and depends only on the Hessian structure (D, g). If the points p and q are sufficiently close, then $\mathcal{D}(p, q)$ is approximated by

$$\mathcal{D}(p, q) \doteqdot \frac{1}{2} \sum_{ij} g_{ij}(p)(x^i(q) - x^i(p))(x^j(q) - x^j(p)).$$

Let $\mathcal{D}'(p, q)$ be the divergence for the dual Hessian structure $(D', g = D'd\varphi')$ of the Hessian structure $(D, g = Dd\varphi)$. Then

$$\mathcal{D}'(p, q) = \mathcal{D}(q, p).$$

In fact, it follows from $\varphi' = \sum_i x^i x_i' - \varphi$ and $\dfrac{\partial \varphi'}{\partial x_i'} = x^i$ that

$$\mathcal{D}'(p,q) = \sum_i (x_i'(q) - x_i'(p))\frac{\partial \varphi'}{\partial x_i'}(q) - (\varphi'(q) - \varphi'(p))$$

$$= \sum_i (x^i(p) - x^i(q))x_i'(p) - (\varphi(p) - \varphi(q))$$

$$= \mathcal{D}(q,p).$$

Example 2.11. Let $(\Delta_n^*, D^*, g^* = D^* d\varphi^*)$ be the Hessian domain of Example 2.8, and let $\mathcal{D}^*$ be the divergence for the Hessian structure. Upon substituting for $\varphi^* = \sum_{i=1}^{n+1} x_i^* \log x_i^* - 1$ and $\dfrac{\partial \varphi^*}{\partial x_i^*} = \log \dfrac{x_i^*}{x_{n+1}^*}$ into the expression $\mathcal{D}^*(p,q) = \sum_{i=1}^{n} (x_i^*(q) - x_i^*(p))\dfrac{\partial \varphi^*}{\partial x_i^*}(q) - \varphi^*(q) + \varphi^*(p)$ we have

$$\mathcal{D}^*(p,q) = \sum_{i=1}^{n+1} x_i^*(p) \log \frac{x_i^*(p)}{x_i^*(q)}.$$

Example 2.12. Let $\left(\mathbf{R}^n, D, g = Dd\left(\dfrac{1}{2}\sum_i (x^i)^2\right)\right)$ be the Euclidean space. The divergence is then given by

$$\mathcal{D}(p,q) = \frac{1}{2}\sum_i (x^i(p) - x^i(q))^2,$$

that is, $\mathcal{D}(p,q)$ is half of the square of the Euclidean distance between p and q.

In view of the above Example 2.12, the following theorem is regarded as an extension of Pythagoras's Theorem in Euclidean space [Amari and Nagaoka (2000)].

Theorem 2.3. *Let $(\Omega, D, g = Dd\varphi)$ be a Hessian domain and let p, q and $r \in \Omega$. If σ, the geodesic for D connecting q and p, and σ', the geodesic for the dual connection D' of D connecting q and r, are orthogonal at q, then we have*

$$\mathcal{D}(p,r) = \mathcal{D}(p,q) + \mathcal{D}(q,r).$$

Proof. Using an affine coordinate system $\{x^1, \cdots, x^n\}$ with respect to D, we may suppose

$$x^i(\sigma(t)) = x^i(q) + (x^i(p) - x^i(q))t.$$

In the same way we have

$$x'_j(\sigma'(t)) = x'_j(q) + (x'_j(r) - x'_j(q))t,$$

where $\{x'_1 = \partial\varphi/\partial x^1, \cdots, x'_n = \partial\varphi/\partial x^n\}$ is an affine coordinate system with respect to D'. By the definition of the divergence and $g\left(\dfrac{\partial}{\partial x^i}, \dfrac{\partial}{\partial x'_j}\right) = \delta^j_i$ we obtain

$$
\begin{aligned}
\mathcal{D}(p,r) &- \mathcal{D}(p,q) - \mathcal{D}(q,r) \\
&= \sum_i (x^i(r) - x^i(p))x'_i(r) - (\varphi(r) - \varphi(p)) \\
&\quad - \Big\{ \sum_i (x^i(q) - x^i(p))x'_i(q) - (\varphi(q) - \varphi(p)) \Big\} \\
&\quad - \Big\{ \sum_i (x^i(r) - x^i(q))x'_i(r) - (\varphi(r) - \varphi(q)) \Big\} \\
&= - \sum_i (x^i(p) - x^i(q))(x'_i(r) - x'_i(q)) \\
&= -g_q(\dot\sigma(0), \dot\sigma'(0)) = 0.
\end{aligned}
$$

$\square$

2.5   Codazzi structures

We first assert that the Codazzi equation for a pair of a flat connection and a Riemannian metric is restated as follows.

Lemma 2.3. *Let D be a torsion-free connection and let g be a Riemannian metric. Let us define a new connection D' by*

$$Xg(Y, Z) = g(D_X Y, Z) + g(Y, D'_X Z).$$

Then the following conditions (1)-(3) are equivalent.

(1) *The connection D' is torsion-free.*
(2) *The pair (D, g) satisfies the Codazzi equation,*

$$(D_X g)(Y, Z) = (D_Y g)(X, Z).$$

(3) *Let ∇ be the Levi-Civita connection for g, and let $\gamma_X Y = \nabla_X Y - D_X Y$. Then g and γ satisfy*

$$g(\gamma_X Y, Z) = g(Y, \gamma_X Z).$$

If the pair (D, g) satisfies the Codazzi *equation, then the pair (D', g) also satisfies this equation and*

$$D' = 2\nabla - D,$$

$$(D_X g)(Y, Z) = 2g(\gamma_X Y, Z).$$

Proof. By the definition of D', it follows that

$$(D_X g)(Y, Z) = Xg(Y, Z) - g(D_X Y, Z) - g(Y, D_X Z)$$
$$= g(D'_X Y - D_X Y, Z).$$

Hence

$$(D_X g)(Y, Z) - (D_Y g)(X, Z) = g(D'_X Y - D'_Y X - [X, Y], Z).$$

This implies that (1) and (2) are equivalent. Since $(D_X g)(Y, Z) = g(\gamma_X Y, Z) + g(Y, \gamma_X Z)$ and $\gamma_X Y = \gamma_Y X$, we obtain

$$(D_X g)(Y, Z) - (D_Y g)(X, Z) = g(\gamma_Z X, Y) - g(X, \gamma_Z Y).$$

Hence (2) and (3) are also equivalent. If (D, g) satisfies the Codazzi equation, then it follows from (3) that

$$g(D'_X Y, Z) = Xg(Y, Z) - g(Y, D_X Z)$$
$$= g(\nabla_X Y, Z) + g(Y, \gamma_X Z)$$
$$= g((\nabla_X + \gamma_X)Y, Z).$$

This shows that $D' = 2\nabla - D$. Hence $(D'_X g)(Y, Z) = -(D_X g)(Y, Z)$, which implies that (D', g) satisfies the Codazzi equation. Furthermore, we have $(D_X g)(Y, Z) = g(D'_X Y - D_X Y, Z) = 2g(\gamma_X Y, Z)$. $\square$

Proposition 2.1 asserts that a pair (D, g) of a flat connection D and a Riemannian metric g on M is a Hessian structure if and only if it satisfies the Codazzi equation. In view of this fact, the notion of Hessian structures was generalized by [Delanoe (1989)] as follows.

Definition 2.8. A pair (D, g) of a torsion-free connection D and a Riemannian metric g on M is called a **Codazzi structure** if it satisfies Codazzi equation,

$$(D_X g)(Y, Z) = (D_Y g)(X, Z).$$

A manifold M equipped with a Codazzi structure (D, g) is said to be a **Codazzi manifold**, and is denoted by (M, D, g).

For a Codazzi structure (D, g) the connection D' defined by

$$Xg(Y, Z) = g(D_X Y, Z) + g(Y, D'_X Z)$$

is called the **dual connection** of D with respect to g, and the pair (D', g) is said to be the **dual Codazzi structure** of (D, g).

Let M be a manifold with a torsion-free connection D and let TM be the tangent bundle over M with canonical projection $\pi : TM \longrightarrow M$. For a local coordinate system $\{x^1, \cdots, x^n\}$ on M we set $q^i = x^i \circ \pi$ and $r^i = dx^i$. Then $\{q^1, \cdots, q^n, r^1, \cdots, r^n\}$ forms a local coordinate system on TM. A tangent vector

$$A = \sum_i a^i \left(\frac{\partial}{\partial q^i}\right)_z + \sum_i a^{n+i} \left(\frac{\partial}{\partial r^i}\right)_z$$

at $z = \displaystyle\sum_i z^i \left(\frac{\partial}{\partial x^i}\right)_{\pi(z)} \in TM$ is said to be a **horizontal vector** if it satisfies the following conditions,

$$a^{n+i} + \sum_{j,k} \Gamma^i_{jk}(\pi(z)) a^j z^k = 0,$$

where Γ^i_{jk} are the Christoffel symbols for D [Dom]. The set of all horizontal vectors $\mathcal{H}_z$ at $z \in TM$ is called the horizontal subspace at z. Then we have

$$T_z TM = \mathrm{Ker}\pi_{*z} \oplus \mathcal{H}_z, \qquad \dim \mathcal{H}_z = n.$$

Let T^*M represent the cotangent vector bundles over M with canonical projection $\pi^* : T^*M \longrightarrow M$. For a local coordinate system $\{x^1, \cdots, x^n\}$ on M we set $q^i = x^i \circ \pi^*$ and $p_i = \dfrac{\partial}{\partial x^i}$. Then the $2n$-tuple $\{q^1, \cdots, q^n, p_1, \cdots, p_n\}$ forms a local coordinate system on T^*M. The canonical symplectic form on T^*M is given by

$$\rho = \sum_{i=1}^n dq^i \wedge dp_i.$$

A Riemannian metric g on M induces a mapping

$$g : z \in TM \longrightarrow g(z, \ ) \in T^*M.$$

We denote by g_{*z} the differential of the mapping at z. Then we have the following theorem due to [Delanoe (1989)].

Theorem 2.4. *Let D be a torsion-free connection and let g be a Riemannian metric on M. Then the following conditions are equivalent.*

(1) *(D, g) is a Codazzi structure.*
(2) *$g_*(\mathcal{H})$ is a Lagrangian subspace with respect to the canonical symplectic form ρ on T^*M.*

Proof. The mapping $g : TM \longrightarrow T^*M$ is expressed by

$$g : (q^1, \cdots, q^n, r^1, \cdots, r^n) \longrightarrow (q^1, \cdots, q^n, \sum_k g_{1k} r^k, \cdots, \sum_k g_{nk} r^k),$$

and so

$$g_* \left(\frac{\partial}{\partial q^i} \right) = \frac{\partial}{\partial q^i} + \sum_{j,k} \frac{\partial g_{jk}}{\partial q^i} r^k \frac{\partial}{\partial p_j},$$

$$g_* \left(\frac{\partial}{\partial r^i} \right) = \sum_j g_{ij} \frac{\partial}{\partial p_j}.$$

Since $\left\{ A_i = \dfrac{\partial}{\partial q^i} - \sum_{k,l} \Gamma_{ik}^l r^k \dfrac{\partial}{\partial r^l} \;\middle|\; 1 \le i \le n \right\}$ is a basis of the horizon-

tal subspace $\mathcal{H}$, $\{ A_i^* = g_*(A_i) \mid \le i \le n \}$ forms a basis of $g_*(\mathcal{H})$. Upon introducing $D_i = D_{\partial/\partial x^i}$, we have

$$A_i^* = g_* \left(\frac{\partial}{\partial q^i} \right) - \sum_{k,l} \Gamma_{ik}^l r^k g_* \left(\frac{\partial}{\partial r^l} \right)$$

$$= \frac{\partial}{\partial q^i} + \sum_{j,k} \frac{\partial g_{jk}}{\partial q^i} r^k \frac{\partial}{\partial p_j} - \sum_{j,k,l} \Gamma_{ik}^l r^k g_{lj} \frac{\partial}{\partial p_j}$$

$$= \frac{\partial}{\partial q^i} + \sum_{j,k} \left(D_i g_{jk} + \sum_l \Gamma_{ij}^l g_{lk} \right) r^k \frac{\partial}{\partial p_j}.$$

Hence $\left\{ A_1^*, \cdots, A_n^*, \dfrac{\partial}{\partial p_1}, \cdots, \dfrac{\partial}{\partial p_n} \right\}$ is a basis of each tangent space of T^*M. Note that ρ vanishes on the space spanned by $\left\{ \dfrac{\partial}{\partial p_1}, \cdots, \dfrac{\partial}{\partial p_n} \right\}$. Since

$$dq^s(A_i^*) = \delta_i^s,$$

$$dp_s(A_i^*) = \left(D_i g_{sk} + \sum_l \Gamma_{is}^l g_{lk} \right) r^k,$$

we have

$$\rho(A_i^*, A_j^*) = \sum_s (dq^s \wedge dp_s)(A_i^*, A_j^*)$$

$$= \sum_s \left\{ dq^s(A_i^*) dp_s(A_j^*) - dq^s(A_j^*) dp_s(A_i^*) \right\}$$

$$= \sum_{k,s} \left\{ \delta_i^s \left(D_j g_{sk} + \sum_l \Gamma_{js}^l g_{lk} \right) - \delta_j^s \left(D_i g_{sk} + \sum_l \Gamma_{is}^l g_{lk} \right) \right\} r^k$$

$$= \sum_k \left(D_j g_{ik} - D_i g_{jk} \right) r^k.$$

Therefore ρ vanishes on $g_*(\mathcal{H})$ if and only if

$$D_j g_{ik} - D_i g_{jk} = 0.$$

Hence (D, g) is a Codazzi structure if and only if $g_*(\mathcal{H})$ is a Lagrangian subspace with respect to ρ. $\qquad\square$

Definition 2.9. A Codazzi structure (D, g) is said to be of a **constant curvature** c if the curvature tensor R_D of D satisfies

$$R_D(X, Y)Z = c\{g(Y, Z)X - g(X, Z)Y\},$$

where c is a constant real number [Kurose (1990)].

Proposition 2.8. *Let (D, g) be a Codazzi structure and let (D', g) be the dual Codazzi structure.*

(1) *Denoting by R_D and $R_{D'}$ the curvature tensors of D and D' respectively we have*

$$g(R_D(X, Y)Z, W) + g(Z, R_{D'}(X, Y)W) = 0.$$

(2) *If (D, g) is a Codazzi structure of constant curvature c, then (D', g) is also a Codazzi structure of constant curvature c.*

The proof is straightforward, and so we do not present it here.

By Proposition 2.1 we have the following proposition.

Proposition 2.9. *A Codazzi structure (D, g) is of constant curvature 0 if and only if (D, g) is a Hessian structure.*

The above Proposition 2.9 is generalized as follows.

Proposition 2.10. *Let (D, g) be a Codazzi structure of constant curvature. Then g is locally expressed by*

$$g = Dd\varphi + \frac{\varphi}{n-1}\mathrm{Ric}_D, \qquad n = \dim M,$$

where Ric_D is the Ricci tensor of D and φ is a local function (cf. [Nomizu and Simon (1992)]).

Proof. For the proof of this proposition the reader may refer to the above literature. $\qquad\square$

Chapter 3

Curvatures for Hessian structures

In section **2.1** we introduced the difference tensor $\gamma = \nabla - D$ on a Hessian manifold (M, D, g). The covariant differential $Q = D\gamma$ of γ is called the Hessian curvature tensor for (D, g). It reflects the properties of the Hessian structure (D, g), and performs a variety of important roles. The Hessian curvature tensor is analogous to the Riemannian curvature tensor for a Kählerian metric. Indeed, denoting by g^T the Kählerian metric on the tangent bundle TM over M induced by g, the Riemannian curvature tensor for g^T may be related to the Hessian curvature tensor (Proposition 3.3). Using D and the volume element of g, we define the second Koszul form β, which is related to the Ricci tensor for g^T (Proposition 3.5). This suggests that the second Koszul form plays an important role similar to the Ricci tensor for a Kählerian metric. We also define the Hessian sectional curvature, which is similar to the holomorphic sectional curvature for a Kählerian metric, and construct Hessian manifolds of constant Hessian sectional curvature.

3.1 Hessian curvature tensors and Koszul forms

In this section we define the Hessian curvature tensor and the Koszul forms α and β for a Hessian structure (D, g). These tensors play important roles in Hessian geometry similar to that of the Riemannian curvature tensor and the Ricci tensor in Kählerian geometry.

Definition 3.1. Let (D, g) be a Hessian structure and let $\gamma = \nabla - D$ be the difference tensor between the Levi-Civita connection ∇ for g and D. A tensor field Q of type (1,3) defined by the covariant differential

$$Q = D\gamma$$

of γ is said to be the **Hessian curvature tensor** for (D, g). The components $Q^i{}_{jkl}$ of Q with respect to an affine coordinate system $\{x^1, \cdots, x^n\}$ are given by

$$Q^i{}_{jkl} = \frac{\partial \gamma^i{}_{jl}}{\partial x^k}.$$

Proposition 3.1. Let $g_{ij} = \dfrac{\partial^2 \varphi}{\partial x^i \partial x^j}$. Then we have

(1) $Q_{ijkl} = \dfrac{1}{2} \dfrac{\partial^4 \varphi}{\partial x^i \partial x^j \partial x^k \partial x^l} - \dfrac{1}{2} g^{rs} \dfrac{\partial^3 \varphi}{\partial x^i \partial x^k \partial x^r} \dfrac{\partial^3 \varphi}{\partial x^j \partial x^l \partial x^s}.$

(2) $Q_{ijkl} = Q_{kjil} = Q_{klij} = Q_{ilkj} = Q_{jilk}.$

Proof. By Proposition 2.2, we obtain

$$\begin{aligned}
Q_{ijkl} &= g_{ir} \frac{\partial \gamma^r{}_{jl}}{\partial x^k} = \frac{\partial (g_{ir} \gamma^r{}_{jl})}{\partial x^k} - \frac{\partial g_{ir}}{\partial x^k} \gamma^r{}_{jl} \\
&= \frac{\partial \gamma_{ijl}}{\partial x^k} - \frac{1}{2} g^{rs} \frac{\partial g_{ir}}{\partial x^k} \frac{\partial g_{sj}}{\partial x^l} \\
&= \frac{1}{2} \frac{\partial^4 \varphi}{\partial x^i \partial x^j \partial x^k \partial x^l} - \frac{1}{2} g^{rs} \frac{\partial^3 \varphi}{\partial x^i \partial x^k \partial x^r} \frac{\partial^3 \varphi}{\partial x^j \partial x^l \partial x^s},
\end{aligned}$$

which proves (1). Assertion (2) follows directly from (1). □

Proposition 3.2. Let R be the Riemannian curvature tensor for g. Then

$$R_{ijkl} = \frac{1}{2}(Q_{ijkl} - Q_{jikl}).$$

Proof. From Propositions 2.2, 2.3 and 3.1 it follows that

$$Q_{ijkl} = \frac{1}{2} \frac{\partial^4 \varphi}{\partial x^i \partial x^j \partial x^k \partial x^l} - 2g^{rs} \gamma_{ikr} \gamma_{jls},$$

whereby

$$\begin{aligned}
Q_{ijkl} - Q_{jikl} &= -2g^{rs} \gamma_{ikr} \gamma_{jls} + 2g^{rs} \gamma_{jkr} \gamma_{ils} \\
&= 2(\gamma_{jkr} \gamma^r{}_{il} - \gamma_{jlr} \gamma^r{}_{ik}) \\
&= 2R_{ijkl}.
\end{aligned}$$

□

From the above proposition it may be seen that the Hessian curvature tensor Q carries more detailed information than the Riemannian curvature tensor R.

The following Proposition 3.3 suggests that the Hessian curvature tensor plays a similar role to that of the Riemannian curvature tensor in Kählerian geometry. In Proposition 2.6 we proved that a Hessian structure (D, g) on M induces a Kählerian structure (J, g^T) on the tangent bundle TM. With the same notation as section 2.2, we have the following proposition.

Proposition 3.3. *Let R^T be the Riemannian curvature tensor on the Kählerian manifold (TM, J, g^T). Then we have*

$$R^T_{i\bar{j}k\bar{l}} = \frac{1}{2}Q_{ijkl} \circ \pi.$$

Proof. We introduce $F = 4\varphi \circ \pi$, and by expression (2.3) we obtain

$$g^T_{i\bar{j}} = \frac{\partial^2 F}{\partial z^i \partial \bar{z}^j}.$$

Making use of a formula in [Kobayashi and Nomizu (1963, 1969)](II,p157), we have

$$
\begin{aligned}
R^T_{i\bar{j}k\bar{l}} &= \frac{\partial^2 g^T_{i\bar{j}}}{\partial z^k \partial \bar{z}^l} - \sum_{p,q} g^{T\bar{p}q}\frac{\partial g^T_{i\bar{p}}}{\partial z^k}\frac{\partial g^T_{q\bar{j}}}{\partial \bar{z}^l} \\
&= \frac{\partial^4 F}{\partial z^i \partial \bar{z}^j \partial z^k \partial \bar{z}^l} - \sum_{p,q}(g^{pq} \circ \pi)\frac{\partial^3 F}{\partial z^k \partial z^i \partial \bar{z}^p}\frac{\partial^3 F}{\partial \bar{z}^j \partial \bar{z}^l \partial z^q} \\
&= \frac{1}{4}\left\{\frac{\partial^4 \varphi}{\partial x^i \partial x^j \partial x^k \partial x^l} - \sum_{p,q}g^{pq}\frac{\partial^3 \varphi}{\partial x^i \partial x^k \partial x^p}\frac{\partial^3 \varphi}{\partial x^j \partial x^l \partial x^q}\right\} \circ \pi \\
&= \frac{1}{2}Q_{ijkl} \circ \pi.
\end{aligned}
$$

$\square$

Definition 3.2. Let v be the volume element of g. We define a closed 1-form α and a symmetric bilinear form β by

$$D_X v = \alpha(X)v,$$
$$\beta = D\alpha.$$

The forms α and β are called **the first Koszul form** and **the second Koszul form** for a Hessian structure (D, g) respectively.

Proposition 3.4. *We have*

(1) $\alpha(X) = \text{Tr } \gamma_X.$

(2) $\alpha_i = \dfrac{1}{2}\dfrac{\partial \log \det[g_{kl}]}{\partial x^i} = \gamma^r{}_{ri}.$

(3) $\beta_{ij} = \dfrac{\partial \alpha_i}{\partial x^j} = \dfrac{1}{2}\dfrac{\partial^2 \log \det[g_{kl}]}{\partial x^i \partial x^j} = Q^r{}_{rij} = Q_{ij}{}^r{}_r.$

Proof.   Since $v = (\det[g_{ij}])^{\frac{1}{2}} dx^1 \wedge \cdots \wedge dx^n$ and $\gamma_{\partial/\partial x^i} dx^k = -\sum_l \gamma^k_{il} dx^l$

we have

$$
\begin{aligned}
\alpha_i v &= \alpha\left(\frac{\partial}{\partial x^i}\right) v = D_{\partial/\partial x^i} v \\
&= \frac{\partial}{\partial x^i} (\det[g_{ij}])^{\frac{1}{2}} dx^1 \wedge \cdots \wedge dx^n \\
&= \frac{\partial}{\partial x^i} \log(\det[g_{ij}])^{\frac{1}{2}} v,
\end{aligned}
$$

$$
\begin{aligned}
\alpha_i v &= D_{\partial/\partial x^i} v = (D - \nabla)_{\partial/\partial x^i} v = -\gamma_{\partial/\partial x^i} v \\
&= -(\det[g_{ij}])^{\frac{1}{2}} \sum_k dx^1 \wedge \cdots \wedge \gamma_{\partial/\partial x^i} dx^k \wedge \cdots \wedge dx^n \\
&= \gamma^k_{ik} v.
\end{aligned}
$$

These imply (1) and (2). Assertion (3) follows from (2) and Definition 3.1. $\square$

Lemma 3.1. *Let α and β be the* Koszul *forms for a Hessian structure* (D, g), *and let α' and β' be the* Koszul *forms for the dual Hessian structure* (D', g). *Then we have*

$$ \alpha' = -\alpha, \qquad \beta' = \beta - 2\nabla\alpha. $$

Proof.   We have

$$
\begin{aligned}
\alpha'(X) &= D'_X v = (2\nabla_X - D_X) v = -D_X v = -\alpha(X) v, \\
\beta' &= \beta - 2\nabla\alpha.
\end{aligned}
$$
$\square$

Proposition 3.5. *Let $R^T_{i\bar{j}}$ be the* Ricci *tensor on the Kählerian manifold* (TM, J, g^T). *Then we have*

$$ R^T_{i\bar{j}} = -\frac{1}{2}\beta_{ij} \circ \pi. $$

Proof.   By Propositions 3.3, 3.4 and a formula in [Kobayashi and Nomizu (1963, 1969)](II,p157), we obtain

$$
\begin{aligned}
R^T_{i\bar{j}} &= -g^{p\bar{q}} R^T_{i\bar{q}p\bar{j}} = -\frac{1}{2}(g^{pq} \circ \pi)(Q_{iqpj} \circ \pi) = -\frac{1}{2}(g^{pq} Q_{ijpq}) \circ \pi \\
&= -\frac{1}{2} Q_{ij}{}^q{}_q \circ \pi = -\frac{1}{2}\beta_{ij} \circ \pi.
\end{aligned}
$$
$\square$

The above proposition suggests that the second Koszul form β plays a similar role to that of the Ricci tensor in Kählerian geometry. Accordingly, it is appropriate to introduce the following definition, which originated in [Cheng and Yau (1982)].

Definition 3.3. If a Hessian structure (D, g) satisfies the condition

$$\beta = \lambda g, \quad \lambda = \frac{\beta^i{}_i}{n},$$

then the Hessian structure is said to be **Einstein-Hessian**.

The following theorem then follows from Proposition 3.5 and expression (2.3).

Theorem 3.1. *Let (D, g) be a Hessian structure on M and let (J, g^T) the Kählerian structure on the tangent bundle TM induced by (D, g). Then the following conditions (1) and (2) are equivalent.*

(1) (D, g) *is Einstein-Hessian.*
(2) (J, g^T) *is Einstein-Kählerian.*

Example 3.1. Let us find the Koszul forms for Hessian domains given in Example 2.2. Let α and β be the Koszul forms for a Hessian structure (D, g) on Ω, and let α' and β' be the Koszul forms for the dual Hessian structure (D', g).

(1) Let $\Omega = \left\{ x \in \mathbf{R}^n \mid x^n > \frac{1}{2} \sum_{i}^{n-1} (x^i)^2 \right\}$ and $\varphi = -\log\left\{ x^n - \frac{1}{2} \sum_{i=1}^{n-1} (x^i)^2 \right\}$.
Then

$$\det \left[\frac{\partial^2 \varphi}{\partial x^i \partial x^j} \right] = f^{-n-1}, \quad \text{where } f = x^n - \frac{1}{2} \sum_{i=1}^{n-1} (x^i)^2.$$

Hence

$$\alpha = -\frac{n+1}{2} d\log f = \frac{n+1}{2} d\varphi, \quad \beta = \frac{n+1}{2} Dd\varphi = \frac{n+1}{2} g.$$

By Lemma 3.1 we have

$$\alpha' = -\alpha = \frac{n+1}{2} d\log f = -\frac{n+1}{2} d\log(-x'_n),$$

$$\beta' = \frac{n+1}{2} \left(\frac{dx'_n}{x'_n} \right)^2, \quad \text{where } x'_i = \frac{\partial \varphi}{\partial x^i}.$$

(2) Let $\Omega = \mathbf{R}^n$ and $\varphi = \log \left(1 + \sum_{i=1}^{n} e^{x^i} \right)$. Then

$$\det \left[\frac{\partial^2 \varphi}{\partial x^i \partial x^j} \right] = \frac{e^{x^1} \cdots e^{x^n}}{f^{n+1}}, \quad \text{where } f = \sum_{i=1}^{n} e^{x^i} + 1.$$

Hence

$$\alpha = \frac{1}{2} \left\{ \sum_{i=1}^{n} dx^i - (n+1) d \log f \right\},$$

$$\beta = D\alpha = -\frac{n+1}{2} Dd \log f = -\frac{n+1}{2} g.$$

Upon introducing $x_i' = \partial \varphi / \partial x^i$, we have

$$f = \frac{1}{1 - \sum_{k=1}^{n} x_k'}, \quad x^i = \log \frac{x_i'}{1 - \sum_{k=1}^{n} x_k'}.$$

By Lemma 3.1 we further obtain

$$\alpha' = -\alpha = -\frac{1}{2} d \log \left\{ \left(1 - \sum_{k=1}^{n} x_k' \right)^n x_1' \cdots x_n' \right\},$$

$$\beta' = \frac{n}{2} \left(d \log \left(1 - \sum_{k=1}^{n} x_k' \right) \right)^2 + \frac{1}{2} \sum_{k=1}^{n} (d \log x_k')^2.$$

(3) Let

$$\Omega = \left\{ x \in \mathbf{R}^n \mid x^n > \left(\sum_{i=1}^{n-1} (x^i)^2 \right)^{\frac{1}{2}} \right\}, \quad \varphi = -\log \left\{ (x^n)^2 - \sum_{i=1}^{n-1} (x^i)^2 \right\}.$$

Then $\det \left[\frac{\partial^2 \varphi}{\partial x^i \partial x^j} \right] = \left(\frac{2}{f} \right)^n$, where $f = (x^n)^2 - \sum_{i=1}^{n-1} (x^i)^2$. Hence

$$\alpha = -\frac{n}{2} d \log f = \frac{n}{2} d\varphi, \quad \beta = \frac{n}{2} Dd\varphi = \frac{n}{2} g.$$

Introducing $x_i' = \frac{\partial \varphi}{\partial x^i}$ and $f' = (x_n')^2 - \sum_{i=1}^{n-1} (x_i')^2$, we have $f = \frac{4}{f'}$. The Legendre transform of φ is given by $\varphi' = -\varphi - 2 = -\log f' + \log 4 - 2$. By the same way as above, we have $\det \left[\frac{\partial^2 \varphi'}{\partial x_i' \partial x_j'} \right] = \left(\frac{2}{f'} \right)^n$. Hence

$$\alpha' = -\frac{n}{2} d \log f' = \frac{n}{2} d\varphi', \quad \beta' = \frac{n}{2} D' d\varphi' = \frac{n}{2} g.$$

3.2 Hessian sectional curvature

A Kählerian manifold of constant holomorphic sectional curvature is called a complex space form. It is known that a simply connected, complete, complex space form is holomorphically isometric to the complex projective space $P^n(\mathbf{C})$, the complex Euclidean space $\mathbf{C}^n$, or the complex hyperbolic space $H^n(\mathbf{C})$ according to whether the sectional curvature is positive, zero or negative respectively [Kobayashi and Nomizu (1963, 1969)](II, Theorem 7.8, 7.9). In this section we define a Hessian sectional curvature on a Hessian manifold (M, D, g) corresponding to a holomorphic sectional curvature on a Kählerian manifold, and construct Hessian domains of constant Hessian sectional curvature.

Definition 3.4. Let Q be a Hessian curvature tensor on a Hessian manifold (M, D, g). We define an endomorphism $\hat{Q}$ on the space of symmetric contravariant tensor fields of degree 2 by

$$\hat{Q}(\xi)^{ik} = Q^i{}_j{}^k{}_l \xi^{jl}.$$

The endomorphism $\hat{Q}$ is symmetric with respect to the inner product $\langle , \rangle$ induced by the Hessian metric g. In fact, by Proposition 3.1 we have

$$\langle \hat{Q}(\xi), \eta \rangle = Q^i{}_j{}^k{}_l \xi^{jl} \eta_{ik} = Q_{ijkl} \xi^{jl} \eta^{ik}$$
$$= Q_{jilk} \eta^{ik} \xi^{jl} = (Q^j{}_i{}^l{}_k \eta^{ik}) \xi_{jl} = \langle \xi, \hat{Q}(\eta) \rangle.$$

Definition 3.5. Let $\xi_x \neq 0$ be a symmetric contravariant tensor field of degree 2. We put

$$q(\xi_x) = \frac{\langle \hat{Q}(\xi_x), \xi_x \rangle}{\langle \xi_x, \xi_x \rangle},$$

and call it the **Hessian sectional curvature** for ξ_x.

Definition 3.6. If $q(\xi_x)$ is a constant c for all symmetric contravariant tensor field $\xi_x \neq 0$ of degree 2 and for all $x \in M$, then (M, D, g) is said to be a **Hessian manifold of constant Hessian sectional curvature** c.

Proposition 3.6. *The Hessian sectional curvature of $(M, D, \tilde{g})$ is a constant c if and only if*

$$Q_{ijkl} = \frac{c}{2}(g_{ij}g_{kl} + g_{il}g_{kj}).$$

Proof. Suppose that the Hessian sectional curvature is a constant c. Since $\hat{Q}$ is symmetric and $\langle \hat{Q}(\xi_x), \xi_x \rangle = c\langle \xi_x, \xi_x \rangle$ for all ξ_x, we have

$$\hat{Q}(\xi_x) = c\xi_x.$$

Introducing

$$T^i{}_j{}^k{}_l = Q^i{}_j{}^k{}_l - \frac{c}{2}(\delta^i_j \delta^k_l + \delta^i_l \delta^k_j),$$

we have

$$T^i{}_j{}^k{}_l = T^k{}_j{}^i{}_l = T^i{}_l{}^k{}_j, \qquad T^i{}_j{}^k{}_l \xi^{jl}_x = 0.$$

Hence

$$0 = T^i{}_j{}^k{}_l(a^j b^l + a^l b^j)(c_i d_k + c_k d_i) = 4T_{ijkl}a^i b^j c^k d^l$$

for all tangent vectors a^i, b^i, c^i, d^i at x. This implies $T^i{}_j{}^k{}_l = 0$ and

$$Q_{ijkl} = \frac{c}{2}(g_{ij}g_{kl} + g_{il}g_{kj}).$$

Conversely, if the above relations hold, it is easy to see that the Hessian sectional curvature is constant with value c. □

Corollary 3.1. *The following conditions* (1) *and* (2) *are equivalent.*

(1) *The Hessian sectional curvature of* (M, D, g) *is a constant c.*
(2) *The holomorphic sectional curvature of* (TM, J, g^T) *is a constant* $-c$.

Proof. The holomorphic sectional curvature of the Kähler manifold (TM, J, g^T) is a constant $-c$ if and only if

$$R^T{}_{i\bar{j}k\bar{l}} = \frac{1}{2}c(g_{i\bar{j}}g_{k\bar{l}} + g_{i\bar{l}}g_{\bar{j}k})$$

cf. [Kobayashi and Nomizu (1963, 1969)](II, p169)D It therefore follows from Proposition 3.3 and $g_{i\bar{j}} = g_{ij} \circ \pi$ that assertions (1) and (2) are equivalent. □

This suggests that the notion of Hessian sectional curvature corresponds to that of holomorphic sectional curvature for Kählerian manifolds.

Corollary 3.2. *Suppose that a Hessian manifold* (M, D, g) *is a space of constant Hessian sectional curvature c. Then the Riemannian manifold* (M, g) *is a space form of constant sectional curvature* $-\dfrac{c}{4}$.

Proof. By Propositions 3.2 and 3.6 we have

$$R_{ijkl} = -\frac{c}{4}(g_{ik}g_{jl} - g_{il}g_{jk}).$$

This implies that the Riemannian manifold (M, g) is a space form of constant sectional curvature $-\dfrac{c}{4}$. $\qquad\square$

Corollary 3.3. *If the Hessian sectional curvature of (M, D, g) is a constant c, then the Hessian structure (D, g) is Einstein-Hessian and*

$$\beta = \frac{(n+1)c}{2}g.$$

Proof. The above assertion follows from Proposition 3.4 (3) and Proposition 3.6. $\qquad\square$

Definition 3.7. We define a tensor field W of type $(1, 3)$ by

$$W^i{}_{jkl} = Q^i{}_{jkl} - \frac{1}{n+1}(\delta^i_j\beta_{kl} + \delta^i_l\beta_{kj}).$$

Theorem 3.2. *A Hessian sectional curvature is a constant if and only if $W = 0$.*

Proof. Suppose that the Hessian sectional curvature is a constant c. By Proposition 3.6 and Corollary 3.3 we have

$$\begin{aligned} W^i{}_{jkl} &= Q^i{}_{jkl} - \frac{1}{n+1}(\delta^i_j\beta_{kl} + \delta^i_l\beta_{kj}) \\ &= Q^i{}_{jkl} - \frac{c}{2}(\delta^i_j g_{kl} + \delta^i_l g_{kj}) = 0. \end{aligned}$$

Conversely, suppose $W = 0$. Then

$$Q_{ijkl} = \frac{1}{n+1}(g_{ij}\beta_{kl} + g_{il}\beta_{kj}).$$

From this expression, together with $Q_{ijkl} = Q_{klij}$, it follows that

$$g_{ij}\beta_{kl} + g_{il}\beta_{kj} = g_{kl}\beta_{ij} + g_{kj}\beta_{il}.$$

Multiplying both sides of this formula by g^{ij} and contracting i, j, we have

$$\beta_{kl} = \frac{\beta^r{}_r}{n}g_{kl}.$$

Hence

$$Q_{ijkl} = \frac{\beta^r{}_r}{n(n+1)}(g_{ij}g_{kl} + g_{il}g_{kj}),$$

and by Proposition 3.6 the Hessian sectional curvature is a constant. $\qquad\square$

The above theorem suggests that the tensor field W plays a similar role to that of the projective curvature tensor on a Kählerian manifold.

Theorem 3.3. *We have*

$$\operatorname{Tr} \hat{Q}^2 \geq \frac{2}{n(n+1)} (\operatorname{Tr} \hat{\beta})^2,$$

where $\operatorname{Tr} \hat{\beta} = \beta^r{}_r$. *The equality holds if and only if the Hessian sectional curvature is a constant.*

Proof. Following the notation of Proposition 3.6, we introduce

$$T^i{}_j{}^k{}_l = Q^i{}_j{}^k{}_l - \frac{\operatorname{Tr} \hat{\beta}}{n(n+1)} (\delta^i_j \delta^k_l + \delta^i_l \delta^k_j),$$

and then

$$T^i{}_j{}^k{}_l T_i{}^j{}_k{}^l = Q^i{}_j{}^k{}_l Q_i{}^j{}_k{}^l - \frac{2\operatorname{Tr} \hat{\beta}}{n(n+1)} Q_i{}^j{}_k{}^l (\delta^i_j \delta^k_l + \delta^i_l \delta^k_j)$$

$$+ \frac{\operatorname{Tr} \hat{\beta}}{n(n+1)} 2(\delta^i_j \delta^k_l + \delta^i_l \delta^k_j)(\delta^j_i \delta^l_k + \delta^l_i \delta^j_k)$$

$$= \operatorname{Tr} \hat{Q}^2 - \frac{2\operatorname{Tr} \hat{\beta}}{n(n+1)} (Q_i{}^i{}_k{}^k + Q_i{}^j{}_j{}^i) + \frac{2(\operatorname{Tr} \hat{\beta})^2}{n(n+1)}$$

$$= \operatorname{Tr} \hat{Q}^2 - \frac{2}{n(n+1)} (\operatorname{Tr} \hat{\beta})^2.$$

This ensures

$$\operatorname{Tr} \hat{Q}^2 \geq \frac{2}{n(n+1)} (\operatorname{Tr} \hat{\beta})^2,$$

and the equality holds if and only if

$$Q^i{}_j{}^k{}_l = \frac{\operatorname{Tr} \hat{\beta}}{n(n+1)} (\delta^i_j \delta^k_l + \delta^i_l \delta^k_j). \qquad \square$$

Now let us construct a Hessian domain $(\Omega, D, g = Dd\varphi)$ in $\mathbf{R}^n$ of constant Hessian sectional curvature c.

Proposition 3.7. *The following Hessian domains are examples of spaces of constant Hessian sectional curvature* 0.

(1) *The Euclidean space* $\left(\mathbf{R}^n, D, g = Dd\left(\frac{1}{2} \sum_{i=1}^n (x^i)^2 \right) \right) D$

(2) $\left(\mathbf{R}^n, D, g = Dd\left(\sum_{i=1}^n e^{x^i} \right) \right).$

Proof. (1) Since $g_{ij} = \delta_{ij}$ we have $\gamma^i{}_{jk} = 0C$ and so the Hessian curvature tensor Q vanishes identically.

(2) Putting $\varphi = \sum_{i=1}^{n} e^{x^i}$ and $\eta_i = \dfrac{\partial \varphi}{\partial x^i} = e^{x^i}$ we have

$$[g_{ij}] = [\delta_{ij}\eta_i].$$

Let us denote by γ_i and γ^i matrices whose (j, l) components are γ_{ijl} and $\gamma^i{}_{jl}$ respectively. Then

$$\gamma_i = \frac{1}{2}\frac{\partial}{\partial x^i}[g_{jl}] = \frac{1}{2}[\delta_{jl}g_{ij}],$$

hence

$$\gamma^i = \frac{1}{2}[\delta_{jl}\delta^i_j].$$

We denote by $Q^i{}_k$ a matrix whose (j, l) components are $Q^i{}_{jkl}$. Then

$$Q^i{}_k = [Q^i{}_{jkl}] = \left[\frac{\partial \gamma^i{}_{jl}}{\partial x^k}\right] = \frac{\partial}{\partial x^k}\gamma^i = 0.$$

Thus the Hessian sectional curvature vanishes identically. □

Proposition 3.8. *Let c be a positive real number and let*

$$\Omega = \left\{(x^1, \cdots, x^n) \in \mathbf{R}^n \mid x^n > \frac{1}{2}\sum_{i=1}^{n-1}(x^i)^2\right\},$$

$$\varphi = -\frac{1}{c}\log\left\{x^n - \frac{1}{2}\sum_{i=1}^{n-1}(x^i)^2\right\}.$$

Then $(\Omega, D, g = Dd\varphi)$ is a Hessian domain of constant Hessian sectional curvature c. The Riemannian manifold (Ω, g) is isometric to a hyperbolic space form $(\mathbf{H}(-\dfrac{c}{4}), g)$ of constant sectional curvature $-\dfrac{c}{4}$, where

$$\mathbf{H} = \{(y^1, \cdots, y^{n-1}, y^n) \in \mathbf{R}^n \mid y^n > 0\},$$

$$g = \frac{1}{(y^n)^2}\left\{\sum_{i=1}^{n-1}(dy^i)^2 + \frac{4}{c}(dy^n)^2\right\}.$$

Proof. Introducing $\eta_i = \dfrac{\partial \varphi}{\partial x^i}$, we have

$$\eta_i = \begin{cases} \dfrac{1}{c}x^i\left\{x^n - \dfrac{1}{2}\sum_{r=1}^{n-1}(x^r)^2\right\}^{-1}, & 1 \le i \le n-1, \\[3mm] -\dfrac{1}{c}\left\{x^n - \dfrac{1}{2}\sum_{r=1}^{n-1}(x^r)^2\right\}^{-1}, & i = n, \end{cases}$$

$$[g_{ij}] = \begin{bmatrix} c\eta_i\eta_j - \delta_{ij}\eta_n & c\eta_i\eta_n \\ c\eta_n\eta_j & c\eta_n\eta_n \end{bmatrix}.$$

From the above expressions we can express g in the form

$$g = c\left\{ \sum_{i=1}^{n-1} \eta_i dx^i + \eta_n dx^n \right\}^2 - \eta_n \sum_{i=1}^{n-1} (dx^i)^2$$

$$= \frac{1}{2}(d\log f)^2 + \frac{1}{2} \sum_{i=1}^{n-1} (dx^i)^2,$$

where $f = x^n - \frac{1}{2} \sum_{i=1}^{n-1} (x^i)^2$. Upon further introducing

$$y^i = \begin{cases} c^{-\frac{1}{2}} x^i, \ 1 \le i \le n-1, \\[2mm] f^{\frac{1}{2}}, \qquad i = n, \end{cases}$$

then

$$\eta_i = \begin{cases} c^{-\frac{1}{2}} y^i (y^n)^{-2}, \ 1 \le i \le n-1, \\[2mm] -c^{-1}(y^n)^{-2}, \qquad i = n, \end{cases}$$

$$x^i = \begin{cases} c^{\frac{1}{2}} y^i, \qquad\qquad\qquad 1 \le i \le n-1, \\[2mm] (y^n)^2 + \frac{c}{2} \sum_{i=1}^{n-1} (y^i)^2, \qquad i = n, \end{cases}$$

$$-\infty < y^i < +\infty, \qquad 1 \le i \le n-1,$$
$$y^n > 0.$$

Using these expressions we can express g in the form

$$g = \sum_{i,j=1}^{n} \frac{\partial \eta_i}{\partial x^j} dx^i dx^j = \sum_{i=1}^{n-1} d\eta_i dx^i + d\eta_n dx^n$$

$$= \sum_{i=1}^{n-1} d\{c^{-\frac{1}{2}} y^i (y^n)^{-2}\} d(c^{\frac{1}{2}} y^i) + d\left\{ -\frac{1}{c}\frac{1}{(y^n)^2} \right\} d\left\{ (y^n)^2 + \frac{c}{2} \sum_{i=1}^{n-1} (y^i)^2 \right\}$$

$$= \frac{1}{(y^n)^2} \left\{ \sum_{i=1}^{n-1} (dy^i)^2 + \frac{4}{c} (dy^n)^2 \right\}.$$

Hence g is positive definite and $(D, g = Dd\varphi)$ is a Hessian structure on Ω. With reference to the above expression for g, we deduce that (Ω, g) is isometric to a hyperbolic space form $(\mathbf{H}(-\frac{c}{4}), g)$ of constant sectional curvature $-\dfrac{c}{4}$. Let γ_i and γ^i be matrices whose (j,l) components are γ_{ijl} and $\gamma^i{}_{jl}$ respectively. Then

$$\gamma_i = \frac{1}{2}\frac{\partial}{\partial x^i}[g_{jl}]$$

$$= \frac{1}{2}\begin{bmatrix} cg_{ij}\eta_l + cg_{il}\eta_j - \delta_{jl}g_{in} & cg_{ij}\eta_n + cg_{in}\eta_j \\[4pt] cg_{in}\eta_l + cg_{il}\eta_n & 2cg_{in}\eta_n \end{bmatrix},$$

and so

$$\gamma^i = \frac{1}{2}\begin{bmatrix} c\delta^i_j\eta_l + c\delta^i_l\eta_j - \delta_{jl}\delta^i_n & c\delta^i_j\eta_n + c\delta^i_n\eta_j \\[4pt] c\delta^i_n\eta_l + c\delta^i_l\eta_n & 2c\delta^i_n\eta_n \end{bmatrix}.$$

Denoting by $Q^i{}_k$ a matrix whose (j,l) components are $Q^i{}_{jkl}$, we have

$$Q^i{}_k = [Q^i{}_{jkl}] = \left[\frac{\partial \gamma^i{}_{jl}}{\partial x^k}\right] = \frac{\partial}{\partial x^k}\gamma^i$$

$$= \frac{c}{2}\begin{bmatrix} \delta^i_j g_{kl} + \delta^i_l g_{kj} & \delta^i_j g_{kn} + \delta^i_n g_{kj} \\[4pt] \delta^i_n g_{kl} + \delta^i_l g_{kn} & 2\delta^i_n g_{kn} \end{bmatrix},$$

that is

$$Q^i{}_{jkl} = \frac{c}{2}(\delta^i_j g_{kl} + \delta^i_l g_{kj}).$$

Hence the Hessian sectional curvature of $(\Omega, D, g = D^2\varphi)$ is the constant c. $\qquad\square$

Proposition 3.9. *Let c be a negative real number and let*

$$\varphi = -\frac{1}{c}\log\left(\sum_{i=1}^{n} e^{x^i} + 1\right).$$

Then $(\mathbf{R}^n, D, g = Dd\varphi)$ is a Hessian domain of constant Hessian sectional curvature c. The Riemannian manifold $(\mathbf{R}^n, g)$ is isometric to a domain in a sphere defined by

$$\sum_{i=1}^{n+1}(y^i)^2 = -\frac{4}{c},$$
$$y^1 > 0,\ y^2 > 0,\ \cdots,\ y^{n+1} > 0.$$

Proof.   We introduce

$$\eta_i = \frac{\partial \varphi}{\partial x^i} = -\frac{1}{c}e^{x^i}\Big(\sum_{j=1}^n e^{x^j} + 1\Big)^{-1},$$

$$y_i = \begin{cases} 2(\eta_i)^{\frac{1}{2}}, & 1 \le i \le n, \\[2mm] 2\big\{-c\big(\sum_{j=1}^n e^{x^j} + 1\big)\big\}^{-\frac{1}{2}}, & i = n+1. \end{cases}$$

Then

$$y_1^2 + \cdots + y_n^2 + y_{n+1}^2 = -\frac{4}{c}, \qquad y_i > 0,$$

$$dy_{n+1} = -\sum_{i=1}^n \frac{y_i}{y_{n+1}}dy_i,$$

$$x^i = 2\log\frac{y_i}{y_{n+1}} - \log(-c).$$

With these expressions we have

$$g = \sum_{i,j=1}^n \frac{\partial \eta_i}{\partial x^j}dx^i dx^j = \frac{1}{4}\sum_{i,j=1}^n \frac{\partial y_i^2}{\partial x^j}dx^i dx^j$$

$$= \frac{1}{2}\sum_{i=1}^n y_i\Big(\sum_{j=1}^n \frac{\partial y_i}{\partial x^j}dx^j\Big)dx^i = \frac{1}{2}\sum_{i=1}^n (y_i dy_i)dx^i$$

$$= \sum_{i=1}^n (y_i dy_i)d\log\frac{y_i}{y_{n+1}} = \sum_{i=1}^n (y_i dy_i)\frac{y_{n+1}}{y_i}\frac{y_{n+1}dy_i - y_i dy_{n+1}}{y_{n+1}^2}$$

$$= \sum_{i=1}^n (dy_i)^2 - \Big(\sum_{i=1}^n \frac{y_i}{y_{n+1}}dy_i\Big)dy_{n+1}$$

$$= \sum_{i=1}^n (dy_i)^2 + (dy_{n+1})^2.$$

Hence g is positive definite and $(D, g = Dd\varphi)$ is a Hessian structure. With reference to the above expression for g, we deduce that the Riemannian manifold $(\mathbf{R}^n, g)$ is isometric to a domain in the sphere $\sum_{i=1}^{n+1} y_i^2 = -\frac{4}{c}$ defined by $y_1 > 0,\ y_2 > 0, \cdots, y_{n+1} > 0$. Since $g_{ij} = \delta_{ij}\eta_i + c\eta_i\eta_j$, using the same notation in the proof of Proposition 3.8 we have

$$\gamma_i = \frac{1}{2}\frac{\partial}{\partial x^i}[g_{jl}] = \frac{1}{2}[\delta_{jl}g_{ij} + c(g_{ij}\eta_l + g_{il}\eta_j)],$$

$$\gamma^i = \frac{1}{2}[\delta_{jl}\delta_j^i + c(\delta_j^i\eta_l + \delta_l^i\eta_j)].$$

Thus

$$Q^i{}_k = [Q^i{}_{jkl}] = \left[\frac{\partial \gamma^i{}_{jl}}{\partial x^k}\right] = \frac{\partial \gamma^i}{\partial x^k}$$

$$= \left[\frac{c}{2}(\delta^i_j g_{kl} + \delta^i_l g_{kj})\right],$$

that is

$$Q^i{}_{jkl} = \frac{c}{2}(\delta^i_j g_{kl} + \delta^i_l g_{kj}).$$

This means that $(\mathbf{R}^n, D, g = Dd\varphi)$ is a Hessian domain of constant Hessian sectional curvature c. □

Chapter 4

Regular convex cones

In Chapters 2 and 3 we surveyed the general theory and fundamental results of Hessian structures. In this chapter we give a brief exposition on regular convex cones. We will first show that regular convex cones admit canonical Hessian structures. The study of regular convex cones is the origin of the geometry of Hessian structures. I.I. Pyateckii-Shapiro studied realizations of homogeneous bounded domains by considering Siegel domains in connection with automorphic forms [Pyatetskii-Shapiro (1969)]. For a problem proposed by [Cartan (1935)] he constructed a 4-dimensional non-symmetric affine homogeneous Siegel domain which is an example of a non-symmetric homogeneous bounded domain [Pyatetskii-Shapiro (1959)]. Regular convex cones are used to define Siegel domains. In section **4.1** we define a characteristic function ψ of a regular convex cone Ω, and show that $g = Dd \log \psi$ is a Hessian metric on Ω invariant under affine automorphisms of Ω. The Hessian metric $g = Dd \log \psi$ is called the canonical Hessian metric on Ω. In section **4.2** we prove that if Ω is a homogeneous self dual cone, then the gradient mapping is a symmetry with respect to the canonical Hessian metric, and Ω is a symmetric homogeneous Riemannian manifold. Furthermore, we give a bijective correspondence between homogeneous self dual cones and compact Jordan algebras [Koecher (1962)][Vinberg (1960)].

4.1 Regular convex cones

In this section we outline the fundamental theoretical results for regular convex cones, such as dual cones, characteristic functions and canonical Hessian structures.

Definition 4.1.

(1) A subset S in $\mathbf{R}^n$ is said to be a **convex set** if a segment of a line joining any two points in S is contained in S.

(2) A subset S in $\mathbf{R}^n$ is said to be a **cone** with vertex 0 if, for any x in S and any positive real number λ, λx belongs to S.

Suppose that $\mathbf{R}^n$ is equipped with an inner product (x, y). The polar set S^0 of any subset S in $\mathbf{R}^n$ is defined by

$$S^o = \{y \in \mathbf{R}^n \mid (y, x) \leq 1, \ for \ all \ x \in S\}.$$

Theorem 4.1. *Suppose that S is a closed convex set containing the origin 0. Then*

$$(S^o)^o = S.$$

Proof. It is clear that S is contained in $(S^o)^o$. For the proof that S contains $(S^o)^o$, it suffices to show for any $x_0 \notin S$ there exists y_0 satisfying

$$(x_0, y_0) > 1, \quad (x, y_0) \leq 1$$

for all $x \in S$. Let $x_1 \in S$ be a point realizing the minimal distance from x_0 to S, then

$$\|x - x_0\| \geq \|x_1 - x_0\|$$

for any $x \in S$, where $\|x\| = (x, x)^{1/2}$. Since $\lambda x + (1 - \lambda)x_1 \in S$ for all $x \in S$ and $0 \leq \lambda \leq 1$, we have

$$\|\lambda x + (1 - \lambda)x_1 - x_0\|^2 \geq \|x_1 - x_0\|^2,$$

that is

$$\lambda^2 \|x - x_1\|^2 + 2\lambda(x - x_1, x_1 - x_0) \geq 0,$$

and so

$$(x - x_1, x_1 - x_0) \geq 0.$$

Therefore

$$(x, x_0 - x_1) \leq (x_1, x_0 - x_1),$$

for all $x \in S$, and so substituting x for 0, we have

$$(x_1, x_0 - x_1) \geq 0.$$

Since $(x_0 - x_1, x_0 - x_1) > 0$, there exists μ such that

$$0 \leq (x_1, x_0 - x_1) < \mu < (x_0, x_0 - x_1).$$

From these inequalities we have

$$(x, x_0 - x_1) \leq (x_1, x_0 - x_1) < \mu < (x_0, x_0 - x_1),$$

for any $x \in S$. Therefore the selection $y_0 = \dfrac{1}{\mu}(x_0 - x_1)$ satisfies the required condition. $\qquad\square$

Let C be a convex cone. Then the set

$$C - C = \{x - y \mid x, y \in C\}$$

is the minimal subspace containing C. The set

$$C \cap -C$$

is the maximal subspace contained in C, where $-C = \{-x \mid x \in C\}$. We define

$$C^\sharp = \{y \in \mathbf{R}^n \mid (x, y) \geq 0, \ for \ all \ x \in C\}.$$

Then $C^\sharp$ is a closed convex cone and it is straightforward to deduce that

$$C^\sharp = -C^o.$$

From this we have immediately the following corollary.

Corollary 4.1. *Let C be a closed convex cone. Then*

$$(C^\sharp)^\sharp = C.$$

Corollary 4.2. *Let C be a closed convex cone. Then*

$$(C^\sharp - C^\sharp)^\perp = C \cap -C.$$

Proof. $x \in (C^\sharp - C^\sharp)^\perp$ if and only if $(x, y) = 0$ for all $y \in C^\sharp$, which is equivalent to x and $-x \in (C^\sharp)^\sharp = C$. □

Proposition 4.1. *Let C be a closed convex cone. Then*

$$int(C^\sharp) = \{y \mid (y, x) > 0, \ for \ all \ x \neq 0 \in C\},$$

where $int(C^\sharp)$ is the set of interior points of $C^\sharp$.
The following conditions are equivalent.

(1) C *contains no full straight lines.*
(2) $C \cap -C = \{0\}D$
(3) $int(C^\sharp) \neq \emptyset D$

Proof. Put $B = \{y \mid (y, x) > 0, \ for \ all \ x \neq 0 \in C\}$. Then

$$B = \{y \mid (y, x) > 0, \ for \ all \ x \in C \cap S^{n-1}\},$$

where S^{n-1} is the unit sphere with centre 0. Since $C \cap S^{n-1}$ is compact, B is an open set. Therefore B is contained in the interior of $C^\sharp$. Let y be an inner point of $C^\sharp$. If $\|u\|$ is small enough, then $y + u \in C^\sharp$, and so $(x, y) + (x, u) = (x, y + u) \geq 0$ for all $x \neq 0 \in C$. Hence $(x, y) > 0$, and $y \in B$. Thus the interior of $C^\sharp$ is contained in B. Therefore $int(C^\sharp) = B$. It is obvious that (1) and (2) are equivalent. By corollary 4.2, the condition (2) holds if and only if $C^\sharp - C^\sharp = \mathbf{R}^n$, which is equivalent to (3). □

Definition 4.2.

(1) A convex domain not containing any full straight line is said to be a
    regular convex domain.
(2) An open convex cone not containing any full straight line is called a
    regular convex cone.

Proposition 4.2. *Let Ω be a regular convex domain. Then the following
assertions hold.*

(1) *There exists an affine coordinate system $\{y^1, \cdots, y^n\}$ on $\mathbf{R}^n$ such that*

$$y^i(p) > 0, \quad for \ all \ p \in \Omega, \quad where \ i = 1, \cdots, n.$$

(2) *The tube domain $T_\Omega = \mathbf{R}^n + \sqrt{-1}\Omega \subset \mathbf{C}^n$ over Ω is holomorphically
    isomorphic to a bounded domain in $\mathbf{C}^n$.*

Proof.   Without loss of generality, we may assume that Ω contains the
origin o. Let K be the set of all points p such that $tp \in \Omega$ for all $t > 0$.
Then K is a closed convex cone not containing any full straight line with
vertex o. Since, by Proposition 4.1 $int(K^\sharp) = \{y \mid (y, x) > 0, \ x \neq 0 \in K\}$ is
a non-empty open set, we can choose a basis $\{v_1, \cdots, v_n\}$ of $\mathbf{R}^n$ contained
in $int(K^\sharp)$. We show that there exists a positive real number m such that
$(v_i, p) > -m$ for all $p \in \Omega$. Suppose that such a positive number does not
exist, then we can find a sequence of points $\{p_n\}$ such that $(v_i, p_n) \to -\infty$.
Let S be a small sphere centred at o such that $\{p_n\}$ lie outside of S, and
$S \subset \Omega$. Let q_n be a point of intersection between S and the line segment
joining p_n and o, and let $p_n = \lambda_n q_n$, $\lambda_n > 0$. Selecting, if necessary, a
subsequence of p_n, we may assume that $q_n \to q_0 \in S$. Since $(v_i, q_n) \to$
(v_i, q_0) and $(v_i, p_n) = \lambda_n(v_i, q_n) \to -\infty$, it follows that $\lambda_n \to \infty$. Suppose
that there exists a positive number $t_0 > 0$ such that $t_0 q_0$ is contained
in the boundary of Ω. Since $\bar{\Omega}$ is a convex set we have $s q_0 \in \Omega$ for all
$0 \leq s < t_0$. For a fixed $t_1 > t_0$ the point $t_1 q_0$ is an exterior point of Ω.
Choosing a sufficiently large n, we have $\lambda_n > t_1$. Since Ω is a convex set and
$\lambda_n q_n = p_n \in \Omega$, it follows that $t_1 q_n \in \Omega$. Hence $t_1 q_0 \in \bar{\Omega}$. However, this
contradicts that $t_1 q_0$ is an exterior point of Ω. Thus we know $t q_0 \in \Omega$ for
all $t > 0$, that is, $q_0 \in K$. This implies $(v_i, q_0) > 0$, and so $(v_i, q_n) > 0$ for a
sufficiently large n. This means $(v_i, p_n) = \lambda_n(v_i, q_n) > 0$, which contradicts
the limiting behaviour $(v_i, p_n) \to -\infty$.

(1) Taking a basis $\{u_1, \cdots, u_n\}$ of $\mathbf{R}^n$ such that $(u_i, v_j) = \delta_{ij}$, we define a
function y^i by $y^i : p = \sum_k p^k u_k \longrightarrow p^i + m$. Then $\{y^1, \cdots, y^n\}$ is an affine

coordinate system on $\mathbf{R}^n$ satisfying $y^i(p) > 0$ for all $p \in \Omega$.

(2) We define a holomorphic coordinate system on $T_\Omega = \mathbf{R}^n + \sqrt{-1}\Omega$ by $z^1 = x^1 + \sqrt{-1}y^1, \cdots, z^n = x^n + \sqrt{-1}y^n$. Applying Cayley transformations to all the components z^j, mapping upper half-planes to unit disks, we derive the result that T_Ω is holomorphically isomorphic to a bounded domain. $\square$

Theorem 4.2. *Let Ω be a regular convex cone and let $\bar{\Omega}$ be the closure of Ω. We set*

$$\Omega^\star = \{y \mid (y, x) > 0, \ for \ all \ x \neq 0 \in \bar{\Omega}\}.$$

Then

(1) $\Omega^\star$ *is a regular convex cone.*

(2) $(\Omega^\star)^\star = \Omega$.

Proof. Since Ω is a convex set not containing any full straight line, its closure $\bar{\Omega}$ also does not contain any full straight line and

$$\bar{\Omega} \cap -\bar{\Omega} = \{0\}.$$

By Proposition 4.1, we have $\Omega^\star = int((\bar{\Omega})^\sharp) \neq \emptyset$D Since Ω is an open set, by Corollary 4.1 and 4.2 we have $(\bar{\Omega})^\sharp \cap -(\bar{\Omega})^\sharp = (((\bar{\Omega})^\sharp)^\sharp - ((\bar{\Omega})^\sharp)^\sharp)^\perp = (\bar{\Omega} - \bar{\Omega})^\perp = \bar{\Omega} \cap -\bar{\Omega} = \{0\}$D Therefore $\Omega^\star$ is a regular convex cone. Furthermore, we have $(\Omega^\star)^\star = int(((\bar{\Omega})^\sharp)^\sharp) = int(\bar{\Omega}) = \Omega$. $\square$

Definition 4.3. The regular convex cone $\Omega^\star$ is called the **dual cone** of Ω with respect to the inner product $(\ ,\ )$.

Let $\mathbf{R}_n^*$ be the dual vector space of $\mathbf{R}^n$. We denote by $\langle x, y^* \rangle$ the value $y^*(x)$ of $y^* \in \mathbf{R}_n^*$ at $x \in \mathbf{R}^n$. If we replace the inner product (x, y) on $\mathbf{R}^n$ by the pairing $\langle x, y^* \rangle$, by the same method as the proof of Theorem 4.2, we can prove the following theorem.

Theorem 4.3. *Let Ω be a regular convex cone in $\mathbf{R}^n$. We define a subset Ω^* of $\mathbf{R}_n^*$ by*

$$\Omega^* = \{y^* \in \mathbf{R}_n^* \mid \langle x, y^* \rangle > 0, \ x \neq 0 \in \bar{\Omega}\}.$$

Then

(1) Ω^* *is a regular convex cone in $\mathbf{R}_n^*$.*

(2) *Identifying the dual space of $\mathbf{R}_n^*$ with $\mathbf{R}^n$ we have $(\Omega^*)^* = \Omega$.*

Definition 4.4. The regular convex cone Ω^* is said to be the **dual cone** of Ω in $\mathbf{R}_n^*$.

Let Ω be a regular convex cone. We define a function ψ on Ω by

$$\psi(x) = \int_{\Omega^*} e^{-\langle x, x^* \rangle} dx^*, \tag{4.1}$$

where dx^* is a D^*-parallel volume element in $\mathbf{R}_n^*$. We show that the integral defining $\psi(x)$ is convergent. For a fixed $x \in \Omega$ we define a hyperplane P_t in $\mathbf{R}_n^*$ by $P_t = \{x^* \in \mathbf{R}_n^* \mid \langle x, x^* \rangle = t\}$. Then the set $P_t \cap \bar{\Omega}^*$ is bounded with respect to the Euclidean norm $\| \cdot \|$ on $\mathbf{R}_n^*$. Suppose that the set is unbounded, then there exists a sequence of points $\{p_k^*\}$ in $P_t \cap \bar{\Omega}^*$ such that $\lim_{k \to \infty} \|p_k^*\| = \infty$ and $\lim_{k \to \infty} \dfrac{p_k^*}{\|p_k^*\|} = p^*$. Then $p^* \neq 0 \in \bar{\Omega}^*$ and so $\langle x, p^* \rangle > 0$ since $x \in (\Omega^*)^*$. However, we also have $\langle x, p^* \rangle = \lim_{k \to \infty} \langle x, \dfrac{p_k^*}{\|p_k^*\|} \rangle = \lim_{k \to \infty} \dfrac{t}{\|p_k^*\|} = 0$. This is a contradiction. Hence $P_t \cap \bar{\Omega}^*$ is bounded, and the volume

$$v(t) = \int_{P_t \cap \Omega^*} dx_t^*$$

is finite, where dx_t^* is the volume element on P_t induced by dx^*. Since $P_t = tP(1)$, we have $v(t) = t^{n-1}v(1)$, and so

$$\psi(x) = \int_{\Omega^*} e^{-\langle x, x^* \rangle} dx^* = \int_0^\infty e^{-t} \left(\int_{P_t \cap \Omega^*} dx_t^* \right) dt$$

$$= v(1) \int_0^\infty e^{-t} t^{n-1} dt = (n-1)! v(1) < \infty.$$

Definition 4.5. The function $\psi(x)$ is said to be the **characteristic function** of a regular convex cone Ω.

Proposition 4.3. *Let x_0 be a boundary point of a regular convex cone Ω. Then*

$$\lim_{x \to x_0} \psi(x) = \infty.$$

Proof. Let $\{p_k\}$ be a sequence of points in Ω converging to x_0. Then $F_k(x^*) = e^{-\langle p_k, x^* \rangle}$ is uniformly convergent to $F_0(x^*) = e^{-\langle x_0, x^* \rangle}$ on any compact subset of $\mathbf{R}_n^*$. Hence

$$\liminf_{k \to \infty} \psi(p_k) = \liminf_{k \to \infty} \int_{\Omega^*} F_k(x^*) dx^*$$

$$\geq \int_{\Omega^*} \liminf_{k \to \infty} F_k(x^*) dx^* = \int_{\Omega^*} F_0(x^*) dx^*.$$

Since $(\Omega^*)^* = \Omega$, there exists $x_0^* \neq 0 \in \bar{\Omega}^*$ such that $\langle x_0, x_o^* \rangle = 0$. Let K be a closed ball contained in Ω^*, and let

$$L = K + \{\lambda x_0^* \mid \lambda > 0\} \subset \Omega^*.$$

We set $c = \min_{x^* \in K} F_0(x^*)$. Then $c > 0$ and $c = \min_{x^* \in L} F_0(x^*)$. Therefore

$$\int_{\Omega^*} F_0(x^*)dx^* \geq \int_L F_0(x^*)dx^* \geq c \int_L dx^* = \infty.$$

Hence

$$\lim_{k \to \infty} \psi(p_k) = \infty.$$

$\square$

An endomorphism s of $\mathbf{R}^n$ preserving a regular convex cone Ω is said to be a **linear automorphism** of Ω. The dual endomorphism s^* of a linear automorphism s of Ω is a linear automorphism of the dual cone Ω^*. For a linear automorphism s of Ω, changing variables to $y^* = s^* x^*$ we have

$$\psi(sx) = \int_{\Omega^*} e^{-\langle sx, x^* \rangle} dx^* = \int_{\Omega^*} e^{-\langle x, s^* x^* \rangle} dx^*$$
$$= \int_{\Omega^*} e^{-\langle x, y^* \rangle} \frac{1}{\det s} dy^*.$$

Thus

$$\psi(sx) = \frac{\psi(x)}{\det s}, \tag{4.2}$$

and the following proposition has been proved.

Proposition 4.4. *Let dx be a D-parallel volume element on $\mathbf{R}^n$. Then $\psi(x)dx$ is a volume element on Ω invariant under linear automorphisms of Ω.*

Proposition 4.5. *The Hessian of $\log \psi$ with respect to D,*

$$Dd \log \psi = \sum_{i,j} \frac{\partial^2 \log \psi}{\partial x^i \partial x^j} dx^i dx^j,$$

is positive definite on Ω.

Proof. Let $x + ta$ be a straight line passing through $x \in \Omega$ in the direction $a \in \mathbf{R}^n$. Then it follows from definition (4.1) that

$$\frac{d}{dt}\psi(x + ta) = -\int_{\Omega^*} e^{-\langle x+ta, x^* \rangle} \langle a, x^* \rangle dx^*,$$
$$\frac{d^2}{dt^2}\psi(x + ta)) = \int_{\Omega^*} e^{-\langle x+ta, x^* \rangle} \langle a, x^* \rangle^2 dx^*.$$

Put $F(x^*) = e^{-\langle x, x^* \rangle/2}$ and $G(x^*) = e^{-\langle x, x^* \rangle/2} \langle a, x^* \rangle$. Then by Schwartz's inequality we have

$$
\begin{aligned}
(Dd \log \psi)(a, a) &= \left. \frac{d^2}{dt^2} \right|_{t=0} \log \psi(x + ta)) \\
&= -\frac{1}{\psi^2(x)} \left(\left. \frac{d}{dt} \right|_{t=0} \psi(x + ta) \right)^2 + \frac{1}{\psi(x)} \left. \frac{d^2}{dt^2} \right|_{t=0} \psi(x + ta(t)) \\
&= \frac{1}{\psi^2(x)} \left\{ \int_{\Omega^*} F^2 dx^* \int_{\Omega^*} G^2 dx^* - \left(\int_{\Omega^*} FG dx^* \right)^2 \right\} > 0.
\end{aligned}
$$

$\square$

Definition 4.6. We call the metric $g = Dd \log \psi$ and the pair (D, g) the **canonical Hessian metric** and the **canonical Hessian structure** on a regular convex cone Ω respectively.

Let s be a linear automorphism of Ω. From $\log \psi(sx) = \log \psi(x) - \log \det s$, it follows that

$$
d(\log \psi \circ s) = d(\log \psi - \log \det s) = d \log \psi,
$$

that is, $d \log \psi$ is invariant under s. Therefore $Dd \log \psi$ is also invariant under s. Thus we have proved the following proposition.

Proposition 4.6. *The canonical Hessian metric $g = D \log \psi$ on a regular convex cone Ω is invariant under linear automorphisms of Ω.*

Let ι be the gradient mapping for a regular convex cone $(\Omega, D, g = Dd \log \psi)$. For a straight line $x + ta$ passing through $x \in \Omega$ in direction $a \in \mathbf{R}^n$ we have

$$
(d\psi)(a) = \left. \frac{d}{dt} \right|_{t=0} \psi(x + ta) = -\int_{\Omega^*} e^{-\langle x, x^* \rangle} \langle a, x^* \rangle dx^*,
$$

and so

$$
(d\psi)_x = -\int_{\Omega^*} e^{-\langle x, x^* \rangle} x^* dx^*.
$$

Hence

$$
\iota(x) = -(d \log \psi)_x = -\frac{(d\psi)_x}{\psi(x)} = \frac{\int_{\Omega^*} e^{-\langle x, x^* \rangle} x^* dx^*}{\int_{\Omega^*} e^{-\langle x, x^* \rangle} dx^*}.
$$

Using the same notation as in the proof of the convergence of $\psi(x)$, we have

$$
\iota(x) = \frac{\int_0^\infty e^{-t} \left(\int_{P_t \cap \Omega^*} x^* dx_t^* \right) dt}{\int_0^\infty e^{-t} \left(\int_{P_t \cap \Omega^*} dx_t^* \right) d}.
$$

Applying this expression together with

$$\int_{P_t \cap \Omega^*} x^* dx_t^* = \left(\frac{t}{n}\right)^n \int_{P_n \cap \Omega^*} x^* dx_n^*,$$

$$\int_{P_t \cap \Omega^*} dx_t^* = \left(\frac{t}{n}\right)^{n-1} \int_{P_n \cap \Omega^*} dx_n^*,$$

we obtain

$$\iota(x) = \frac{\int_0^\infty e^{-t} t^n dt \int_{P_n \cap \Omega^*} x^* dx_n^*}{n \int_0^\infty e^{-t} t^{n-1} dt \int_{P_n \cap \Omega^*} dx_n^*} = \frac{\int_{P_n \cap \Omega^*} x^* dx_n^*}{\int_{P_n \cap \Omega^*} dx_n^*},$$

and so $\iota(x)$ is the center of gravity of $P_n \cap \Omega^*$. Since $P_n \cap \Omega^*$ is convex we have $\iota(x) \in P_n \cap \Omega^*$, that is,

$$\iota(x) \in \Omega^*, \quad \langle x, \iota(x) \rangle = n. \tag{4.3}$$

Proposition 4.7. *The gradient mapping ι is a bijection from Ω to Ω^*.*

Proof. It suffices to show that for any $x^* \in \Omega^*$ there exists a unique $x \in \Omega$ such that $\iota(x) = x^*$. Let Q be a hyperplane defined by

$$Q = \{z \in \mathbf{R}^n \mid \langle z, x^* \rangle = n\}.$$

Suppose that there exists $x \in \Omega$ such that $\iota(x) = x^*$, then by (4.3) it follows that $x \in Q$ and

$$-(d \log \psi)_x (z - x) = \langle z - x, \iota(x) \rangle = \langle z, x^* \rangle - n = 0$$

for all $z \in Q$. Hence ψ restricted on $Q \cap \Omega$ attains its relative minimum at x. Conversely, if ψ attains its relative minimum at x on $Q \cap \Omega$, then

$$0 = (d \log \psi)_x (z - x) = \langle z - x, -\iota(x) \rangle = \langle z, -\iota(x) \rangle + n$$
$$= \langle z, -\iota(x) \rangle + \langle z, x^* \rangle$$

for all $z \in Q$. Hence $\iota(x) = x^*$. Therefore it is enough to show that ψ restricted on $Q \cap \Omega$ has a unique minimum. The characteristic function ψ attains its minimum on a bounded closed set $Q \cap \bar{\Omega}$, and, by Proposition 4.3, ψ grows without limit approaching the boundary of Ω. Therefore ψ attains its minimum on $Q \cap \Omega$. If ψ attains its minimum at different points x_1 and $x_2 \in Q \cap \Omega$, then for the middle point x_3 of the segment joining x_1 and x_2, it follows from the convexity of ψ that $\psi(x_3) < \psi(x_1) = \psi(x_2)$. This is a contradiction. $\qquad \square$

Proposition 4.8. *For a linear automorphism s of a regular convex cone Ω we have*

$$\iota(sx) = s^{*-1} \iota(x), \quad \text{for all } x \in \Omega,$$

where s^* is the dual endomorphism of s.

Proof. For $a \in \mathbf{R}^n$ it follows from equation (4.2) that

$$\langle a, \iota(sx) \rangle = -(d \log \psi)_{sx}(a) = -\left. \frac{d}{dt} \right|_{t=0} \log \psi(sx + ta)$$

$$= -\left. \frac{d}{dt} \right|_{t=0} \log \psi(x + ts^{-1}a) = -(d \log \psi)_x(s^{-1}a) = \langle s^{-1}a, \iota(x) \rangle$$

$$= \langle a, s^{*-1}\iota(x) \rangle. \qquad \square$$

A regular convex cone Ω is said to be **homogeneous** if for any two points x and $y \in \Omega$ there exists a linear automorphism s such that $y = sx$.

Proposition 4.9. *Let Ω be a homogeneous regular convex cone. Then*

(1) *The dual cone Ω^* is homogeneous.*
(2) *Let ψ_Ω and ψ_{Ω^*} be the characteristic functions of Ω and Ω^* respectively. Then $\psi_\Omega(x)\psi_{\Omega^*}(\iota(x))$ is a constant for all $x \in \Omega$.*
(3) *The* Legendre *transform of $\log \psi_\Omega$ coincides with $\log \psi_{\Omega^*}$ except for a constant.*

Proof. Let x^* and y^* be any two points in Ω^*. By Proposition 4.7 there exist x and $y \in \Omega$ such that $x^* = \iota(x)$ and $y^* = \iota(y)$. By the homogeneity, Ω admits a linear automorphism s such that $y = sx$. From Proposition 4.8 we further have that $x^* = \iota(x) = \iota(s^{-1}y) = s^*\iota(y) = s^*y^*$. Thus Ω^* is homogeneous. For any $x \in \Omega$, and any linear automorphism s, we have

$$\psi_\Omega(sx)\psi_{\Omega^*}(\iota(sx)) = \psi_\Omega(sx)\psi_{\Omega^*}(s^{*-1}\iota(x))$$
$$= \frac{\psi_\Omega(x)\psi_{\Omega^*}(\iota(x))}{\det s \det s^{*-1}}$$
$$= \psi_\Omega(x)\psi_{\Omega^*}(\iota(x)).$$

This implies that $\psi_\Omega(x)\psi_{\Omega^*}(\iota(x))$ is a constant c because Ω is homogeneous. The Legendre transform of $\varphi = \log \psi_\Omega$ is given by

$$\varphi^*(\iota(x)) = -\langle x, \iota(x) \rangle - \varphi(x).$$

On the other hand from assertion (2) it follows that $\varphi(x) = \log \psi_\Omega(x) = -\log \psi_{\Omega^*}(\iota(x)) + \log c$. Hence $\varphi^* = \log \psi_{\Omega^*} - n - \log c$. $\qquad \square$

Proposition 4.10. *Let Ω be a homogeneous regular convex cone. Then the Koszul forms α and β for the canonical Hessian structure $(D, g = Dd\psi)$ are expressed by*

(1) $\alpha = d \log \psi$.
(2) $\beta = g$.

Proof. The volume element v determined by g is invariant under the linear automorphism group G of Ω. Moreover, by Proposition 4.4, the volume element $\psi(x)dx$ is also G-invariant. Hence there exists a constant $c \neq 0$ such that

$$v = c\psi(x)dx.$$

Therefore we have

$$\alpha(X)v = D_X v = c(d\psi)(X)dx = (d\log\psi)(X)v. \qquad \square$$

Let B be a strictly bounded domain in $\mathbf{R}^{n-1}$ and let $\Omega_B = \{(tx, t) \mid x \in B, t \in \mathbf{R}^+\}$ be the cone over B. If the boundary ∂B of B is C^2-class, then it follows from [Sasaki (1985)] that the canonical Hessian metric on Ω_B is complete. The author doesn't know whether this metric is complete on Ω_B with less smooth boundary.

We mention here that there is another natural projectively invariant Hessian metric on any regular convex cone Ω given by [Cheng and Yau (1982)].

Theorem 4.4. *Let Ω be a regular convex cone. Then there exists a convex function φ satisfying*

$$\det\left[\frac{\partial^2\varphi}{\partial x^i \partial x^j}\right] = e^{2\varphi},$$

$$\varphi \longrightarrow \infty \quad at \; \partial\Omega,$$

and that the Einstein-Hessian metric $Dd\varphi$ is complete and invariant under linear automorphism of Ω.

4.2 Homogeneous self-dual cones

In this section we show that a homogeneous self-dual cone is symmetric with respect to the canonical Hessian metric and give a bijective correspondence between homogeneous self-dual cones and compact Jordan algebras. Let Ω be a regular convex cone in $\mathbf{R}^n$, and let $\Omega^\star$ be the dual cone of Ω with respect to an inner product (x, y) ;

$$\Omega^\star = \{y \in \mathbf{R}^n \mid (y, z) > 0, \; for \; all \; z \neq 0 \in \bar{\Omega}\}.$$

Theorem 4.5. *Let Ω be a regular convex cone in $\mathbf{R}^n$ with canonical Hessian structure $(D, g = D\log\psi)$, and let $\Omega^\star$ be the dual cone with respect to an inner product (x, y). Identifying $\mathbf{R}_n^*$ with $\mathbf{R}^n$ by the inner product, the gradient mapping $\iota : \Omega \to \Omega^\star$ has a fixed point.*

Proof. By the identification of $\mathbf{R}_n^*$ with $\mathbf{R}^n$ we have

$$(\iota(x), z) = -\sum_i \frac{\partial \log \psi(x)}{\partial x^i} z^i, \quad z \in \mathbf{R}^n.$$

Put $S^{n-1} = \{x \in \mathbf{R}^n \mid (x,x) = n\}$ and suppose that $\log \psi$ attains its minimum at o on a compact set $S^{n-1} \cap \bar{\Omega}$. By Proposition 4.3 it follows that $o \in S^{n-1} \cap \Omega$. Identifying the tangent space of Ω at o with $\mathbf{R}^n$, the tangent space of $S^{n-1} \cap \bar{\Omega}$ at o coincides with $\{a \in \mathbf{R}^n \mid (a, o) = 0\}$. Hence

$$0 = \sum_i \frac{\partial \log \psi(o)}{\partial x^i} a^i = (-\iota(o), a)$$

for all $a \in \mathbf{R}^n$ such that $(a, o) = 0$. Thus $\iota(o) = \lambda\, o$, where $\lambda \in \mathbf{R}$. On the other hand, by (4.3) we have $n = (o, \iota(o)) = \lambda n$. Hence $\iota(o) = o$. $\qquad\square$

Corollary 4.3. *The set $\Omega \cap \Omega^*$ is not an empty set* [Ochiai (1966)].

Definition 4.7. A regular convex cone Ω is said to be a **self-dual cone** if there exists an inner product such that $\Omega = \Omega^*$.

Theorem 4.6. *Let Ω be a homogeneous self-dual regular convex cone with canonical Hessian structure $(D, g = Dd \log \psi)$. Then the gradient mapping $\iota : \Omega \longrightarrow \Omega$ is a symmetry with respect to the canonical Hessian metric g. That is,*

(1) ι *is an isometry.*
(2) ι^2 *is the identity mapping.*
(3) ι *has an isolated fixed point.*

Thus (Ω, g) is a Riemannian symmetric space.

Proof. Let Ω be a self-dual cone with respect to an inner product (x, y). Identifying $\mathbf{R}_n^*$ with $\mathbf{R}^n$ by the inner product, we have $D^* = D$ and $\Omega^* = \Omega^* = \Omega$. Let φ^* be the Legendre transform of $\varphi = \log \psi$. Then $\varphi^* = \varphi - n - \log c$ in the proof of Proposition 4.9. Hence

$$g^* = D^* d\varphi^* = Dd\varphi = g.$$

The assertions (1) and (2) follow from the above result and Proposition 2.7. Let o be a fixed point in Theorem 4.5. For $a = (a^1, \cdots, a^n) \in \mathbf{R}^n$ we set $X_a = \sum_i a^i \frac{\partial}{\partial x^i}$. Identifying the tangent space of Ω at o with $\mathbf{R}^n$, we have

$$g(X_a, X_b) = (D_{X_a} d \log \psi)(X_b) = X_a((d \log \psi)(X_b))$$
$$= X_a(-\iota, b) = -(\iota_*(X_a), b).$$

Hence

$$g_o(a,b) = -((\iota_*)_o(a), b).$$

This shows that $(\iota_*)_o$ is symmetric with respect to (x,y) and negative definite. Moreover $(\iota_*)_o^2$ is the identity mapping by (2). These results together imply $(\iota_*)_o = -I$, hence o is an isolated fixed point of ι. $\qquad\square$

Remark 4.1. Let ψ be the characteristic function of Ω in Theorem 4.6. For positive numbers μ and ν we set $\hat\psi = \mu\psi^\nu$. Then $\hat g = Dd\log\hat\psi$ is positive definite. The gradient mapping $\hat\iota$ for the Hessian structure $(D, \hat g)$ is given by $\hat\iota = \nu\iota$ and is a symmetry with respect to $\hat g$. Since $\iota(\kappa o) = \dfrac{1}{\kappa}o$ for $\kappa > 0$ the isolated fixed point $\hat o$ of $\hat\iota$ is given by $\hat o = \sqrt{\nu}o$.

Let Ω be a homogeneous regular convex cone. The set of all linear automorphisms $G(\Omega)$ of Ω is a closed subgroup of $GL(n, \mathbf{R})$. The identity component G of $G(\Omega)$ is also a closed subgroup of $GL(n, \mathbf{R})$ and acts transitively on Ω. By [Vinberg (1963)] we have:

(i) An isotropy subgroup of G at any point of Ω is a maximal compact subgroup of G.

(ii) G coincides with the connected component of a certain algebraic subgroup of $GL(n, \mathbf{R})$.

For the proof see Lemma 10.1 and Proposition 10.2.

Theorem 4.7. *Let Ω be a homogeneous self-dual regular convex cone with respect to an inner product. We denote by G the identity component of the group of all linear automorphisms of Ω, and by K the isotropy subgroup of G at the isolated fixed point o of the gradient mapping ι. Then*

(1) *G is self-adjoint with respect to the inner product.*

(2) *Let $\mathfrak{g}$ be the Lie algebra of G and let $\mathfrak{m} = \{X \in \mathfrak{g} \mid {}^tX = X\}$ where ${}^tX$ is the adjoint endomorphism of $X \in \mathfrak{g}$. Then the mapping*

$$X \in \mathfrak{m} \longrightarrow (\exp X)o \in \Omega$$

is a diffeomorphism.

(3) *$\Omega = G/K$ is a symmetric space (Definition 9.4)D*

Proof. Since Ω is a self-dual cone, the adjoint linear transformation ${}^ts$ of $s \in G$ is contained in G. The mapping defined by

$$\sigma : s \to {}^ts^{-1}$$

is an involutive automorphism of G. By Proposition 4.8, we have $(\iota s \iota)(x) = {}^t s^{-1} \iota(\iota(x)) = \sigma(s)(x)$ for $s \in G$ and $x \in \Omega$. Hence

$$\sigma(s) = \iota s \iota.$$

Let $\mathfrak{g}$ be the Lie algebra of G and let σ_* be the differential of σ. Then

$$\sigma_*(X) = -{}^t X.$$

For the proof of this theorem we require the following lemma due to [Murakami (1952)].

Lemma 4.1. *Let G be the identity component of an algebraic subgroup of $GL(n, \mathbf{R})$ and let $\mathfrak{g} \subset \mathfrak{gl}(n, \mathbf{R})$ be the Lie algebra of G. Suppose that G is self-adjoint with respect to an inner product. Then*

(1) *Let $\mathfrak{k} = \{X \in \mathfrak{g} \mid {}^t X = -X\}$ and $\mathfrak{m} = \{X \in \mathfrak{g} \mid {}^t X = X\}$. Then*

$$\mathfrak{g} = \mathfrak{k} + \mathfrak{m}, \quad [\mathfrak{k}, \mathfrak{k}] \subset \mathfrak{k}, \quad [\mathfrak{k}, \mathfrak{m}] \subset \mathfrak{m}, \quad [\mathfrak{m}, \mathfrak{m}] \subset \mathfrak{k}.$$

(2) *A subgroup given by $K = \{s \in G \mid {}^t s^{-1} = s\}$ is a maximal compact subgroup of G.*

(3) *A mapping defined by $(X, k) \in \mathfrak{m} \times K \to (\exp X)k \in G$ is a diffeomorphism.*

The isotropy subgroup H of G at the fixed point o of the gradient mapping ι of Ω is a maximal compact subgroup of G. By Lemma 4.1 (2) the group $K = \{s \in G \mid {}^t s^{-1} = s\}$ is a maximal compact subgroup of G. For $s \in K$ we have $\iota(so) = (\iota s \iota)(o) = {}^t s^{-1} o = so$. Hence so is a fixed point of ι. Since o is an isolated fixed point of ι, we have $so = o$ for s sufficiently near the identity, and so $s \in H$. Because K is generated by a neighbourhood of the identity, we have $K \subset H$. Since H and K are both maximal compact subgroups of G, we have $H = K$. Since any element $s \in G$ has a unique expression $s = (\exp X)k$ $(X \in \mathfrak{m}, \ k \in K)$ by the above lemma, and $\Omega = Go = (\exp \mathfrak{m})o$, the mapping $X \in \mathfrak{m} \longrightarrow (\exp X)o \in \Omega$ is a diffeomorphism. Assertion (3) follows from Lemma 4.1 and $\sigma_*(X) = -{}^t X$.

$\square$

Theorem 4.8. *Let Ω be a homogeneous regular convex cone. Then the following conditions (1) and (2) are equivalent.*

(1) *Ω is a self-dual cone.*

(2) *There exists a self-adjoint linear Lie group which acts transitively on Ω.*

Proof. By Theorem 4.7, assertion (2) follows from (1). Suppose that Ω admits a linear Lie group G which acts transitively on Ω and is self-adjoint with respect to an inner product. Let $\Omega^\star$ be the dual cone of Ω with respect to the inner product. By Corollary 4.3. there exists an element $o \in \Omega \cap \Omega^\star$. Since G and ${}^tG$ act transitively on Ω and $\Omega^\star$ respectively, we have $\Omega = Go = {}^tGo = \Omega^\star$. $\square$

Example 4.1. Let V be the vector space of all real symmetric matrices of degree n, and let Ω be the set of all positive definite matrices in V. Then Ω is a regular convex cone in V. We define a representation f of $GL(n, \mathbf{R})$ on V by $f(a)(x) = ax^ta$ where $a \in GL(n, \mathbf{R})$ and $x \in V$. Then $G = f(GL(n, \mathbf{R}))$ acts transitively on Ω and is self-adjoint with respect to the inner product $(x, y) = \mathrm{Tr}\ xy$. By Theorem 4.7 the cone Ω is self-dual with respect to the inner product. Let ψ be the characteristic function of Ω. Then we have

$$\psi(x) = (\det x)^{-\frac{n+1}{2}} \psi(e),$$

where e is the unit matrix. Hence the canonical Hessian metric is given by

$$g = -\frac{n+1}{2} Dd \log \det x.$$

Example 4.2. Let us consider $\mathbf{R}^n$ as the space of all column vectors of degree n. Let J be a matrix of degree n by given by

$$J = \begin{bmatrix} -I_{n-1} & 0 \\ 0 & 1 \end{bmatrix},$$

where I_{n-1} is the unit matrix of degree $n - 1$. The **Lorentz cone** Ω in $\mathbf{R}^n$ is defined by

$$\Omega = \{x = [x^i] \in \mathbf{R}^n \mid {}^txJx > 0,\ x^n > 0\}.$$

We denote $x = [x^i] \in \mathbf{R}^n$ by $x = \begin{bmatrix} x' \\ x^n \end{bmatrix}$ where ${}^tx' = [x^1, \cdots, x^{n-1}]$. Suppose that Ω contains a full straight line $a + sb$, where $a \in \Omega$ and $b \in \mathbf{R}^n$. Since $a^n + sb^n > 0$ for all s, it follows that $b^n = 0$. Hence ${}^t(a + sb)J(a + sb) = -({}^tb'b')s^2 + 2({}^tbJa)s + {}^taJa$. However, this contradicts the inequality ${}^t(a+sb)J(a+sb) > 0$ for all s. We show the convexity of Ω. Note that $x \in \Omega$ if and only if $(x^n)^2 > {}^tx'x'$ and $x^n > 0$. For any $x, y \in \Omega$ and $0 < \lambda < 1$ we have $\{(1-\lambda)x^n + \lambda y^n\}^2 \geq (1-\lambda)^2\ {}^tx'x' + 2\lambda(1-\lambda)({}^tx'x'{}^ty'y')^{1/2} + \lambda^2\ {}^ty'y' \geq {}^t\{(1 - \lambda)x' + \lambda y'\}\{(1 - \lambda)x' + \lambda y'\}$, and so $(1 - \lambda)x + \lambda y \in \Omega$. Thus Ω is a regular convex cone. We claim that the group $\mathbf{R}^+SO(n - 1, 1)$ acts transitively on Ω. Note that $e_n = {}^t[0, \cdots, 0, 1] \in \Omega$. Let $x \in \Omega$ and set $y =$

$({}^txJx)^{-\frac{1}{2}}x$. Then $(y^n)^2 - {}^ty'y' = 1$. There exists $s' \in SO(n-1)$ such that $s'y' = {}^t[0,\cdots,0,({}^ty'y')^{\frac{1}{2}}]$. Putting $s = \begin{bmatrix} s' & 0 \\ 0 & 1 \end{bmatrix}$, we have $s \in SO(n-1,1)$ and $sy = {}^t[0,\cdots,0,({}^ty'y')^{\frac{1}{2}},y^n]$. Introducing

$$r' = \begin{bmatrix} y^n & -({}^ty'y')^{\frac{1}{2}} \\ -({}^ty'y')^{\frac{1}{2}} & y^n \end{bmatrix}, \quad r = \begin{bmatrix} I_{n-2} & 0 \\ 0 & r' \end{bmatrix},$$

we have $r \in SO(n-1,1)$ and $({}^txJx)^{-\frac{1}{2}}rsx = e_n \in \Omega$. Thus $\mathbf{R}^+SO(n-1,1)$ acts transitively on Ω. Since the group $\mathbf{R}^+SO(n-1,1)$ is self-adjoint with respect to the inner product $(x,y) = {}^txy$, by Theorem 4.7 we know that Ω is a self-dual cone with respect to the inner product. Let ψ be the characteristic function of Ω. From $({}^txJx)^{-\frac{1}{2}}rsx = e_n$ it follows that $\psi(e_n) = \psi(({}^txJx)^{-\frac{1}{2}}rsx) = ({}^txJx)^{\frac{n}{2}}\psi(x)$, that is,

$$\psi(x) = ({}^txJx)^{-\frac{n}{2}}\psi(e_n).$$

The canonical Hessian metric is therefore expressed by

$$g = -\frac{n}{2}Dd\log {}^txJx.$$

We shall now give a brief survey of Jordan algebras. For details, the reader may refer to [Braun and Koecher (1966)][Faraut and Korányi (1994)][Koecher (1962)].

Let $\mathbf{A}$ be an algebra over $\mathbf{R}$ with multiplication $a * b$. We set

$$[a * b * c] = a * (b * c) - (a * b) * c.$$

Definition 4.8. A commutative algebra $\mathbf{A}$ over $\mathbf{R}$ is said to be a **Jordan algebra** if it satisfies

$$[a^2 * b * a] = 0,$$

where $a^2 = a * a$.

Let $\mathbf{A}$ be a Jordan algebra. For $a,b,c,d \in \mathbf{A}$ and $\lambda,\mu \in \mathbf{R}$ we expand the equation $[(a+\lambda b+\mu d)^2 * c * (a+\lambda b+\mu d)] = 0$ and compare coefficients of the term $\lambda\mu$ to obtain

$$(a * b) * (c * d) + (b * d) * (c * a) + (d * a) * (c * b)$$
$$= ((a * b) * c) * d + ((b * d) * c) * a + ((d * a) * c) * b.$$

Denoting by L_a the multiplication by $a \in \mathbf{A}$, the above equation is equivalent to

$$L_{(a*b)*c} = L_{a*b}L_c + L_{c*a}L_b + L_{c*b}L_a - L_aL_cL_b - L_bL_cL_a.$$

By the same method we have

$$L_{a*(b*c)} = L_{(c*b)*a} = L_{c*b}L_a + L_{c*a}L_b + L_{a*b}L_c - L_cL_aL_b - L_bL_aL_c.$$

It follows from these formulae that

$$L_{[a*b*c]} = [[L_a, L_c], L_b].$$

Conversely, we have the following lemma due to [Vinberg (1960)].

Lemma 4.2. *Let* **A** *be a finite-dimensional commutative algebra over* **R** *satisfying the following conditions.*

(1) *The bilinear form given by* $(a, b) = \operatorname{Tr} L_{a*b}$ *is non-degenerate.*
(2) $L_{[a*b*c]} = [[L_a, L_c], L_b].$

Then **A** *is a Jordan algebra.*

Proof. By condition (2) we have

$$\operatorname{Tr} L_{[a*b*c]} = 0,$$

for all a, b and $c \in \mathbf{A}$. By a straightforward calculation we obtain

$$[a^2 * b * a] * x = [(a^2 * b) * a * x] + [a^2 * b * (a * x)] - [a^2 * (b * a) * x]$$
$$+ \frac{2}{3}[a * a * [b * a * x]] - \frac{1}{3}[b * a^3 * x].$$

Hence we have

$$([a^2 * b * a], x) = \operatorname{Tr} L_{[a^2*b*a]*x} = 0,$$

for all $x \in \mathbf{A}$. Therefore, by condition (1), we obtain

$$[a^2 * b * a] = 0.$$

$\square$

Definition 4.9. A Jordan algebra **A** is said to be **semisimple** (resp. **compact semisimple**) if the bilinear form given by $(a, b) = \operatorname{Tr} L_{a*b}$ on **A** is non-degenerate (resp. positive definite).

Example 4.3. Let **A** be an associative algebra with multiplication ab. We define a new multiplication on **A** by

$$a * b = \frac{1}{2}(ab + ba).$$

Then it is easy to see that **A** is a Jordan algebra with multiplication $a * b$.

Example 4.4. We denote $a = [a^1, \cdots, a^n] \in \mathbf{R}^n$ by $a = [a', a^n]$ where $a' = [a^1, \cdots, a^{n-1}]$. For a real symmetric matrix S' of degree $n-1$ we define a multiplication on $\mathbf{R}^n$ by

$$a * b = [a^n b' + b^n a', a^n b^n + a' S'\, {}^t b'],$$

where $a = [a', a^n]$ and $b = [b', b^n] \in \mathbf{R}^n$. Since $L_a = a^n I + L_{[a',0]}$ and $L_{a^2} = \{(a^n)^2 + a' S'\, {}^t a'\} I + 2a^n L_{[a',0]}$ where I is the identity mapping on $\mathbf{R}^n$, we have $[L_a, L_{a^2}] = 0$. Hence $\mathbf{R}^n$ is a Jordan algebra with respect to this multiplication. Since $\operatorname{Tr} L_a = n a^n$ we have $\operatorname{Tr} L_{a*b} = n(a^n b^n + a' S'\, {}^t b')$. Therefore, if S' is positive definite, then the inner product given by $(a, b) = \operatorname{Tr} L_{a*b}$ is positive definite, and so the Jordan algebra is compact semisimple.

Now, let us return to the situation of Theorem 4.7. For $X = [a^i_j] \in \mathfrak{g}$ the vector field X^* induced by $\exp(-tX)$ is expressed by

$$X^* = -\sum_{i,j} a^i_j x^j \frac{\partial}{\partial x^i}.$$

We identify the tangent space at the fixed point o of the gradient mapping ι with $\mathbf{R}^n$. Then a mapping $q : \mathfrak{g} \longrightarrow \mathbf{R}^n$ which assigns $X \in \mathfrak{g}$ to the value X^*_o of X^* at o is a surjective linear mapping. The kernel $q^{-1}(o)$ coincides with the Lie algebra $\mathfrak{k}$ of K and the restriction $q|_{\mathfrak{m}}$ of q to $\mathfrak{m}$ is a linear isomorphism from $\mathfrak{m}$ onto $\mathbf{R}^n$. For $a \in \mathbf{R}^n$ we introduce

$$X_a = q|_{\mathfrak{m}}^{-1}(a) \in \mathfrak{m},$$

and define a multiplication on $\mathbf{R}^n$ by

$$a * b = -X_a b. \tag{4.4}$$

We may then derive the following lemma.

Lemma 4.3.

(1) $a * b = b * a.$

(2) $o * a = a * o = a.$

(3) $[[X_a, X_b], X_c] = X_{a*(c*b)-(a*c)*b}.$

Proof. We set $X_a = [a^i_j]$ and $X_b = [b^i_j]$. Then we have $D_{X^*_a} X^*_b = \sum_{i,j,k} b^i_j a^j_k x^k \frac{\partial}{\partial x^i}$. Hence

$$(D_{X^*_a} X^*_b)_o = (X_b X_a) o = b * a.$$

Since $[X_a, X_b] \in \mathfrak{k}$ we have

$$a * b - b * a = (D_{X_b^*} X_a^*)_o - (D_{X_a^*} X_b^*)_o$$
$$= -[X_a^*, X_b^*]_o = -[X_a, X_b]_o^* = 0,$$

which implies (1). Assertion (2) follows from $a = (X_a^*)_o = -X_a o = a * o$. Since

$$[[X_a, X_b], X_c] \in \mathfrak{m},$$
$$[[X_a, X_b], X_c]_o^* = [[X_a, X_b]^*, X_c^*]_o = [X_a, X_b](X_c^*)_o - X_c[X_a, X_b]_o^*$$
$$= [X_a, X_b]c = a * (b * c) - b * (c * a),$$

we obtain

$$[[X_a, X_b], X_c] = X_{a*(c*b)-(a*c)*b}.$$ $\square$

Lemma 4.4.

(1) $(d \log \psi)_o(a) = \operatorname{Tr} X_a$.
(2) $(D d \log \psi)_o(a, b) = -\operatorname{Tr} X_{a*b}$.

Proof. By (4.2) together with $\det \exp X_a = \exp \operatorname{Tr} X_a$, we have

$$(d \log \psi)_o(a) = (X_a^* \log \psi)(o) = \frac{d}{dt}\Big|_{t=0} \log \psi((\exp -tX_a)o)$$
$$= \frac{d}{dt}\Big|_{t=0} \log(\det(\exp tX_a)\psi(o)) = \operatorname{Tr} X_a.$$

Since $d \log \psi$ is G-invariant, $(d \log \psi)(X_a^*)$ is a constant. Hence, from $(D_{X_b^*} X_a^*)_o = a * b$, we have

$$(D d \log \psi)_a(a, b) = (D_{X_b^*} d \log \psi)(X_a^*)(o)$$
$$= (X_b^*)_o((d \log \psi)(X_a^*)) - (d \log \psi)_o((D_{X_b^*} X_a^*)_o)$$
$$= -(d \log \psi)_o(a * b) = -\operatorname{Tr} X_{a*b}.$$ $\square$

Theorem 4.9. *Let Ω be a homogeneous self-dual regular convex cone in $\mathbf{R}^n$. We define an operation of multiplication $a * b$ on $\mathbf{R}^n$ by (4.4). Then the algebra $\mathbf{R}^n$ with multiplication $a * b$ is a compact semisimple Jordan algebra, and*

$$\Omega = \left\{ \exp a = \sum_{n=0}^{\infty} \frac{1}{n!} a^n \mid a \in \mathbf{R}^n \right\},$$

where $a^n = \overbrace{a * \cdots * a}^{n \text{ terms}}$.

Proof. By Lemmata 4.2, 4.3 and 4.4, we know that the algebra $\mathbf{R}^n$ with multiplication $a * b = -X_a b$ is a compact semisimple Jordan algebra. Since

$$(\exp - X_a)o = \sum_{n=0}^{\infty} \frac{1}{n!} (-X_a)^n o = \sum_{n=0}^{\infty} \frac{1}{n!} a^n = \exp a,$$ by Theorem 4.7 and

Lemma 4.3 we have

$$\Omega = \{(\exp X)o \mid X \in \mathfrak{m}\} = \left\{ \exp a = \sum_{n=0}^{\infty} \frac{1}{n!} a^n \mid a \in \mathbf{R}^n \right\}.$$

□

It is known that the converse of Theorem 4.9 also holds.

Theorem 4.10. *Let* $\mathbf{A}$ *be a finite-dimensional compact semisimple Jordan algebra over* $\mathbf{R}$*. Then a set* $\Omega(\mathbf{A})$ *defined by*

$$\Omega(\mathbf{A}) = \left\{ \exp a = \sum_{n=0}^{\infty} \frac{1}{n!} a^n \mid a \in \mathbf{A} \right\}$$

is a homogeneous self-dual regular convex cone in $\mathbf{A}$*.*

For the proof of this theorem the reader may refer to [Faraut and Korányi (1994)][Koecher (1962)].

Theorem 4.11. *Let* Ω *be a homogeneous self-dual regular convex cone in* $\mathbf{R}^n$*. Then the multiplication of the Jordan algebra given by equation* (4.4) *is expressed by*

$$(a * b)^i = \sum_{j,k} \gamma^i{}_{jk}(o) a^j b^k, \tag{4.5}$$

where $\gamma^i{}_{jk}(o)$ *are the values of the components of the difference tensor* $\gamma = \nabla - D$ *at the fixed point* o *of the gradient mapping.*

Proof. We set $a = [a^i]$ and $X_a = [a_j^i]$. By definition we have $a * b = -X_a b = -\sum_j a_j^i b^j \left(\frac{\partial}{\partial x^i} \right)_o$ D Upon setting $A_{X_a^*} = \mathcal{L}_{X_a^*} - D_{X_a^*}$, where $\mathcal{L}_{X_a^*}$

is the Lie derivative with respect to X_a^*, we have $A_{X_a^*} \frac{\partial}{\partial x^j} = -D_{\frac{\partial}{\partial x^j}} X_a^* = \sum_k a_j^k \frac{\partial}{\partial x^k}$. Hence

$$(A_{X_a^*} X_b^*)_o = -\sum_{j,k} a_j^k b^j \left(\frac{\partial}{\partial x^k} \right)_o = a * b.$$

Since X_a^* is a Killing vector field with respect to g, we have

$$0 = (\mathcal{L}_{X_a^*} g) \left(\frac{\partial}{\partial x^k}, \frac{\partial}{\partial x^l} \right) = X_a^* g_{kl} - g \left(\left[X_a^*, \frac{\partial}{\partial x^k} \right], \frac{\partial}{\partial x^l} \right) - g \left(\frac{\partial}{\partial x^k}, \left[X_a^*, \frac{\partial}{\partial x^l} \right] \right).$$

This implies

$$\sum_{i,j} \gamma_{ikl} a^i_j x^j = -\frac{1}{2} \Big(\sum_i a^i_k g_{il} + \sum_i a^i_l g_{ik} \Big),$$

and so we have

$$\gamma_{X^*_a} \frac{\partial}{\partial x^j} = -\sum_{i,k,p} \gamma^k_{ij} a^i_p x^p \frac{\partial}{\partial x^k} = \frac{1}{2} \Big(\sum_{i,k} a^{ik} g_{ij} \frac{\partial}{\partial x^k} + \sum_{i,k} a^i_j \delta^k_i \frac{\partial}{\partial x^k} \Big)$$

$$= \sum_k a^k_j \frac{\partial}{\partial x^k} = A_{X^*_a} \frac{\partial}{\partial x^j}.$$

and

$$\gamma_{X^*_a} = A_{X^*_a}.$$

Finally, we have therefore

$$a * b = (A_{X^*_a} X^*_b)_o = (\gamma_{X^*_a} X^*_b)_o = \sum_{i,j,k} \gamma^k_{ij}(o) a^i b^j \Big(\frac{\partial}{\partial x^k} \Big)_o.$$

$\square$

Theorem 4.12. *Let Ω be a homogeneous self-dual regular convex cone. Then the difference tensor $\gamma = \nabla - D$ is ∇-parallel. In particular, the curvature tensor of ∇ is ∇-parallel, that is, (Ω, g) is a Riemannian symmetric space.*

Proof. Since X^*_a is an infinitesimal affine transformation with respect to D and ∇, we have

$$[\mathcal{L}_{X^*_a}, D_{X^*_b}] = D_{\mathcal{L}_{X^*_a} X^*_b}, \quad [\mathcal{L}_{X^*_a}, \nabla_{X^*_b}] = \nabla_{\mathcal{L}_{X^*_a} X^*_b},$$

and so

$$[\mathcal{L}_{X^*_a}, \gamma_{X^*_b}] = \gamma_{\mathcal{L}_{X^*_a} X^*_b}.$$

Introducing $A^{\nabla}_{X^*_a} = \mathcal{L}_{X^*_a} - \nabla_{X^*_a}$, we have $\gamma_{X^*_a} = \nabla_{X^*_a} - D_{X^*_a} = -A^{\nabla}_{X^*_a} + A_{X^*_a}$. In the proof of Theorem 4.11 we showed that $A_{X^*_a} = \gamma_{X^*_a}$. Hence $A^{\nabla}_{X^*_a} = 0$, that is, $\nabla_{X^*_a} = \mathcal{L}_{X^*_a}$, and so

$$[\nabla_{X^*_a}, \gamma_{X^*_b}] = [\mathcal{L}_{X^*_a}, \gamma_{X^*_b}] = \gamma_{\mathcal{L}_{X^*_a} X^*_b} = \gamma_{\nabla_{X^*_a} X^*_b}.$$

It follows from Lemma 2.2 that γ is ∇-parallel and the curvature tensor of ∇ is also ∇-parallel. $\square$

Let us consider two typical examples along the lines of the above argument.

Example 4.5. Let Ω be the set of all positive definite real symmetric matrices of degree n. Then, by Example 4.1, Ω is a homogeneous self-dual regular convex cone. We employ the same notation as in this previous example. We proved the group $G = f(GL(n, \mathbf{R}))$ is self-adjoint with respect to the inner product $(x, y) = \mathrm{Tr}\, xy$ and acts transitively on Ω. Put $\hat{\psi}(x) = (\det x)^{-1}$, then $\hat{g} = Dd \log \hat{\psi}$ is a G-invariant Hessian metric. Denoting by X_{ij} the cofactor of the (i, j) component of a matrix x, we have

$$\frac{\partial \log(\det x)^{-1}}{\partial x^{ij}} = \begin{cases} -(\det x)^{-1} X_{ii} \\ -2(\det x)^{-1} X_{ij}, \quad i < j. \end{cases}$$

Let $\hat{\imath}$ be the gradient mapping with respect to $\hat{\psi}$. Then

$$(\hat{\imath}(x), a) = -(d \log \hat{\psi})_x(a)$$

$$= -\frac{d}{dt}\Big|_{t=0} (\log \hat{\psi})(x + ta)$$

$$= -\sum_{i \leq j} \frac{\partial}{\partial x^{ij}} (\log \det x^{-1}) a^{ij}$$

$$= \frac{1}{\det x}\Big\{ \sum_i X_{ii} a^{ii} + 2 \sum_{i<j} X_{ij} a^{ij} \Big\} = \frac{1}{\det x} \sum_{i,j} X_{ij} a^{ij}$$

$$= ((\det x)^{-1} X, a) = (x^{-1}, a).$$

Thus

$$\hat{\imath}(x) = x^{-1},$$

and the isolated fixed point of $\hat{\imath}$ is the unit matrix e. Let f be the differential of $\boldsymbol{f}$, then

$$K = \{\mathbf{f}(a) \mid a \in SO(n)\},$$

$$\mathfrak{k} = \{f(A) \mid {}^t A = -A, \ A \in \mathfrak{gl}(n, \mathbf{R})\},$$

$$\mathfrak{m} = \{f(A) \mid {}^t A = A, \ A \in \mathfrak{gl}(n, \mathbf{R})\},$$

and

$$\Omega = (\exp \mathfrak{m})e = \exp V.$$

For $a \in V$, the vector field X_a induced by $\boldsymbol{f}(\exp(-ta))$ is

$$X_a = -\frac{1}{2} f(a).$$

Hence the multiplication of the corresponding Jordan algebra is given by

$$a * b = \frac{1}{2}(ab + ba).$$

Example 4.6. We consider the Lorentz cone Ω of Example 4.2, and will employ the same notation as in this previous example. We proved that the group $G = \mathbf{R}^+ SO(n-1,1)$ is self-adjoint with respect to the inner product $(x,y) = {}^txy$ and acts transitively on Ω. Set $\hat{\varphi}(x) = -\dfrac{1}{2}\log {}^txJx$. Then $(D, g = Dd\hat{\varphi})$ is a Hessian structure on Ω. The gradient mapping $\hat{\imath}$ for the Hessian structure is given by

$$\hat{\imath}(x) = \frac{1}{{}^txJx}\,{}^t[-x^1,\cdots,-x^{n-1},x^n],$$

and its fixed point is $\hat{o} = {}^t[0,\cdots,0,1]$. Let $\mathfrak{g}$ be the Lie algebra of $G = \mathbf{R}^+ SO(n-1,1)$, in which case we have

$$\mathfrak{g} = \left\{ \begin{bmatrix} A + pI_{n-1} & q \\ {}^tq & p \end{bmatrix} \,\middle|\, {}^tA = -A,\ p \in \mathbf{R},\ q \in \mathbf{R}^{n-1} \right\}.$$

Put $\mathfrak{k} = \{X \in \mathfrak{g} \mid {}^tX = -X\}$ and $\mathfrak{m} = \{X \in \mathfrak{g} \mid {}^tX = X\}$. Then

$$\mathfrak{k} = \left\{ \begin{bmatrix} A & 0 \\ 0 & 0 \end{bmatrix} \,\middle|\, {}^tA = -A \right\},$$

$$\mathfrak{m} = \left\{ \begin{bmatrix} pI_{n-1} & q \\ {}^tq & p \end{bmatrix} \,\middle|\, p \in \mathbf{R},\ q \in \mathbf{R}^{n-1} \right\}.$$

For $X = \begin{bmatrix} pI_{n-1} & q \\ {}^tq & p \end{bmatrix} \in \mathfrak{m}$ we have

$$X^* = -\sum_{i=1}^{n-1}\left(px^i + q^ix^n\right)\frac{\partial}{\partial x^i} - \left(\sum_{i=1}^{n-1}q^ix^i + px^n\right)\frac{\partial}{\partial x^n}.$$

Since

$$X_{\hat{o}}^* = -\sum_{i=1}^{n-1}q^i\left(\frac{\partial}{\partial x^i}\right)_{\hat{o}} - p\left(\frac{\partial}{\partial x^n}\right)_{\hat{o}} = -\,{}^t[q^1,\cdots,q^{n-1},p],$$

for $a = {}^t[a^1,\cdots,a^n] \in \mathbf{R}^n$ we have

$$X_a = -\begin{bmatrix} a^nI_{n-1} & a' \\ {}^ta' & a^n \end{bmatrix}.$$

Hence

$$a * b = -X_a b = -\begin{bmatrix} a^nI_{n-1} & a' \\ {}^ta' & a^n \end{bmatrix}\begin{bmatrix} b' \\ b^n \end{bmatrix} = \begin{bmatrix} a^nb' + b^na' \\ {}^tab \end{bmatrix}.$$

It is known [Faraut and Korányi (1994)][Koecher (1962)] that a finite-dimensional compact semisimple Jordan algebra $\mathbf{A}$ can be decomposed into a direct sum;

$$\mathbf{A} = \mathbf{A}_1 + \cdots + \mathbf{A}_k,$$

where $\mathbf{A}_1, \cdots, \mathbf{A}_k$ are compact simple Jordan algebras. Therefore, by Theorem 4.10, the homogeneous self-dual regular convex cone $\Omega(\mathbf{A})$ corresponding to the Jordan algebra $\mathbf{A}$ may be decomposed in such a way;

$$\Omega(\mathbf{A}) = \Omega(\mathbf{A}_1) + \cdots + \Omega(\mathbf{A}_k).$$

Thus the classification of self-dual homogeneous convex regular cones is reduced to that of compact simple Jordan algebras.

The classification of compact simple Jordan algebras is given by the following theorem [Faraut and Korányi (1994)][Koecher (1962)].

Theorem 4.13. *A finite-dimensional compact simple Jordan algebra over* $\mathbf{R}$ *is isomorphic to one of the followings.*

(1) *The algebra of all real symmetric matrices of degree* n.
(2) *The algebra of all Hermitian matrices of degree* n.
(3) *The algebra of all quaternion Hermitian matrices of degree* n.
(4) *The algebra of all Cayley Hermitian matrices of degree* 3.

The multiplications of the above Jordan algebras are given by

$$a * b = \frac{1}{2}(ab + ba),$$

where the operations of the right of the equation are ordinary matrix operations.

(5) *The algebra* $\mathbf{R}^n$ *with multiplication*

$$a * b = [a^n b' + b^n a', \ a^{\,t}b],$$

where $a = [a', a^{n-1}]$ *and* $b = [b', b^{n-1}] \in \mathbf{R}^n$ (cf. *Example 4.6*).

Chapter 5

Hessian structures and affine differential geometry

The approach of using the concept of affine immersions to study affine differential geometry was proposed by K. Nomizu [Nomizu and Sasaki (1994)]. In section **5.1**, we give a brief survey of affine immersions. In section **5.2**, applying Nomizu's method, we consider level surfaces of the potential function φ of a Hessian domain $(\Omega, D, g = Dd\varphi)$ in $\mathbf{R}^{n+1}$. That is, we develop affine differential geometry of level surfaces by using the gradient vector field E of φ as a transversal vector field. We give a characterization of the potential function φ in terms of affine fundamental forms, shape operators and transversal connection forms for the foliation of φ. In section **5.3** we study the relations between affine differential geometries induced by the gradient vector field E and three connections; D, the Levi-Civita connection ∇ of g, and the dual flat connection D'. We investigate the Laplacian of the gradient mapping and show that an analogy of the affine Bernstein problem proposed by S.S. Chern can be proved.

5.1 Affine hypersurfaces

In this section we give a brief survey of the affine differential geometry of affine hypersurfaces which will be needed in this chapter, for further details the interested reader may refer to [Nomizu and Sasaki (1994)].

Let M be an n-dimensional manifold. A smooth mapping

$$\phi : M \longrightarrow \mathbf{R}^{n+1}$$

is said to be a (hypersurface) **immersion** of M if the differential ϕ_{*p} of ϕ at any point $p \in M$ is injective.

A vector field $\xi : p \in M \longrightarrow \xi_p$ is said to be **transversal** with respect to an immersion $\phi : M \longrightarrow \mathbf{R}^{n+1}$ if it satisfies

$$T_{\phi(p)}\mathbf{R}^{n+1} = \phi_{*p}(T_pM) + \mathbf{R}\xi_p.$$

A pair (ϕ, ξ) of an immersion $\phi : M \longrightarrow \mathbf{R}^{n+1}$ and a transversal vector field ξ is said to be an **affine immersion** of M.

For an immersion $\phi : M \longrightarrow \mathbf{R}^{n+1}$, if a vector field ϕ assigning each point $p \in M$ to the position vector $\phi(p)$ is transversal, then the affine immersion $(\phi, -\phi)$ is said to be a **central affine immersion**.

Let (ϕ, ξ) be an affine immersion of M, and let D be the standard flat connection on $\mathbf{R}^{n+1}$. We denote by

$$D_X \phi_*(Y)$$

the covariant derivative along ϕ induced by D where X and $Y \in \mathfrak{X}(M)$. We decompose $D_X\phi_*(Y)$ into a tangential component to $\phi(M)$ and a component in the direction of ξ, and define a torsion-free affine connection D^M and a symmetric bilinear form h on M by

$$D_X\phi_*(Y) = \phi_*(D_X^M Y) + h(X, Y)\xi. \quad \text{(Gauss formula)} \qquad (5.1)$$

The connection D^M and the form h are said to be the **induced connection** and the **affine fundamental form** induced by an affine immersion (ϕ, ξ) respectively.

We decompose $D_X\xi$ into a tangential component to $\phi(M)$ and a component in the direction ξ, and define a $(1,1)$-tensor field S and a 1-form τ by

$$D_X\xi = -\phi_*(SX) + \tau(X)\xi. \quad \text{(Weingarten formula)} \qquad (5.2)$$

S and τ are called the affine **shape operator** and the **transversal connection form** induced by an affine immersion (ϕ, ξ) respectively.

Definition 5.1. An affine immersion (ϕ, ξ) is said to be **non-degenerate** if the affine fundamental form h is non-degenerate.

We take a fixed D-parallel volume element ω on $\mathbf{R}^{n+1}$. For an affine immersion (ϕ, ξ) we define a volume element θ on M by

$$\theta(X_1, \cdots, X_n) = \omega(\phi_*(X_1), \cdots, \phi_*(X_n), \xi),$$

and call it the **induced volume element** for (ϕ, ξ). Then

$$D_X^M\theta = \tau(X)\theta.$$

Definition 5.2. If $\tau = 0$, that is, $D^M\theta = 0$, then (ϕ, ξ) is called an **equiaffine immersion**.

Proposition 5.1. *Let* (ϕ, ξ) *be an equiaffine immersion of* M. *Then*

(1) *The curvature tensor* R_{DM} *of* D^M *is expressed by*
$$R_{DM}(X, Y)Z = h(Y, Z)SX - h(X, Z)SY. \qquad \text{(Gauss equation)}$$

(2) $(D_X^M h)(Y, Z) = (D_Y^M h)(X, Z).$ (Codazzi equation for h)

(3) $(D_X^M S)(Y) = (D_Y^M S)(X).$ (Codazzi equation for S)

(4) $h(X, SY) = h(SX, Y).$ (Ricci equation)

Theorem 5.1. *Let* (ϕ, ξ') *be a non-degenerate affine immersion. Then there exists a non-degenerate affine immersion* (ϕ, ξ) *satisfying the following conditions.*

(1) *The transversal connection form* τ *vanishes.*
(2) *The induced volume element* θ *coincides with the volume element for the affine fundamental form* h.

The transversal vector field ξ *satisfying these conditions is unique up to its sign.*

Proof. Let h', τ' and θ' be the affine fundamental form, the transversal connection form and the induced volume element for (ϕ, ξ') respectively. Choose a basis $X_1', \cdots, X_n'$ such that
$$\theta'(X_1', \cdots, X_n') = 1,$$
and define a function f' by
$$f' = |\det[h'(X_i', X_j')]|^{\frac{1}{n+2}}.$$

Let Z' be a vector field on M determined by
$$h'(Z', \ ) = -f'\tau' - df'.$$

Then a vector ξ given by
$$\xi = f'\xi' + Z'$$

satisfies the conditions (1) and (2). $\qquad\qquad\square$

Definition 5.3. A non-degenerate affine immersion (ϕ, ξ) satisfying the conditions (1) and (2) of Theorem 5.1 is said to be the **Blaschke immersion**, and ξ is called the **affine normal**.

Let (ϕ, ξ) be a non-degenerate affine immersion. We denote by ∇^M the Levi-Civita connection of the affine fundamental form h, and by γ^M the difference tensor between ∇^M and D^M,

$$\gamma_X^M = \nabla_X^M - D_X^M.$$

We are now in a position to introduce the following proposition.

Proposition 5.2.

(1) $\gamma_X^M Y = \gamma_Y^M X$.

(2) $h(\gamma_X^M Y, Z) = h(Y, \gamma_X^M Z)$.

(3) $h(\gamma_X^M Y, Z) = \dfrac{1}{2}(D_X^M h)(Y, Z)$.

Proof. Assertion (1) follows from the vanishing of the torsion tensors of ∇^M and D^M. By assertion (1) and $\nabla^M h = 0$, we obtain

$$h(\gamma_Z^M Y, X) = -(\gamma_Z^M h)(Y, X) - h(Y, \gamma_Z^M X)$$
$$= (D_Z^M h)(Y, X) - h(Y, \gamma_X^M Z).$$

Using this result together with Proposition 5.1 (2), we have

$$0 = h(\gamma_Z^M Y, X) - h(\gamma_Y^M Z, X)$$
$$= (D_Z^M h)(Y, X) - (D_Y^M h)(Z, X) - h(Y, \gamma_X^M Z) + h(Z, \gamma_X^M Y)$$
$$= h(\gamma_X^M Y, Z) - h(Y, \gamma_X^M Z),$$

which implies (2). Assertion (3) follows from

$$(D_Z^M h)(Y, X) = h(\gamma_Z^M Y, X) + h(Y, \gamma_Z^M X) = 2h(\gamma_Z^M Y, X). \qquad \square$$

Proposition 5.3. Let (ϕ, ξ) be a Blaschke *immersion. Then*

$$\mathrm{Tr}\ \gamma_X^M = 0. \qquad \textbf{(apolarity condition)}$$

Proof. $0 = D_X^M \theta = (-\gamma_X^M + \nabla_X^M)\theta = -\gamma_X^M \theta = -(\mathrm{Tr}\ \gamma_X^M)\theta D \qquad \square$

Let (ϕ, ξ) be an equiaffine immersion. For each point $p \in M$ we define $\nu_p \in \mathbf{R}_{n+1}^*$ by

$$\nu_p(\phi_* X_p) = 0, \quad X_p \in T_p M,$$
$$\nu_p(\xi_p) = 1.$$

Then a mapping defined by

$$\nu : p \in M \longrightarrow \nu_p \in \mathbf{R}_{n+1}^* - \{0\}$$

is said to be the **conormal mapping**. This mapping is an affine immersion. A vector field which assigns each $p \in M$ to a vector ν_p is transversal to $\nu(M)$. Hence the pair $(\nu, -\nu)$ of the affine immersion $\nu : p \in M \longrightarrow \nu_p \in \mathbf{R}^*_{n+1} - \{0\}$ and the transversal vector field $-\nu$ is a central affine immersion. We denote by D^* the standard flat connection on $\mathbf{R}^*_{n+1}$. Let

$$D^*_X \nu_*(Y) = \nu_*(\bar{D}^M_X Y) + \bar{h}(X,Y)(-\nu)$$

be the Gauss formula for the central affine immersion $(\nu, -\nu)$. Then we may derive the following proposition.

Proposition 5.4. *For a non-degenerate equiaffine immersion we have*

(1) $\bar{h}(X,Y) = h(SX,Y)$.
(2) *The connections D^M and $\bar{D}^M$ are dual with respect to the affine fundamental form h (cf. Definition 2.8), that is,*

$$Xh(Y,Z) = h(D^M_X Y, Z) + h(Y, \bar{D}^M_X Z).$$

Proof. We first show $\nu_*(Y)(\xi) = 0$ and $\nu_*(Y)(\phi_*(Z)) = -h(Y,Z)$. Differentiating $\nu(\xi) = 1$ by Y we have

$$0 = Y(\nu(\xi)) = (D^*_Y \nu)(\xi) + \nu(D_Y \xi) = \nu_*(Y)(\xi) + \nu(-SY)$$
$$= \nu_*(Y)(\xi).$$

While, differentiating $\nu(\phi_*(Z)) = 0$ by Y we obtain

$$0 = Y(\nu(\phi_*(Z))) = (D^*_Y \nu)(\phi_*(Z)) + \nu(D_Y \phi_*(Z))$$
$$= \nu_*(Y)(\phi_*(Z)) + h(Y,Z).$$

Therefore, differentiating $\nu_*(Y)(\xi) = 0$ by X, we obtain

$$0 = X(\nu_*(Y)(\xi)) = (D^*_X \nu_*(Y))(\xi) + \nu_*(Y)(D_X \xi)$$
$$= \{\nu_*(\bar{D}_X Y) - \bar{h}(X,Y)\nu\}(\xi) + \nu_*(Y)(-\phi_*(SX))$$
$$= -\bar{h}(X,Y) + h(Y,SX).$$

The proof of assertion (1) is complete. Differentiating $\nu_*(Y)(\phi_*(Z)) = -h(Y,Z)$ by X we have

$$Xh(Y,Z) = -X(\nu_*(Y)(\phi_*(Z)))$$
$$= -(D^*_X \nu_*(Y))(\phi_*(Z)) - \nu_*(Y)(D_X \phi_*(Z))$$
$$= -\{\nu_*(\bar{D}^M_X Y) + \bar{h}(X,Y)(-\nu)\}(\phi_*(Z))$$
$$\quad - \nu_*(Y)(\phi_*(D^M_X Z) + h(X,Z)\xi)$$
$$= h(\bar{D}^M_X Y, Z) + h(Y, D^M_X Z),$$

and so the proof of assertion (2) is also complete. $\qquad\square$

Definition 5.4. For a non-degenerate affine immersion (ϕ, ξ) the Laplacian $\Delta_{(h,D)}\phi$ of $\phi : M \longrightarrow \mathbf{R}^{n+1}$ with respect to h and D is by definition

$$\Delta_{(h,D)}\phi = \sum_{i,j} h^{ij} \left\{ D_{\partial/\partial x^i} \phi_* \left(\frac{\partial}{\partial x^j} \right) - \phi_* \left(\nabla^M_{\partial/\partial x^i} \frac{\partial}{\partial x^j} \right) \right\},$$

where $\{x^1, \cdots, x^n\}$ is a local coordinate system on M.

Theorem 5.2. *Let (ϕ, ξ) be a Blaschke immersion. Then the Laplacian of the conormal mapping $\nu : M \longrightarrow \mathbf{R}^*_{n+1}$ with respect to h and D^* is given by*

$$\Delta_{(h,D^*)}\nu = -\operatorname{Tr} S.$$

Proof. Since $\nabla^M_X = \bar{D}^M_X - \gamma^M_X$ by Proposition 5.4 and Lemma 2.3, it follows that

$$D^*_{\partial/\partial x^i} \nu_* \left(\frac{\partial}{\partial x^j} \right) - \nu_* \left(\nabla^M_{\partial/\partial x^i} \frac{\partial}{\partial x^j} \right)$$

$$= \nu_* \left(\bar{D}^M_{\partial/\partial x^i} \frac{\partial}{\partial x^j} \right) - \bar{h} \left(\frac{\partial}{\partial x^i}, \frac{\partial}{\partial x^j} \right) \nu - \nu_* \left(\bar{D}^M_{\partial/\partial x^i} \frac{\partial}{\partial x^j} - \gamma^M_{\partial/\partial x^i} \frac{\partial}{\partial x^j} \right)$$

$$= -h \left(S \frac{\partial}{\partial x^i}, \frac{\partial}{\partial x^j} \right) \nu + \nu_* \left(\gamma^M_{\partial/\partial x^i} \frac{\partial}{\partial x^j} \right).$$

By the apolarity condition $\sum_{i,j} h^{ij} \gamma^M_{\partial/\partial x^i} \frac{\partial}{\partial x^j} = 0$ we have

$$\Delta_{(h,D^*)}\nu = \sum_{i,j} h^{ij} \left\{ D^*_{\partial/\partial x^i} \nu_* \left(\frac{\partial}{\partial x^j} \right) - \nu_* \left(\nabla^M_{\partial/\partial x^i} \frac{\partial}{\partial x^j} \right) \right\}$$

$$= -(\operatorname{Tr} S)\nu. \qquad \square$$

5.2 Level surfaces of potential functions

Let $(\Omega, D, g = Dd\varphi)$ be a Hessian domain in $\mathbf{R}^{n+1}$. In this section we assume that the Hessian metric $g = Dd\varphi$ is non-degenerate unless otherwise specified.

A non-empty set M given by

$$M = \{x \in \Omega \mid \varphi(x) = c\}$$

is said to be a **level surface** of φ. A level surface M is an n-dimensional submanifold of $\mathbf{R}^{n+1}$ if and only if for all $x \in M$

$$d\varphi_x \neq 0.$$

A vector field E defined by

$$g(X, E) = d\varphi(X)$$

is said to be the **gradient vector field** of φ with respect to g. The gradient vector field E is transversal to M if and only if for all $x \in M$

$$d\varphi(E)(x) \neq 0.$$

Henceforth in this chapter we will assume

(A.1) $d\varphi_x \neq 0$,

(A.2) $d\varphi(E)(x) \neq 0$, for all $x \in \Omega$.

Using the standard flat connection D on $\mathbf{R}^{n+1}$ and the gradient vector field E, we define the induced connection D^M, the affine fundamental form h, the shape operator S and the transversal connection form τ by

$$D_X Y = D_X^M Y + h(X, Y)E, \qquad \text{(Gauss formula)}$$
$$D_X E = -S(X) + \tau(X)E. \qquad \text{(Weingarten formula)}$$

Note that a vector field X along M is tangential to M if and only if

$$d\varphi(X) = 0.$$

Lemma 5.1.

(1) $h = -\dfrac{1}{d\varphi(E)} g.$

(2) $\tau = d \log |d\varphi(E)| D$

Proof. The assertions follow from

$$g(X, Y) = (D_X d\varphi)(Y) = X((d\varphi)(Y)) - d\varphi(D_X Y)$$
$$= -d\varphi(D_X^M Y + h(X, Y)E) = -(d\varphi(E))h(X, Y),$$
$$0 = (d\varphi)(X) = g(X, E) = (D_X d\varphi)(E)$$
$$= X(d\varphi(E)) - d\varphi(D_X E) = X(d\varphi(E)) - \tau(X)d\varphi(E),$$

for X and $Y \in \mathfrak{X}(M)$. □

This lemma implies that level surfaces of the potential function φ are non-degenerate hypersurfaces in the sense of Definition 5.1.

Lemma 5.2. *The following conditions* (1) *and* (2) *are equivalent.*

(1) $\tau = 0$.

(2) $D_E E = \lambda E$.

Under the above conditions we have

$$d(d\varphi(E)) = (\lambda + 1)d\varphi.$$

In particular, $d\varphi(E)$ is a constant on Ω if and only if $\lambda = -1$.

Proof.   For $X \in \mathfrak{X}(M)$ we have

$$(D_E g)(E, X) = E(g(E, X)) - g(D_E E, X) - g(E, D_E X)$$
$$= -g(D_E E, X) - d\varphi(D_E X)$$
$$= -g(D_E E, X) - \{E(d\varphi(X)) - (D_E d\varphi)(X)\}$$
$$= -g(D_E E, X).$$

Alternatively, by the Codazzi equation for (D, g) (Proposition 2.1(2)) and Lemma 5.1 we obtain

$$(D_E g)(E, X) = (D_X g)(E, E) = X(g(E, E)) - 2g(D_X E, E)$$
$$= X(d\varphi(E)) - 2\tau(X)d\varphi(E) = -\tau(X)d\varphi(E).$$

These results together yield $g(D_E E, X) = \tau(X)d\varphi(E)$, which implies that assertions (1) and (2) are equivalent. For the case $\tau = 0$, we have for $X \in \mathfrak{X}(M)$

$$(d(d\varphi(E)))(X) = ((d\varphi(E))\tau)(X) = 0,$$
$$(d(d\varphi(E)))(E) = E(d\varphi(E)) = (D_E d\varphi)(E) + d\varphi(D_E E)$$
$$= ((\lambda + 1)d\varphi)(E).$$

Hence we obtain $d(d\varphi(E)) = (\lambda + 1)d\varphi D$ $\square$

Example 5.1. We denote by g_M the restriction of g to M. Suppose that $\tau = 0$ and $S = kI$ where I is the identity mapping on $\mathfrak{X}(M)$ and k is a constant. Then by Lemma 5.1 $d\varphi(E)$ is a constant. It follows from Proposition 5.1 (1) and (2) and Lemma 5.1 (1) that the pair (D^M, g_M) is the Codazzi structure of constant curvature $-k/d\varphi(E)$.

Theorem 5.3. *Let M be a level surface of φ and let ξ be the affine normal with respect to the D-parallel volume element $\omega = dx^1 \wedge \cdots \wedge dx^{n+1}$ on $\mathbf{R}^{n+1}$. We decompose ξ into the tangential component Z to M and a component in the direction of E,*

$$\xi = Z + \mu E.$$

Then

$$\mu = |\det[g_{ij}]/(d\varphi(E))^{n+1}|^{\frac{1}{n+2}},$$
$$g(X, Z) = (d(\mu d\varphi(E)))(X) \quad for \ X \in \mathfrak{X}(M),$$

and

$$d\log \mid \mu d\varphi(E) \mid = \frac{1}{n+2}(\tau + 2\alpha),$$

where α is the first Koszul form for (D, g).

Proof. According to the procedure in the proof of Theorem 5.1, we shall find the affine normal ξ. We choose $X_1, \cdots, X_n \in \mathfrak{X}(M)$ such that

$$(dx^1 \wedge \cdots \wedge dx^{n+1})(X_1, \cdots, X_n, E) = 1.$$

Putting $\omega_i = g(X_i, \ )$, the volume element v_g determined by g is given by

$$v_g = \frac{\omega_1 \wedge \cdots \wedge \omega_n \wedge d\varphi}{|(d\varphi(E))\det[g(X_i, X_j)]|^{\frac{1}{2}}}.$$

However, v_g may be alternatively expressed

$$v_g = |\det[g_{ij}]|^{\frac{1}{2}} dx^1 \wedge \cdots \wedge dx^{n+1},$$

hence

$$dx^1 \wedge \cdots \wedge dx^{n+1} = \frac{\omega_1 \wedge \cdots \wedge \omega_n \wedge d\varphi}{|(d\varphi(E))\det[g(X_i, X_j)]\det[g_{ij}]|^{\frac{1}{2}}},$$

and so

$$1 = (dx^1 \wedge \cdots \wedge dx^{n+1})(X_1, \cdots, X_n, E)$$
$$= \frac{(d\varphi(E))\det[g(X_i, X_j)]}{|(d\varphi(E))\det[g(X_i, X_j)]\det[g_{ij}]|^{\frac{1}{2}}},$$

and we have

$$|\det[g(X_i, X_j)]| = |(d\varphi(E))^{-1}\det[g_{ij}]|.$$

It follows from the procedure in the proof of Theorem 5.1 that

$$\mu = |\det[h(X_i, X_j)]|^{\frac{1}{n+2}} = |\det[(d\varphi(E))^{-1}g(X_i, X_j)]|^{\frac{1}{n+2}}$$
$$= |(d\varphi(E))^{-n-1}\det[g_{ij}]|^{\frac{1}{n+2}},$$
$$h(X, Z) = -\mu\tau(X) - (d\mu)(X)$$
$$= -\mu(d\log|\mu d\varphi(E)|)(X).$$

Thus

$$g(X, Z) = (d(\mu d\varphi(E)))(X).$$

Since $|\mu d\varphi(E)| = |(d\varphi(E))\det[g_{ij}]|^{\frac{1}{n+2}}$, we have

$$d\log|\mu d\varphi(E)| = \frac{1}{n+2}(\tau + 2\alpha).$$

$\square$

Corollary 5.1. *The gradient vector field* E *is parallel to the affine normal* ξ *if and only if*

$$\alpha = -\frac{1}{2}\tau.$$

Let us now consider characterizations of potential functions in terms of shape operators, transversal connection forms and so on.

Theorem 5.4. *For all level surfaces of* φ, *the conditions*

$$S = -I \quad , \quad \tau = 0, \quad \lambda = 1, \quad D^M h = 0,$$

hold if and only if φ *is a polynomial of degree 2, where* I *is the identity mapping.*

Proof. We identify a vector field $A = \sum_i a^i \dfrac{\partial}{\partial x^i}$ on Ω with a column vector $[a^i]$ and a non-degenerate metric $g = \sum_{i,j} g_{ij} dx^i dx^j$ with a matrix $[g_{ij}]$. Then

$$E = \begin{bmatrix} g_{11} & \cdots & g_{1n+1} \\ \vdots & & \vdots \\ g_{n+11} & \cdots & g_{n+1n+1} \end{bmatrix}^{-1} \begin{bmatrix} \partial\varphi/\partial x^1 \\ \vdots \\ \partial\varphi/\partial x^{n+1} \end{bmatrix}.$$

Suppose that φ is a polynomial of degree 2. Choosing an appropriate affine coordinate system $\{x^1, \cdots, x^{n+1}\}$ we have

$$\varphi(x) = \frac{1}{2}\{(x^1)^2 + \cdots + (x^p)^2 - (x^{p+1})^2 - \cdots - (x^{n+1})^2\} + k$$

where k is a constant. Denoting by I_r the unit matrix of degree r we have

$$[g_{ij}] = \begin{bmatrix} I_p & 0 \\ 0 & -I_{n-p+1} \end{bmatrix},$$

$$E = \begin{bmatrix} I_p & 0 \\ 0 & -I_{n-p+1} \end{bmatrix} \begin{bmatrix} x^1 \\ \vdots \\ x^p \\ -x^{p+1} \\ \vdots \\ -x^{n+1} \end{bmatrix} = \begin{bmatrix} x^1 \\ \vdots \\ x^p \\ x^{p+1} \\ \vdots \\ x^{n+1} \end{bmatrix} = \sum_i x^i \frac{\partial}{\partial x^i},$$

$$E\varphi = \sum_i x^i \frac{\partial\varphi}{\partial x^i} = 2(\varphi - k),$$

$$D_{\tilde{X}} E = \tilde{X} \quad \text{for } \tilde{X} \in \mathfrak{X}(\Omega).$$

Together these imply

$$S = -I, \quad \tau = 0, \quad \lambda = 1.$$

By Lemma 5.1 we have

$$(D_X^M h)(Y, Z) = -(D_X^M \{(E\varphi)^{-1} g\})(Y, Z) = -(E\varphi)^{-1}(D_X^M g)(Y, Z)$$
$$= -(E\varphi)^{-1}(D_X g)(Y, Z) = 0,$$

for X, Y and $Z \in \mathfrak{X}(M)$.

Conversely, suppose that

$$S = -I, \quad \tau = 0, \quad \lambda = 1, \quad D^M h = 0$$

hold for all level surfaces. Since $D_{\tilde{X}} E = \tilde{X}$ for $\tilde{X} \in \mathfrak{X}(\Omega)$, we have

$$\tilde{X}\varphi = (d\varphi)(\tilde{X}) = g(\tilde{X}, E) = (D_{\tilde{X}} d\varphi)(E) = \tilde{X}(d\varphi(E)) - d\varphi(D_{\tilde{X}} E)$$
$$= \tilde{X}(d\varphi(E)) - \tilde{X}\varphi.$$

Hence $E\varphi - 2\varphi$ is a constant on Ω. We claim

(a) $D_E g = 0$,
(b) $(D_X g)(Y, Z) = 0$ for X, Y and $Z \in \mathfrak{X}(M)$.

Assertion (a) follows from the following equations,

$$(D_E g)(E, E) = E(g(E, E)) - 2g(D_E E, E) = E(E\varphi - 2\varphi) = 0,$$
$$(D_E g)(E, X) = (D_X g)(E, E) = X(g(E, E)) - 2g(D_X E, E)$$
$$= X(E\varphi) = (E\varphi)\tau(X) = 0,$$
$$(D_E g)(X, Y) = (D_X g)(E, Y)$$
$$= X(g(E, Y)) - g(D_X E, Y) - g(E, D_X Y)$$
$$= -g(X, Y) - h(X, Y)d\varphi(E) = 0,$$

for X and $Y \in \mathfrak{X}(M)$. Assertion (b) follows as a consequence of

$$(D_X g)(Y, Z) = (D_X^M g)(Y, Z) = -(D_X^M \{(E\varphi)h\})(Y, Z)$$
$$= -(E\varphi)(\tau(X)h + D_X^M h)(Y, Z) = 0.$$

It follows from (a), (b) and the Codazzi equation (Proposition 2.1(2)) for (D, g) that

$$(D_{\tilde{X}} g)(\tilde{Y}, \tilde{Z}) = 0$$

for all $\tilde{X}, \tilde{Y}$ and $\tilde{Z} \in \mathfrak{X}(\Omega)$. Therefore

$$\frac{\partial^3 \varphi}{\partial x^i \partial x^j \partial x^k} = 0.$$

This implies that φ is a polynomial of degree 2. $\qquad \square$

Example 5.2. (Quadratic Hypersurfaces) Let f be a polynomial of degree 2 given by

$$f(x) = \frac{1}{2}\left\{ \sum_{i=1}^{p}(x^i)^2 - \sum_{j=1}^{n+1-p}(x^{p+j})^2 \right\},$$

and let $M = f^{-1}(c)$. Then $g^0 = Ddf$ is a non-degenerate Hessian metric on $\mathbf{R}^{n+1}$. We denote by g^0_M the restriction of g^0 on M and by ∇^{0M} the Levi-Civita connection of g^0_M. Since

$$g^0 = \sum_{i=1}^{p}(dx^i)^2 - \sum_{j=1}^{n+1-p}(dx^{p+j})^2,$$

the Levi-Civita connection ∇^0 of g^0 coincides with D. Hence

$$\begin{aligned}
0 &= (\nabla^0_X g^0)(Y, Z) = X(g^0(Y,Z)) - g^0(\nabla^0_X Y, Z) - g^0(Y, \nabla^0_X Z) \\
&= X(g^0_M(Y,Z)) - g^0_M(D^M_X Y, Z) - g^0(Y, D^M_X Z) \\
&= (D^M_X g^0_M)(Y, Z),
\end{aligned}$$

for all X, Y and $Z \in \mathfrak{X}(M)$, and so we have $\nabla^{0M} = D^M D$ Since $E = \sum_i x^i \frac{\partial}{\partial x^i}$, $\det[g_{ij}] = (-1)^{n+1-p}$ and $E\varphi = 2c$, by Theorem 5.3 the affine normal ξ is given by $\xi = (2c)^{-\frac{n+1}{n+2}} \sum_i x^i \frac{\partial}{\partial x^i}$.

It follows from $d\varphi(E) = 2c$, $S = -I$ and Example 5.1 that the pair (∇^{0M}, g^0_M) is the Codazzi structure of constant curvature $\dfrac{1}{2c}$. Since the signature of g^0 is $(p, n+1-p)$ and $g^0(E, E) = 2c$, g^0_M is positive definite only when $p = n+1$ and $c > 0$, or $p = n$ and $c < 0$. Hence the pair (M, g^0_M) is a Riemannian manifold of constant curvature $\dfrac{1}{2c}$ if and only if

(i) In case of $c > 0$, M is a sphere defined by

$$\sum_{i=1}^{n+1}(x^i)^2 = 2c.$$

(ii) In case of $c < 0$, M is a level surface of the Lorentz cone defined by

$$(x^{n+1})^2 - \sum_{i=1}^{n}(x^i)^2 = -2c.$$

Let $F(x^1, \cdots, x^n)$ be a smooth function on $\mathbf{R}^n$ such that $\left[\dfrac{\partial^2 F}{\partial x^i \partial x^j}\right] \neq 0$ and let Ω be a domain lying above the graph $x^{n+1} = F(x^1, \cdots, x^n)$;

$$\Omega = \{(x^1, \cdots, x^n, x^{n+1}) \in \mathbf{R}^{n+1} \mid x^{n+1} > F(x^1, \cdots, x^n)\}.$$

We set $f = x^{n+1} - F(x^1, \cdots, x^n)$ and $\varphi = -\log f$. Let us consider the Hessian $g = Dd\varphi$. Putting

$$\partial F = \begin{bmatrix} \partial F/\partial x^1 \\ \vdots \\ \partial F/\partial x^n \end{bmatrix}, \quad \partial^2 F = \left[\frac{\partial^2 F}{\partial x^i \partial x^j}\right],$$

we have

$$[g_{ij}] = \frac{1}{f^2} \begin{bmatrix} f\partial^2 F + \partial F \, {}^t(\partial F) & -\partial F \\ -{}^t(\partial F) & 1 \end{bmatrix},$$

$$\det [g_{ij}] = f^{-n-2} \det \partial^2 F,$$

$$[g_{ij}]^{-1} = f \begin{bmatrix} (\partial^2 F)^{-1} & (\partial^2 F)^{-1}\partial F \\ {}^t\{(\partial^2 F)^{-1}\partial F\} & f \det \left[I_n + \frac{1}{f}(\partial^2 F)^{-1}\partial F \, {}^t(\partial F)\right] \end{bmatrix},$$

$$E = \begin{bmatrix} (\partial^2 F)^{-1} & (\partial^2 F)^{-1}\partial F \\ {}^t\{(\partial^2 F)^{-1}\partial F\} & f \det \left[I_n + \frac{1}{f}(\partial^2 F)^{-1}\partial F \, {}^t(\partial F)\right] \end{bmatrix} \begin{bmatrix} \partial F \\ -1 \end{bmatrix}$$

$$= \begin{bmatrix} 0 \\ -f \end{bmatrix} = -f\frac{\partial}{\partial x^{n+1}}.$$

These expressions imply that $g = Dd\varphi$ is non-degenerate and

$$D_X E = -(Xf)\frac{\partial}{\partial x^{n+1}} = (d\varphi)(X)f\frac{\partial}{\partial x^{n+1}} = 0 \quad for \ X \in \mathfrak{X}(M),$$

$$E\varphi = -f\frac{\partial\varphi}{\partial x^{n+1}} = 1.$$

Thus we have

$$S = 0, \quad \tau = 0, \quad \lambda = -1.$$

Conversely, suppose that $Dd\varphi$ is non-degenerate and the above equations hold for all level surfaces of φ. Then we have

(a) $D_X E = 0$ *for* $X \in \mathfrak{X}(M)$,
(b) $D_E E = -E$.

By Lemma 5.2 and (b) above we know that $d\varphi(E)$ is a constant $k \neq 0$. Put $\tilde{A} = -e^{\frac{\varphi}{k}} E$. It follows from (a) and (b) that

$$D_X \tilde{A} = -(k^{-1} e^{\frac{\varphi}{k}} X\varphi) E - e^{\frac{\varphi}{k}} D_X E = 0 \quad for \ X \in \mathfrak{X}(M),$$
$$D_{\tilde{A}} \tilde{A} = e^{\frac{\varphi}{k}} \left(e^{\frac{\varphi}{k}} E + e^{\frac{\varphi}{k}} D_E E \right) = 0.$$

Thus $\tilde{A}$ is a D-parallel vector field on Ω. Hence we can choose an affine coordinate system $\{x^1, \cdots, x^{n+1}\}$ on $\mathbf{R}^{n+1}$ such that

$$\tilde{A} = \frac{\partial}{\partial x^{n+1}}.$$

Since $\dfrac{\partial}{\partial x^{n+1}} \left(x^{n+1} - e^{-\frac{\varphi}{k}} \right) = 0$ we have

$$x^{n+1} - e^{-\frac{\varphi}{k}} = F(x^1, \cdots, x^n),$$

and hence

$$\varphi = -k \log \left(x^{n+1} - F(x^1, \cdots, x^n) \right).$$

In summary, we have the following theorem.

Theorem 5.5. *For all level surfaces of* φ *the conditions*

$$S = 0, \quad \tau = 0, \quad \lambda = -1$$

hold if and only if, choosing a suitable affine coordinate system $\{x^1, \cdots, x^{n+1}\}$, φ *can be expressed by*

$$\varphi = k \log(x^{n+1} - F(x^1, \cdots, x^n))$$

where $k \neq 0$ *is a constant and* $\det \left[\dfrac{\partial^2 F}{\partial x^i \partial x^j} \right] \neq 0$.

Corollary 5.2. *Let* $F(x^1, \cdots, x^n)$ *be a smooth convex function on* $\mathbf{R}^n$. *For a level surface of the function* $\varphi = \log(x^{n+1} - F(x^1, \cdots, x^n))$ *on the domain* $x^{n+1} > F(x^1, \cdots x^n)$, *the gradient vector field* E *is parallel to the affine normal* ξ *if and only if* F *is a polynomial of degree 2.*

Proof. From the proof of Theorem 5.5 we have

$$\tau = 0, \quad 2\alpha = d \log \det \left[\frac{\partial^2 F}{\partial x^i \partial x^j} \right] + (n+2) d\varphi.$$

Suppose that the gradient vector field E is parallel to the affine normal ξ. By Corollary 5.1 we obtain

$$d \log \det \left[\frac{\partial^2 F}{\partial x^i \partial x^j} \right] = 2\alpha - (n+2)d\varphi = 0$$

on a level surface. Hence $\det \left[\frac{\partial^2 F}{\partial x^i \partial x^j} \right]$ is a positive constant on $\mathbf{R}^n$. Therefore by Theorem 8.6 due to [Cheng and Yau (1986)][Pogorelov (1978)] we know that F is a polynomial of degree 2. Conversely, assume that F is a convex polynomial of degree 2. Then $\det \left[\frac{\partial^2 F}{\partial x^i \partial x^j} \right]$ is a positive constant and $\alpha = \dfrac{n+2}{2} d\varphi$. Again by Corollary 5.1 E is parallel to the affine normal ξ. $\qquad\square$

For a fixed point $p \in \mathbf{R}^{n+1}$, and a positive constant c, an affine transformation of $\mathbf{R}^{n+1}$ defined by $x \longrightarrow c(x-p) + p$ is said to be a dilation at p. Let H be a vector field H induced by a 1-parameter transformation group of dilations $x \longrightarrow e^t(x-p) + p$. Then

$$H = \sum_i (x^i - x^i(p)) \frac{\partial}{\partial x^i}.$$

Theorem 5.6. *For all level surfaces of φ the conditions*

$$S = I, \quad \tau = 0, \quad \lambda = -1$$

hold if and only if $d\varphi$ is invariant under a 1-parameter transformation group of dilations.

Proof. Suppose that the conditions $S = I$, $\tau = 0$ and $\lambda = -1$ hold for all level surfaces. Then

$$D_X E = -X,$$

for $X \in \mathfrak{X}(\Omega)$. Introducing $K = \sum_i x^i \frac{\partial}{\partial x^i}$, we have

$$D_X(E + K) = 0,$$

and $E + K$ is therefore a D-parallel vector field, and so

$$E + K = \sum_i p^i \frac{\partial}{\partial x^i},$$

where p^i are constants. Let H be the vector field induced by the 1-parameter transformation group of dilations at the point $p = [p^i]$ and let

$\mathcal{L}_H$ be the Lie differentiation by H. It follows from Lemma 5.2 and $H = -E$ that

$$\mathcal{L}_H d\varphi = -\mathcal{L}_E d\varphi = -(d\iota_E + \iota_E d)d\varphi = -d(d\varphi(E)) = 0,$$

where ι_E is the interior product operator by E. Therefore $d\varphi$ is invariant by the 1-parameter group of dilations at p. Conversely, suppose that $d\varphi$ is invariant under a 1-parameter group of dilations at p. Let H be the vector field induced by the 1-parameter group. Then

$$0 = \mathcal{L}_H d\varphi = (d\iota_H + \iota_H d)d\varphi = d(d\varphi(H)),$$

which implies that $d\varphi(H)$ is a constant. Therefore

$$g(X, -H) = -(D_X d\varphi)(H) = -X(d\varphi(H)) + (d\varphi)(D_X H)$$
$$= (d\varphi)(X)$$

for $X \in \mathfrak{X}(\Omega)$, which gives us that

$$E = -H,$$
$$D_X E = -D_X H = -X.$$

Hence

$$S = I, \quad \tau = 0, \quad \lambda = -1. \qquad \square$$

Corollary 5.3. *Let Ω be a regular convex cone and let ψ be the characteristic function of Ω. Then for all level surfaces of $\varphi = \log \psi$, we have*

$$S = I, \quad \tau = 0, \quad \lambda = -1.$$

Example 5.3. Let J be a matrix of degree $n + 1$ given by

$$J = \begin{bmatrix} -I_p & 0 \\ 0 & I_{n+1-p} \end{bmatrix},$$

where $0 \le p \le n$ and I_r is the unit matrix of degree r. We denote by Ω the connected component of the set

$$\{x \in \mathbf{R}^{n+1} \mid {}^t x J x > 0\}$$

containing $e_{n+1} = {}^t[0, \cdots, 0, 1]$. Then Ω is a cone with vertex 0. For any $x \in \Omega$ we set $x_0 = ({}^t x J x)^{-\frac{1}{2}} x$. Since ${}^t x_0 J x_0 = {}^t e_{n+1} J e_{n+1} = 1$, by Witt's theorem there exists $s \in SO(p, n + 1 - p)$ such that $s x_0 = e_{n+1}$. Hence $({}^t x J x)^{-\frac{1}{2}} s x = e_{n+1}$. This means that $\mathbf{R}^+ SO(p, n + 1 - p)$ acts transitively on Ω. We set $f(x) = -\dfrac{1}{2} {}^t x J x$ and $\varphi(x) = \log(-f(x))$. Then

$\det \left[\dfrac{\partial^2 \varphi}{\partial x^i \partial x^j} \right] = \dfrac{(-1)^n}{f^{n+1}}$. Hence $g = Dd\varphi$ is non-degenerate. It follows from $\varphi(e^t x) = \varphi(x) - 2t$ that $d\varphi$ is invariant under a 1-parameter transformation group of dilations $e^t I_{n+1}$. Hence, by Theorem 5.6, for a level surface M of φ we have

$$S = I, \quad \tau = 0, \quad \lambda = -1.$$

In the proof of Theorem 5.6 we proved $E = -H = -\sum_i x^i \dfrac{\partial}{\partial x^i}$, and so we have $d\varphi(E) = -2$. Using this result, together with the relation $S = I$ and Example 5.1, we have that (D^M, g_M) is a Codazzi structure of constant curvature $\dfrac{1}{2}$ (cf. Example 5.2).

5.3 Laplacians of gradient mappings

In this section we study the Laplacian of a gradient mapping, and prove a certain analogy to the affine Bernstein problem proposed by [Chern (1978)]. Let ι be the gradient mapping from a Hessian domain $(\Omega, D, g = Dd\varphi)$ of $\mathbf{R}^{n+1}$ into $(\mathbf{R}^*_{n+1}, D^*)$. The Laplacian of ι with respect to (g, D^*) is given by (cf. Definition 5.4)

$$\Delta_{(g, D^*)} \iota = \sum_{i,j} g^{ij} \left\{ D^*_{\partial/\partial x^i} \iota_* \left(\frac{\partial}{\partial x^j} \right) - \iota_* \left(\nabla_{\partial/\partial x^i} \frac{\partial}{\partial x^j} \right) \right\}.$$

A vector field $X^*_x = \sum_i \xi^*_i(x) \left(\dfrac{\partial}{\partial x^*_i} \right)_{\iota(x)}$ along ι is identified with a 1-form $\sum_i \xi^*_i dx^i$. Since $\iota_*(\tilde{X}) = -\sum_{i,j} g_{ij} \xi^i \dfrac{\partial}{\partial x^*_j}$ for $\tilde{X} = \sum_i \xi^i \dfrac{\partial}{\partial x^i}$, by the above identification, the vector field $\iota_*(\tilde{X})$ along ι is considered as a 1-formG

$$\iota_*(\tilde{X}) = -g(\tilde{X}, \ ).$$

By Theorem 2.2 and Proposition 3.4 we have

$$\Delta_{(g, D^*)} \iota = \iota_* \left\{ \sum_{i,j} g^{ij} (D' - \nabla)_{\partial/\partial x^i} \frac{\partial}{\partial x^j} \right\}$$

$$= \iota_* \left\{ \sum_{i,j} g^{ij} (\nabla - D)_{\partial/\partial x^i} \frac{\partial}{\partial x^j} \right\} = \iota_* \left(\sum_k g^{ij} \gamma^k_{ij} \frac{\partial}{\partial x^k} \right)$$

$$= \iota_* \left(\sum_k \alpha^k \frac{\partial}{\partial x^k} \right) = -\sum_i \left(\alpha_i \circ \iota^{-1} \right) \frac{\partial}{\partial x^*_i},$$

and so we have proved the following proposition.

Proposition 5.5. *The Laplacian of the gradient mapping ι with respect to (g, D^*) is expressed by*

$$\Delta_{(g,D^*)}\iota = -\alpha.$$

By Proposition 3.4 and Theorem 8.6 due to [Cheng and Yau (1986)] [Pogorelov (1978)] we also have the following corollary.

Corollary 5.4. *Let φ be a convex function on $\mathbf{R}^{n+1}$. Then the following conditions (1)-(3) are equivalent.*

(1) ι *is harmonic* ; $\Delta_{(g,D^*)}\iota = 0$.
(2) $\alpha = 0$.
(3) φ *is a polynomial of degrre 2.*

In the previous section 5.2 we studied affine differential geometry of level surfaces of φ using a pair (D, E) of the flat connection D and the gradient vector field E. In this section we study level surfaces using pairs (D', E) and (∇, E) where D' and ∇ are the flat dual connection of D and the Levi-Civita connection of g respectively.

We denote by D'^M, h', S' and τ' the induced connection, the affine fundamental form, the shape operator and the transversal connection form with respect to $(D', -E)$ respectively, and have the following relations

$$D'_X Y = D'^M_X Y + h'(X,Y)(-E),$$
$$D'_X(-E) = -S'(X) + \tau'(X)(-E).$$

This immersion coincides with the central affine immersion $(\iota, -\iota)$ from M to $\mathbf{R}^*_{n+1}$. In fact, by Theorem 2.2 we have

$$\begin{aligned}
D^*_X \iota(Y) &= \iota_*(D'_X Y) = \iota_*\{D'^M_X Y + h'(X,Y)(-E)\}\\
&= \iota_*(D'^M_X Y) + h'(X,Y)\iota_*(-E)\\
&= \iota_*(D'^M_X Y) + h'(X,Y)(-\iota).
\end{aligned}$$

Lemma 5.3.

(1) $h'(X,Y) = h(S(X),Y)$.
(2) $S' = I$.
(3) $\tau' = 0$.

Proof. By Theorem 2.2 and Lemma 5.1 we have

$$
\begin{aligned}
0 = Xg(E,Y) &= g(D_X E, Y) + g(E, D'_X Y) \\
&= \{h(S(X), Y) - h'(X, Y)\} d\varphi(E), \\
0 = Xg(Y, E) &= g(D_X Y, E) + g(Y, D'_X E) \\
&= -g(X, Y) + g(S'(X), Y),
\end{aligned}
$$

for $X, Y \in \mathfrak{X}(M)$. These expressions prove assertions (1) and (2). Assertion (3) follows from

$$
\tau(X) = (d \log |d\varphi(E)|)(X) = \frac{1}{d\varphi(E)} X g(E, E)
$$

$$
= \frac{1}{d\varphi(E)} \{g(D_X E, E) + g(E, D'_X E)\} = \tau(X) + \tau'(X).
$$

$\square$

We denote by ∇^M, h_∇, S_∇ and τ_∇ the induced connection, the affine fundamental form, the shape operator and the transversal connection form with respect to (∇, E) respectively, and have the following relations

$$
\nabla_X Y = \nabla_X^M Y + h_\nabla(X, Y) E,
$$
$$
\nabla_X E = -S_\nabla(X) + \tau_\nabla(X) E.
$$

Since $g(\nabla_X Y, Z) = g(\nabla_X^M Y, Z)$ for X, Y and $Z \in \mathfrak{X}(M)$, the induced connection ∇^M is the Levi-Civita connection of the restriction g_M of g on M. From $\nabla = \frac{1}{2}(D + D')$ and Lemma 5.3 we obtain the following lemma.

Lemma 5.4.

(1) $\nabla^M = \frac{1}{2}(D^M + D'^M)$.

(2) $S_\nabla = \frac{1}{2}(S - I)$.

(3) $h_\nabla(X, Y) = -h(S_\nabla X, Y)$.

(4) $\tau_\nabla = \frac{1}{2}\tau$.

Proposition 5.6. *Let $\gamma = \nabla - D$ be the difference tensor. Then we have*

(1) $\operatorname{Tr} \gamma_E = \frac{1}{2}(\operatorname{Tr} S - \lambda + n + 1)$.

(2) *If $\tau = 0$, then*

$$
\gamma_E X = \frac{1}{2}(X + S(X)) \ \text{ for } X \in \mathfrak{X}(M), \qquad \gamma_E E = \frac{1}{2}(1 - \lambda)E.
$$

Proof. By Proposition 2.2 and Lemmata 5.2 and 5.4 we have

$$\gamma_E X = \gamma_X E = (\nabla_X - D_X)E$$
$$= (-S_\nabla + S)(X) + (\tau_\nabla - \tau)(X)E$$
$$= \frac{1}{2}(I + S)(X) - \frac{1}{2}\tau(X)E,$$

$$g(\gamma_E E, E) = \frac{1}{2}(D_E g)(E, E) = \frac{1}{2}\{g(E, E) - g(D_E E, E)\}$$
$$= g(\frac{1}{2}(1 - \lambda)E, E),$$

$$g(\gamma_E E, X) = g(E, \gamma_E X) = g(E, \gamma_X E)$$
$$= -\frac{1}{2}\tau(X)d\varphi(E).$$

These expressions together prove assertions (1) and (2). $\qquad\square$

Corollary 5.5.

(1) $\operatorname{Tr} S = 2\alpha(E) - n - 1 + \lambda.$

(2) $\operatorname{Tr} S_\nabla = \alpha(E) - n + \frac{1}{2}(\lambda - 1).$

Proof. The assertions follow from Proposition 3.4, 5.6 and Lemma 5.4. $\square$

Theorem 5.7. *Let ι_M and g_M be the restriction of ι and g to M respectively. We denote by $\Delta_{(g_M, D^*)}\iota_M$ the Laplacian of ι_M with respect to (g_M, D^*). Then we have*

(1) $\Delta_{(g_M, D^*)}\iota_M(X) = -\left(\alpha + \frac{1}{2}\tau\right)(X)$ *for* $X \in \mathfrak{X}(M).$

(2) $\Delta_{(g_M, D^*)}\iota_M(E) = -\operatorname{Tr} S.$

Proof. Let $\{X_1, \cdots, X_n\}$ be locally independent vector fields on M. Then

$$\Delta_{(g_M, D^*)}\iota_M = \sum_{i,j=1}^{n} g_M^{ij}\{D_{X_i}^* \iota_{M*}(X_j) - \iota_{M*}(\nabla_{X_i}^M X_j)\},$$

where $[g_M^{ij}] = [g_M(X_i, X_j)]^{-1}$. By Theorem 2.2 and Lemma 5.4 we obtain

$$\Delta_{(g_M, D^*)}\iota_M = \iota_* \left\{ \sum_{i,j=1}^{n} g_M^{ij}(D'_{X_i} X_j - \nabla_{X_i}^M X_j) \right\}$$

$$= \iota_* \left\{ \sum_{i,j=1}^{n} g_M^{ij}(D' - \nabla)_{X_i} X_j + \sum_{i,j=1}^{n} g_M^{ij} h_\nabla(X_i, X_j)E \right\}$$

$$= \iota_* \left\{ \sum_{i,j=1}^{n} g_M^{ij}\gamma_{X_i} X_j + \frac{1}{2}\sum_{i,j=1}^{n} g_M^{ij} h(X_i - SX_i, X_j)E \right\}.$$

We set $\gamma_{X_i} X_j = \sum_{k=1}^{n+1} \gamma^k_{ij} X_k$ where $X_{n+1} = E$. By Lemma 5.1 and Proposition 5.6, we have for $1 \le i, j \le n$,

$$
\begin{aligned}
g(\gamma_{X_i} X_j, E) &= g(X_j, \gamma_{X_i} E) = g(X_j, \gamma_E X_i) \\
&= g\left(X_j, \frac{1}{2}(X_i + S X_i) - \frac{1}{2}\tau(X_i)E\right) \\
&= \frac{1}{2}\{g(X_i, X_j) + g(S X_i, X_j)\},
\end{aligned}
$$

and so

$$
\gamma^{n+1}_{ij} = -\frac{1}{2}\{h(X_i, X_j) + h(S X_i, X_j)\}.
$$

This implies

$$
\begin{aligned}
\sum_{i,j=1}^{n} g_M^{ij} \gamma_{X_i} X_j &= \sum_{i,j=1}^{n} g_M^{ij} \left(\sum_{k=1}^{n} \gamma^k_{ij} X_k + \gamma^{n+1}_{ij} E \right) \\
&= \sum_{i,j,k=1}^{n} g_M^{ij} \gamma^k_{ij} X_k + \frac{1}{2}\left\{ \sum_{i,j=1}^{n} g_M^{ij} h(-X_i - S X_i, X_j) \right\} E.
\end{aligned}
$$

Hence

$$
\begin{aligned}
\Delta_{(g_M, D^*)} \iota_M &= \iota_* \left\{ \sum_{i,j,k=1}^{n} g^{ij} \gamma^k_{ij} X_k - \sum_{i,j=1}^{n} g^{ij} h(S X_i, X_j) E \right\} \\
&= \iota_* \left\{ \sum_{i,j,k=1}^{n} g^{ij} \gamma^k_{ij} X_k + \frac{\operatorname{Tr} S}{d\varphi(E)} E \right\} \\
&= \iota_*(\tilde{X}),
\end{aligned}
$$

where

$$
\tilde{X} = \sum_{i,j,k=1}^{n} g^{ij} \gamma^k_{ij} X_k + \frac{\operatorname{Tr} S}{d\varphi(E)} E.
$$

The first term of the right side of the above equation may be reduced as

$$
\begin{aligned}
\sum_{i,j,k=1}^{n} g^{ij} \gamma^k_{ij} X_k &= \sum_{k=1}^{n} \left\{ \sum_{i,j=1}^{n+1} g^{ij} \gamma^k_{ij} - g^{n+1n+1} \gamma^k_{n+1n+1} \right\} X_k \\
&= \sum_{k=1}^{n} \alpha^k X_k - g^{n+1n+1} \sum_{k=1}^{n} \gamma^k_{n+1n+1} X_k \\
&= \sum_{k=1}^{n+1} \alpha^k X_k - \alpha^{n+1} E - g^{n+1n+1}(\gamma_E E - \gamma^{n+1}_{n+1n+1} E) \\
&= \sum_{k=1}^{n+1} \alpha^k X_k - \frac{1}{d\varphi(E)} \gamma_E E + \left(\frac{1}{d\varphi(E)} \gamma^{n+1}_{n+1n+1} - \alpha^{n+1} \right) E.
\end{aligned}
$$

Thus

$$\tilde{X} = \sum_{k=1}^{n+1} \alpha^k X_k - \frac{1}{d\varphi(E)} \gamma_E E + \left\{ \frac{1}{d\varphi(E)} \left(\operatorname{Tr} S + \gamma_{n+1 n+1}^{n+1} \right) - \alpha^{n+1} \right\} E.$$

By Proposition 5.6 we have

$$g(\tilde{X}, Y) = \alpha(Y) + \frac{1}{2}\tau(Y) \quad for \ Y \in \mathfrak{X}(M),$$

and

$$g(\tilde{X}, E) = \operatorname{Tr} S.$$

These together imply (1) and (2).   □

Corollary 5.6. *The following conditions* (1)-(3) *are equivalent.*

(1) ι_M *is harmonic with respect to* (g_M, D^*); $\ \Delta_{(g_M, D^*)} \iota_M = 0.$

(2) $\operatorname{Tr} S = 0$ *and* $\alpha = -\dfrac{1}{2}\tau.$

(3) $\operatorname{Tr} S = 0$ *and* E *is parallel to the affine normal.*

Proof. It follows from Corollary 5.1 and Theorem 5.7 that (1), (2) and (3) are equivalent.   □

The trace of the shape operator of the Blaschke immersion is said to be the **affine mean curvature**. An affine hypersurface is called **affine minimal** if the affine mean curvature vanishes identically. It follows from Theorem 5.2 that an affine hypersurface is affine minimal if and only if the conormal mapping is harmonic.

S.S.Chern proposed the following problem analogous to the Bernstein problem in Euclidean geometry [Chern (1978)].

Affine Bernstein Problem *If the graph*

$$x^{n+1} = F(x^1, \cdots, x^n)$$

of a convex function $F(x^1, \cdots, x^n)$ *on* $\mathbf{R}^n$ *is affine minimal, that is, in other words, if the conormal mapping is harmonic, then is the graph an elliptic paraboloid?*

This conjecture has been confirmed when $n = 2$ by [Trudinger and Wang (2000)].

If we replace the affine normal ξ by the gradient vector field E, the conormal mapping ν is replaced by the gradient mapping ι_M. Then the analogous problem as above can be proved as follows.

Corollary 5.7. *Let* $F(x^1, \cdots, x^n)$ *be a convex function on* $\mathbf{R}^n$ *and let* Ω *be a domain in* $\mathbf{R}^{n+1}$ *defined by*

$$\Omega = \{(x^1, \cdots, x^n, x^{n+1}) \in \mathbf{R}^{n+1} \mid x^{n+1} > F(x^1, \cdots, x^n)\}.$$

We set

$$\varphi(x^1, \cdots, x^n, x^{n+1}) = \log(x^{n+1} - F(x^1, \cdots, x^n)).$$

Then $(D, g = Dd\varphi)$ *is a Hessian structure on* Ω *(Theorem 5.5). For a level surface* M *of* φ *the following conditions are equivalent.*

(1) *The gradient mapping* ι_M *is harmonic with respect to* (g_M, D^*).

(2) M *is an elliptic paraboloid.*

Proof. If M is an elliptic paraboloid, then F is a convex polynomial of degree 2. It follows from the proof of Theorem 5.5 and Corollary 5.2 that

$$S = 0, \quad \tau = 0, \quad \alpha = \frac{n+2}{2} d\varphi = 0.$$

Hence we have $\Delta_{(g_M, D^*)}\iota_M = 0$ by Theorem 5.7. Conversely, suppose $\Delta_{(g_M, D^*)}\iota_M = 0$. Then

$$\alpha(X) = -\frac{1}{2}\tau(X) = 0,$$

for all $X \in \mathfrak{X}(M)$. We obtained in the proof of Corollary 5.2

$$d \log \det \left[\frac{\partial^2 F}{\partial x^i \partial x^j} \right] = 2\alpha - (n+2)d\varphi = 0,$$

on $\mathfrak{X}(M)$. This implies that $\det \left[\dfrac{\partial^2 F}{\partial x^i \partial x^j} \right]$ is a positive constant on $\mathbf{R}^n$. Hence, by Theorem 8.6, F is a polynomial of degree 2. $\qquad\square$

Let ∇^* be the Levi-Civita connection of the Hessian metric $g^* = D^* d\varphi^*$ on $\Omega^* = \iota(\Omega)$ (cf. Proposition 2.7).

Theorem 5.8. *The Laplacian* $\Delta_{(g_M, \nabla^*)}\iota_M$ *of* ι_M *with respect to* (g_M, ∇^*) *is given by*

$$\Delta_{(g_M, \nabla^*)}\iota_M = \frac{\operatorname{Tr} S_\nabla}{d\varphi(E)}\iota = \frac{\operatorname{Tr} S - n}{2d\varphi(E)}\iota.$$

Proof. Let $\{X_1, \cdots, X_n\}$ be locally independent vector fields on M and let $[g^{ij}]$ be the inverse matrix of $[g(X_i, X_j)]$. Since $\iota : (\Omega, g) \longrightarrow (\Omega^*, g^*)$ is an isometry (cf. Proposition 2.7), by Lemmata 5.1 and 5.4 we have

$$\Delta_{(g_M, \nabla^*)}\iota_M = \sum_{i,j} g^{ij}\{\nabla^*_{X_i}\iota_*(X_j) - \iota_*(\nabla^M_{X_i}X_j)\}$$

$$= \sum_{i,j} g^{ij}\iota_*(\nabla_{X_i}X_j - \nabla^M_{X_i}X_j) = \sum_{i,j} g^{ij}h_\nabla(X_i, X_j)\iota_*(E)$$

$$= \Big\{\frac{1}{d\varphi(E)}\sum_{i,j} g^{ij}g(S_\nabla X_i, X_j)\Big\}\iota$$

$$= \frac{\operatorname{Tr} S_\nabla}{d\varphi(E)}\iota. \qquad \qquad \square$$

Corollary 5.8. *Suppose that $d\varphi$ is invariant under a 1-parameter group of dilations. Then we have*

$$\Delta_{(g_M, \nabla^*)}\iota_M = 0.$$

Proof. By Theorem 5.6 we have $S = I$ and so $\operatorname{Tr} S = n$. Hence our assertion follows from Theorem 5.8. $\qquad \square$

Corollary 5.9. *Suppose that a Hessian metric g on a Hessian domain $(\Omega, D, g = Dd\varphi)$ is positive definite and that $d\varphi$ is invariant under a 1-parameter group of dilations. Then each level surface is a minimal surface of the Riemannian manifold (Ω, g).*

Corollary 5.10. *Let Ω be a regular convex cone and let $g = Dd\log\psi$ be the canonical Hessian metric. Then each level surface of the characteristic function ψ is a minimal surface of the Riemannian manifold (Ω, g).*

Example 5.4. Let Ω be a regular convex cone consisting of all positive definite symmetric matrices of degree n. Then $(D, g = -Dd\log\det x)$ is a Hessian structure on Ω (cf. Example 4.1), and each level surface of $\det x$ is a minimal surface of the Riemannian manifold $(\Omega, g = -Dd\log\det x)$.

Example 5.5. Let $(\Omega, D, g = Dd\varphi)$ be a Hessian domain in $\mathbf{R}^{n+1}$. The Laplacian $\Delta_g\varphi$ of the potential $\varphi : \Omega \longrightarrow \mathbf{R}$ with respect to g is given by

$$\Delta_g\varphi = \sum_{i,j} g^{ij}\Big\{\frac{\partial}{\partial x^i}\Big(\varphi_*\Big(\frac{\partial}{\partial x^j}\Big)\Big) - \varphi_*\Big(\nabla_{\partial/\partial x^i}\frac{\partial}{\partial x^j}\Big)\Big\}$$

$$= \sum_{i,j} g^{ij}\Big(\frac{\partial^2\varphi}{\partial x^i\partial x^j} - \sum_k \gamma^k_{ij}\frac{\partial\varphi}{\partial x^k}\Big)$$

$$= n + 1 - \alpha(E).$$

In the case of Ω being a regular convex cone with canonical Hessian metric $g = Dd \log \psi$, we have $\alpha(E) = \dfrac{1}{2}(\mathrm{Tr}\, S - \lambda + n + 1)$ by Proposition 5.6 and $S = I$, $\lambda = -1$ by Corollary 5.3, and so $\alpha(E) = n + 1$. This proves that the function $\log \psi$ is harmonic,

$$\Delta_g \log \psi = 0.$$

Chapter 6

Hessian structures and information geometry

Let $\mathcal{P} = \{p(x; \lambda) \mid \lambda \in \Lambda\}$ be a smooth family of probability distributions parametrized by $\lambda \in \Lambda$. Then $\mathcal{P}$, identified with a domain Λ, admits a Riemannian metric given by the Fisher information matrix and a pair of dual connections with respect to the metric. S. Amari, H. Nagaoka among others proposed *Information Geometry*, which aims to study smooth families of probability distributions from the viewpoint of dual connections [Amari and Nagaoka (2000)]. In the case when a pair of dual connections is flat, the structures considered by this subject are Hessian structures. It is known that many important smooth families of probability distributions, for example normal distributions, admit Hessian structures. In the manner of [Amari and Nagaoka (2000)], in section **6.1** we introduce the idea of Information Geometry and prove that exponential families of probability distributions carry Hessian structures. In section **6.2** we study a family of probability distributions induced by a linear mapping from a domain in a vector space into the set of all real symmetric positive definite matrices of degree n.

6.1 Dual connections on smooth families of probability distributions

In this section we give a brief survey of dual connections on smooth families of probability distributions. We will give particular attention to flat dual connections on exponential families. For details the reader may refer to [Amari and Nagaoka (2000)].

Definition 6.1. Let $\mathcal{X}$ be a discrete set (countable set) or $\mathbf{R}^m$. A function $p(x)$ on $\mathcal{X}$ is said to be a **probability distribution** if it satisfies the following conditions (1) and (2).

(1) $p(x) \geq 0$ *for* $x \in \mathcal{X}$.

(2) $\sum_{x \in \mathcal{X}} p(x) = 1$, in case $\mathcal{X}$ is a discrete set,

$$\int_{\mathcal{X}} p(x)dx = 1, \text{ in case } \mathcal{X} = \mathbf{R}^m.$$

To simplify the notation, the symbol $\sum_{x \in \mathcal{X}}$ is also denoted by $\int_{\mathcal{X}}$.

The **expectation** of a function $f(x)$ on $\mathcal{X}$ with respect to a probability distribution $p(x)$ is defined by

$$E[f] = \int_{\mathcal{X}} f(x)p(x)dx.$$

In this chapter all the families of probability distributions we consider, $\mathcal{P} = \{p(x; \lambda) \mid \lambda \in \Lambda\}$ on $\mathcal{X}$ parametrized by $\lambda = [\lambda^1, \cdots, \lambda^n] \in \Lambda$, satisfy the following conditions.

   $(\mathbf{P}_1)$ Λ is a domain in $\mathbf{R}^n$.
   $(\mathbf{P}_2)$ $p(x; \lambda)$ is a smooth function with respect to λ.
   $(\mathbf{P}_3)$ The operations of integration with respect to x and differentiation with respect to λ^i are commutative.

Definition 6.2. Let $\mathcal{P} = \{p(x; \lambda) \mid \lambda \in \Lambda\}$ be a family of probability distributions. We set $l_\lambda = l(x; \lambda) = \log p(x; \lambda)$ and denote by E_λ the expectation with respect to $p_\lambda = p(x; \lambda)$. Then a matrix $g = [g_{ij}(\lambda)]$ defined by

$$g_{ij}(\lambda) = E_\lambda \left[\frac{\partial l_\lambda}{\partial \lambda^i} \frac{\partial l_\lambda}{\partial \lambda^j} \right]$$

$$= \int_{\mathcal{X}} \frac{\partial l(x; \lambda)}{\partial \lambda^i} \frac{\partial l(x; \lambda)}{\partial \lambda^j} p(x; \lambda)dx$$

is called the **Fisher information matrix**.

Differentiating both sides of $\int_{\mathcal{X}} p(x; \lambda)dx = 1$ by λ^i and λ^j we have

$$0 = \frac{\partial}{\partial \lambda^i} \int_{\mathcal{X}} p(x; \lambda)dx = \int_{\mathcal{X}} \frac{\partial}{\partial \lambda^i} p(x; \lambda)dx = \int_{\mathcal{X}} \frac{\partial}{\partial \lambda^i} l(x; \lambda)p(x; \lambda)dx,$$

$$0 = \int_{\mathcal{X}} \frac{\partial^2 l(x; \lambda)}{\partial \lambda^j \partial \lambda^i} p(x; \lambda)dx + \int_{\mathcal{X}} \frac{\partial l(x; \lambda)}{\partial \lambda^i} \frac{\partial l(x; \lambda)}{\partial \lambda^j} p(x; \lambda)dx.$$

Therefore

$$E_\lambda \left[\frac{\partial l_\lambda}{\partial \lambda^i} \right] = 0,$$

$$g_{ij}(\lambda) = -E_\lambda \left[\frac{\partial^2 l_\lambda}{\partial \lambda^i \partial \lambda^j} \right].$$

Note that the Fisher information matrix $g = [g_{ij}(\lambda)]$ is positive semi-definite on Λ because

$$\sum_{i,j} g_{ij}(\lambda) c^i c^j = \int_\mathcal{X} \left\{ \sum_i c^i \frac{\partial l(x; \lambda)}{\partial \lambda^i} \right\}^2 p(x; \lambda) dx \geq 0.$$

The following condition concerning the families of probability functions considered here is therefore a natural assumption, and will henceforth be adopted throughout this chapter.

$(\mathbf{P_4})$ The Fisher information matrix $g = [g_{ij}(\lambda)]$ for a family of probability distributions $\mathcal{P} = \{p(x; \lambda) \mid \lambda \in \Lambda\}$ is positive definite on Λ.

By this condition, we may regard the Fisher information matrix $g = [g_{ij}]$ as a Riemannian metric on Λ, and we call it the **Fisher information metric**. Let $\Gamma^i{}_{jk}$ be the Christoffel symbol of the Levi-Civita connection of g, and let $\Gamma_{kij} = \sum_p g_{kp} \Gamma^p{}_{ij}$. Then

$$\Gamma_{kij} = \sum_p g_{kp} \Gamma^p{}_{ij} = \frac{1}{2} \left(\frac{\partial g_{ik}}{\partial \lambda^j} + \frac{\partial g_{jk}}{\partial \lambda^i} - \frac{\partial g_{ij}}{\partial \lambda^k} \right).$$

Differentiating g_{ij} by λ^k, we obtain

$$\frac{\partial g_{ij}}{\partial \lambda^k} = E_\lambda \left[\frac{\partial^2 l_\lambda}{\partial \lambda^k \partial \lambda^i} \frac{\partial l_\lambda}{\partial \lambda^j} \right] + E_\lambda \left[\frac{\partial l_\lambda}{\partial \lambda^i} \frac{\partial^2 l_\lambda}{\partial \lambda^k \partial \lambda^j} \right] + E_\lambda \left[\frac{\partial l_\lambda}{\partial \lambda^i} \frac{\partial l_\lambda}{\partial \lambda^j} \frac{\partial l_\lambda}{\partial \lambda^k} \right],$$

and so

$$\Gamma_{kij} = E_\lambda \left[\frac{\partial^2 l_\lambda}{\partial \lambda^i \partial \lambda^j} \frac{\partial l_\lambda}{\partial \lambda^k} \right] + \frac{1}{2} E_\lambda \left[\frac{\partial l_\lambda}{\partial \lambda^i} \frac{\partial l_\lambda}{\partial \lambda^j} \frac{\partial l_\lambda}{\partial \lambda^k} \right].$$

Let $T_{ijk} = \frac{1}{2} E_\lambda \left[\frac{\partial l_\lambda}{\partial \lambda^i} \frac{\partial l_\lambda}{\partial \lambda^j} \frac{\partial l_\lambda}{\partial \lambda^k} \right]$ and let

$$\Gamma(t)_{kij} = \Gamma_{kij} - t T_{kij}, \qquad \Gamma(t)^i{}_{jk} = \sum_p g^{ip} \Gamma(t)_{pjk}.$$

Since T_{ijk} is a symmetric tensor, $\Gamma(t)^i{}_{jk}$ defines a torsion-free connection $\nabla(t)$, and we have

$$\frac{\partial g_{ij}}{\partial \lambda^k} = \Gamma(t)_{jki} + \Gamma(-t)_{ikj},$$

that is,

$$Xg(Y, Z) = g(\nabla(t)_X Y, Z) + g(Y, \nabla(-t)_X Z).$$

This means that $\nabla(t)$ and $\nabla(-t)$ are dual connections with respect to the Fisher information metric g (Definition 2.8).

Definition 6.3. A family of probability distributions $\mathcal{P} = \{p(x; \theta) \mid \theta \in \Theta\}$ is said to be an **exponential family** if there exist functions $C(x), F_1(x), \cdots, F_n(x)$ on $\mathcal{X}$, and a function $\varphi(\theta)$ on Θ, such that

$$p(x; \theta) = \exp\left\{C(x) + \sum_{i=1}^{n} F_i(x)\theta^i - \varphi(\theta)\right\}.$$

Let $\{p(x; \theta) \mid \theta \in \Theta\}$ be an exponential family. Then

$$\frac{\partial l(x; \theta)}{\partial \theta^i} = F_i(x) - \frac{\partial \varphi}{\partial \theta^i}, \qquad \frac{\partial^2 l(x; \theta)}{\partial \theta^i \partial \theta^j} = -\frac{\partial^2 \varphi}{\partial \theta^i \partial \theta^j},$$

and these expressions imply

$$\Gamma(1)_{kij} = E_\theta\left[\frac{\partial^2 l_\theta}{\partial \theta^i \partial \theta^j}\frac{\partial l_\theta}{\partial \theta^k}\right] = -\frac{\partial^2 \varphi}{\partial \theta^i \partial \theta^j}E_\theta\left[\frac{\partial l_\theta}{\partial \theta^k}\right] = 0.$$

Therefore $\nabla(1)$ is a flat connection and $\{\theta^i\}$ is an affine coordinate system with respect to $\nabla(1)$. Furthermore, we have

$$g_{ij} = -\int_{\mathcal{X}} \frac{\partial^2 l_\theta}{\partial \theta^i \partial \theta^j}p(x; \theta)dx = \frac{\partial^2 \varphi}{\partial \theta^i \partial \theta^j},$$

thus $\left(\nabla(1), g = \left[\dfrac{\partial^2 \varphi}{\partial \theta^i \partial \theta^j}\right]\right)$ is a Hessian structure on Θ. It is known that many important smooth families of probability distributions are exponential families.

Example 6.1. 1-dimensional normal distributions are defined by

$$p(x; \lambda) = \frac{1}{\sqrt{2\pi}\sigma} \exp\left\{-\frac{(x - \mu)^2}{2\sigma^2}\right\},$$

for $x \in \mathcal{X} = \mathbf{R}$ and $\lambda \in \Lambda = \{[\mu, \sigma] \mid \mu \in \mathbf{R}, \ \sigma \in \mathbf{R}^+\}$, where μ is the mean and σ is the standard deviation.

Put

$$F_1(x) = -x^2, \quad F_2(x) = x, \quad \theta^1 = \frac{1}{2\sigma^2}, \quad \theta^2 = \frac{\mu}{\sigma^2},$$

$$\varphi(\theta) = \frac{\mu^2}{2\sigma^2} + \log\sqrt{2\pi}\sigma = \frac{(\theta^2)^2}{4\theta^1} + \frac{1}{2}\log\left(\frac{\pi}{\theta^1}\right).$$

Then for $x \in \mathbf{R}$ and $\theta \in \Theta = \{\theta = [\theta^1, \theta^2] \mid \theta^1 \in \mathbf{R}^+, \theta^2 \in \mathbf{R}\}$ we have

$$p(x; \theta) = \exp\{F_1(x)\theta^1 + F_2(x)\theta^2 - \varphi(\theta)\}.$$

Hence the family of 1-dimensional normal distributions is an exponential family. The Fisher information matrix is given by

$$\left[\frac{\partial^2 \varphi}{\partial\theta^i\partial\theta^j}\right] = \frac{1}{2\theta^1}\begin{bmatrix} \left(\frac{\theta^2}{\theta^1}\right)^2 + \frac{1}{\theta^1} & -\frac{\theta^2}{\theta^1} \\ -\frac{\theta^2}{\theta^1} & 1 \end{bmatrix}.$$

Since

$$\frac{\partial\varphi}{\partial\theta^1} = -\frac{1}{4}\left(\frac{\theta^2}{\theta^1}\right)^2 - \frac{1}{2\theta^1} = -\mu^2 - \sigma^2,$$

$$\frac{\partial\varphi}{\partial\theta^2} = \frac{\theta^2}{2\theta^1} = \mu,$$

the divergence for the Hessian structure $(D, g = Dd\varphi)$ is expressed by

$$\mathcal{D}(p, q) = \sum_{i=1}^{2}(\theta^i(q) - \theta^i(p))\frac{\partial\varphi}{\partial\theta^i}(q) + \varphi(p) - \varphi(q)$$

$$= \frac{1}{2}\left(\frac{1}{\sigma(q)^2} - \frac{1}{\sigma(p)^2}\right)(-\mu(q)^2 - \sigma(q)^2) + \left(\frac{\mu(q)}{\sigma(q)^2} - \frac{\mu(p)}{\sigma(p)^2}\right)\mu(q)$$

$$+ \frac{\mu(p)^2}{2\sigma(p)^2} + \log\sqrt{2\pi}\sigma(p) - \frac{\mu(q)^2}{2\sigma(q)^2} - \log\sqrt{2\pi}\sigma(q)$$

$$= \frac{1}{2}\left\{\frac{(\mu(p) - \mu(q))^2}{\sigma(p)^2} + \left(\frac{\sigma(q)^2}{\sigma(p)^2} - 1\right) + \log\frac{\sigma(p)^2}{\sigma(q)^2}\right\}.$$

Example 6.2. We set

$$\mathcal{X} = \{1, 2, \cdots, n+1\},$$

$$\Lambda = \left\{[\lambda^1, \cdots, \lambda^{n+1}] \in (\mathbf{R}^+)^{n+1} \;\middle|\; \sum_{k=1}^{n+1}\lambda^k = 1\right\}.$$

For $x \in \mathcal{X}$ and $\lambda = [\lambda^1, \cdots, \lambda^{n+1}] \in \Lambda$ we define a probability distribution by

$$p(x; \lambda) = \lambda^x.$$

The members of the family of probability distributions $\{p(x; \lambda) \mid \lambda \in \Lambda\}$ are called **multinomial distributions**. Let $F_i(j) = \delta_{ij}$ where δ_{ij} is the Kronecker's delta. Then

$$p(x; \lambda) = \lambda^1 F_1(x) + \cdots + \lambda^n F(x)_x + \lambda^{n+1} F(x)_{n+1}$$

$$= \exp\left\{F_1(x)\log\lambda^1 + \cdots + F_n(x)\log\lambda^n + F_{n+1}(x)\log\lambda^{n+1}\right\}$$

$$= \exp\left\{F_1(x)\left(\log\lambda^1 - \log\lambda^{n+1}\right) + \cdots\right.$$

$$\left. + F_n(x)\left(\log\lambda^n - \log\lambda^{n+1}\right) + \log\lambda^{n+1}\right\}.$$

Introducing

$$\theta^i = \log\left(\frac{\lambda^i}{\lambda^{n+1}}\right), \quad 1 \le i \le n,$$

$$\varphi(\theta) = -\log\lambda^{n+1} = \log\left(1 + \sum_{i=1}^{n} \exp\theta^i\right),$$

we have

$$p(x;\theta) = \exp\left\{\sum_{i=1}^{n} F_i(x)\theta^i - \varphi(\theta)\right\}, \quad \theta \in \Theta = \mathbf{R}^n.$$

Hence the family of multinomial distributions is an exponential family. Since $\dfrac{\partial\varphi}{\partial\theta^i} = \lambda^i$, the divergence for the Hessian structure is given by

$$
\begin{aligned}
\mathcal{D}(p,q) &= \sum_{i=1}^{n}(\theta^i(q) - \theta^i(p))\frac{\partial\varphi}{\partial\theta^i}(q) - (\varphi(q) - \varphi(p)) \\
&= \sum_{i=1}^{n}\left(\log\frac{\lambda^i(q)}{\lambda^{n+1}(q)} - \log\frac{\lambda^i(p)}{\lambda^{n+1}(p)}\right)\lambda^i(q) \\
&\quad + \log\lambda^{n+1}(q) - \log\lambda^{n+1}(p) \\
&= \sum_{i=1}^{n+1}\lambda^i(q)\log\frac{\lambda^i(q)}{\lambda^i(p)},
\end{aligned}
$$

(cf. Examples 2.2 (4), 2.8, 2.11 and Proposition 3.9).

Example 6.3. Probability distributions given by

$$p(x;\lambda) = e^{-\lambda}\frac{\lambda^x}{x!}, \quad x \in \mathcal{X} = \{0,1,2,\cdots\}, \quad \lambda \in \Lambda = \mathbf{R}^+,$$

are called the **Poisson distributions**. Introducing

$$C(x) = -\log x!, \quad F(x) = x, \quad \theta = \log\lambda,$$
$$\varphi(\theta) = e^\theta, \quad \theta \in \mathbf{R},$$

we have

$$p(x;\theta) = \exp\{C(x) + F(x)\theta - \varphi(\theta)\}, \quad \theta \in \mathbf{R}.$$

Hence the family of Poisson distributions is also an exponential family.

Example 6.4. Let Ω be a regular convex cone and let Ω^* be the dual cone of Ω. We denote by $\psi(\theta)$ the characteristic function on $\Omega^* G$

$$\psi(\theta) = \int_\Omega e^{-\langle x,\theta\rangle}dx.$$

For $x \in \Omega$ and $\theta \in \Omega^*$ we define

$$p(x; \theta) = \frac{e^{-\langle x, \theta \rangle}}{\psi(\theta)} = \exp\{-\langle x, \theta \rangle - \log \psi(\theta)\}.$$

Then $\{p(x; \theta) \mid \theta \in \Omega^*\}$ is an exponential family of probability distributions on Ω parametrized by $\theta \in \Omega^*$. The Fisher information metric coincides with the canonical Hessian metric on Ω^* (cf. Chapter 4).

Example 6.5. A family of probability distributions $\mathcal{P} = \{p(x; \lambda) \mid \lambda \in \Lambda\}$ on $\mathcal{X}$ is called a **mixture family** if it is expressed by

$$p(x; \lambda) = \sum_{i=1}^{n} \lambda^i p_i(x) + \left(1 - \sum_{i=1}^{n} \lambda^i\right) p_{n+1}(x),$$

where $\sum_{i=1}^{n} \lambda^i < 1$, $0 < \lambda^i < 1$, and each $p_i(x)$ is a probability distribution on $\mathcal{X}$. Since

$$\frac{\partial^2 l_\lambda}{\partial \lambda^i \partial \lambda^j} = -\frac{\partial l_\lambda}{\partial \lambda^i} \frac{\partial l_\lambda}{\partial \lambda^j},$$

we obtain

$$\begin{aligned}
\Gamma(-1)_{ij,k} &= \Gamma_{ij,k} + T_{ijk} \\
&= E_\lambda\left[\frac{\partial^2 l_\lambda}{\partial \lambda^i \partial \lambda^j} \frac{\partial l_\lambda}{\partial \lambda^k}\right] + E_\lambda\left[\frac{\partial l_\lambda}{\partial \lambda^i} \frac{\partial l_\lambda}{\partial \lambda^j} \frac{\partial l_\lambda}{\partial \lambda^k}\right] \\
&= 0,
\end{aligned}$$

and so the connection $\nabla(-1)$ is flat and $\{\lambda^1, \cdots, \lambda^n\}$ is an affine coordinate system with respect to $\nabla(-1)$.

The family of multinomial distributions introduced in Example 6.2 is a mixture family, and the Fisher information metric is given by

$$\begin{aligned}
g_{ij}(\lambda) &= \sum_x \frac{p_i(x) - p_{n+1}(x)}{p(x; \lambda)} \frac{p_j(x) - p_{n+1}(x)}{p(x; \lambda)} p(x; \lambda) \\
&= \sum_{k=1}^{n+1} \frac{(\delta_{ik} - \delta_{n+1k})(\delta_{jk} - \delta_{n+1k})}{\lambda^k} \\
&= \frac{1}{\lambda^i} \delta_{ij} + \frac{1}{\lambda^{n+1}}.
\end{aligned}$$

The potential function with respect to $\nabla(-1)$ is expressed by

$$\psi(\lambda) = \sum_{i=1}^{n+1} \lambda^i \log \lambda^i,$$

(cf. Example 2.8).

6.2 Hessian structures induced by normal distributions

Let $\mathbf{S}_n$ be the set of all real symmetric matrices of degree n and let $\mathbf{S}_n^+$ be the subset of $\mathbf{S}_n$ consisting of all positive-definite symmetric matrices. For n column vectors $x \in \mathbf{R}^n$ we define a probability distribution $p(x; \mu, \sigma)$ on $\mathbf{R}^n$ by

$$p(x; \mu, \sigma) = (2\pi)^{-\frac{n}{2}} (\det \sigma)^{-\frac{1}{2}} \exp\left\{ -\frac{{}^t(x-\mu)\sigma^{-1}(x-\mu)}{2} \right\},$$

where $\mu \in \mathbf{R}^n$ and $\sigma \in \mathbf{S}_n^+$. Then $\{p(x; \mu, \sigma) \mid (\mu, \sigma) \in \mathbf{R}^n \times \mathbf{S}_n^+\}$ is a family of probability distributions on $\mathbf{R}^n$ parametrized by (μ, σ), and is called a family of **n-dimensional normal distributions**.

Let Ω be a domain in a finite-dimensional real vector space V, and let ρ be an injective linear mapping from Ω into $\mathbf{S}_n$ satisfying

$(\mathbf{C}_1)$ $\rho(\omega) \in \mathbf{S}_n^+$ for all $\omega \in \Omega$.

We put

$$p(x; \mu, \omega) = (2\pi)^{-\frac{n}{2}} (\det \rho(\omega))^{\frac{1}{2}} \exp\left\{ -\frac{{}^t(x-\mu)\rho(\omega)(x-\mu)}{2} \right\}.$$

Then $\{p(x; \mu, \omega) \mid (\mu, \omega) \in \mathbf{R}^n \times \Omega\}$ is a family of probability distributions on $\mathbf{R}^n$ parametrized by $(\mu, \omega) \in \mathbf{R}^n \times \Omega$, and is called a family of **probability distributions induced by ρ**.

Proposition 6.1. *Let $\{p(x; \mu, \omega) \mid (\mu, \omega) \in \mathbf{R}^n \times \Omega\}$ be a family of probability distributions induced by ρ. Then the family is an exponential family parametrized by $\theta = \rho(\omega)\mu \in \mathbf{R}^n$ and $\omega \in \Omega$. The* Fisher *information metric is a Hessian metric on $\mathbf{R}^n \times \Omega$ with potential function*

$$\varphi(\theta, \omega) = \frac{1}{2}\left\{ {}^t\theta\rho(\omega)^{-1}\theta - \log \det \rho(\omega) \right\}.$$

Proof. Let $\{v^1, \cdots, v^m\}$ be a basis of V. For $x = [x^i]$ and $\mu = [\mu^i] \in \mathbf{R}^n$, and $\omega = \sum_\alpha \omega_\alpha v^\alpha \in \Omega$, we put

$$F^\alpha(x) = -\frac{1}{2} {}^t x \rho(v^\alpha) x, \qquad \theta = \rho(\omega)\mu.$$

Then we have

$$p(x; \mu, \omega) = p(x; \theta, \omega)$$
$$= \exp\left\{ \sum_j \theta_j x^j + \sum_\alpha \omega_\alpha F^\alpha(x) - \varphi(\theta, \omega) - \frac{n}{2}\log 2\pi \right\}.$$

This means that $\{p(x; \theta, \omega)\}$ is an exponential family on $\mathbf{R}^n$ parametrized by $(\theta, \omega) \in \mathbf{R}^n \times \Omega$, and that the Fisher information metric is the Hessian of $\varphi(\theta, \omega)$ with respect to the flat connection on $\mathbf{R}^n \times V$. $\square$

A straightforward calculation shows

$$\frac{\partial \varphi}{\partial \theta_i} = {}^t e^i \rho(\omega)^{-1} \theta,$$

$$\frac{\partial \varphi}{\partial \omega_\alpha} = -\frac{1}{2} \left\{ {}^t \theta \rho(\omega)^{-1} \rho(v^\alpha) \rho(\omega)^{-1} \theta + \mathrm{Tr}\, \rho(\omega)^{-1} \rho(v^\alpha) \right\},$$

$$\left[\frac{\partial^2 \varphi}{\partial \theta_i \partial \theta_j} \right] = \rho(\omega)^{-1},$$

$$\frac{\partial^2 \varphi}{\partial \theta_i \partial \omega_\alpha} = - {}^t e^i \rho(\omega)^{-1} \rho(v^\alpha) \rho(\omega)^{-1} \theta,$$

$$\frac{\partial^2 \varphi}{\partial \omega_\alpha \partial \omega_\beta} = {}^t \theta \rho(\omega)^{-1} \rho(v^\alpha) \rho(\omega)^{-1} \rho(v^\beta) \rho(\omega)^{-1} \theta$$

$$+ \frac{1}{2} \mathrm{Tr}\, \rho(\omega)^{-1} \rho(v^\alpha) \rho(\omega)^{-1} \rho(v^\beta).$$

where e^i is a vector in $\mathbf{R}^n$ whose j-th component is Kronecker's delta δ_{ij}. The Legendre transform φ' of φ is given by

$$\varphi' = \frac{1}{2} \log \det \rho(\omega) - \frac{n}{2}.$$

Proposition 6.2. *Let $\{p(x; \mu, \omega) \mid (\mu, \omega) \in \mathbf{R}^n \times \Omega\}$ be a family of probability distributions induced by ρ. Then the divergence is given by*

$$\mathcal{D}(p, q) = \frac{1}{2} \left\{ {}^t(\mu(p) - \mu(q)) \rho(\omega(p)) (\mu(p) - \mu(q)) + \mathrm{Tr}\, (\rho(\omega(p)) \rho(\omega(q))^{-1}) \right.$$

$$\left. - \log \det(\rho(\omega(p)) \rho(\omega(q))^{-1}) - n \right\}.$$

Proof. Using the above equations we have

$$\mathcal{D}(p, q) = \sum_{i=1}^{n} (\theta_i(q) - \theta_i(p)) \frac{\partial \varphi}{\partial \theta_i}(q) + \sum_{\alpha=1}^{m} (\omega_\alpha(q) - \omega_\alpha(p)) \frac{\partial \varphi}{\partial \omega_\alpha}(q) - (\varphi(q) - \varphi(p))$$

$$= \left\{ {}^t \theta(q) \rho(\omega(q))^{-1} \theta(q) - {}^t \theta(p) \rho(\omega(q))^{-1} \theta(q) \right\}$$

$$- \frac{1}{2} \left\{ {}^t \theta(q) \rho(\omega(q))^{-1} \rho(\omega(q)) \rho(\omega(q))^{-1} \theta(q) + \mathrm{Tr}\, \rho(\omega(q))^{-1} \rho(\omega(q)) \right.$$

$$\left. - {}^t \theta(q) \rho(\omega(q))^{-1} \rho(\omega(p)) \rho(\omega(q))^{-1} \theta(q) - \mathrm{Tr}\, \rho(\omega(q))^{-1} \rho(\omega(p)) \right\}$$

$$- \frac{1}{2} \left\{ {}^t \theta(q) \rho(\omega(q))^{-1} \theta(q) - \log \det \rho(\omega(q)) \right.$$

$$\left. - {}^t \theta(p) \rho(\omega(p))^{-1} \theta(p) + \log \det \rho(\omega(p)) \right\}$$

$$= \frac{1}{2}\Big\{ \, {}^t(\mu(p) - \mu(q))\rho(\omega(p))(\mu(p) - \mu(q)) + \mathrm{Tr}\,(\rho(\omega(p))\rho(\omega(q))^{-1})$$
$$- \log \det(\rho(\omega(p))\rho(\omega(q))^{-1}) - n \Big\}. \qquad \square$$

Example 6.6. Let $\Omega = \mathbf{S}_n^+$ and let $\rho : \Omega \longrightarrow \mathbf{S}_n$ be the inclusion mapping. The probability distributions $\{p(x; \mu, \omega)\}$ induced by ρ is a family of an n-dimensional normal distributions. Then we have

$$\varphi(\theta, \omega) = \frac{1}{2}(\,\theta\omega^{-1}\theta - \log \det \omega).$$

For $\theta = [\theta_i] \in \mathbf{R}^n$, $\omega = [\omega_{ij}] \in \Omega$ and $[\omega^{ij}] = [\omega_{ij}]^{-1}$ we put

$$\eta^i = -\frac{\partial \varphi}{\partial \theta_i} = -\,{}^t e^i \omega^{-1}\theta, \quad \xi^{ij} = -\frac{\partial \varphi}{\partial \omega_{ij}} = \omega^{ij} + \frac{1}{2}\eta^i\eta^j.$$

Then

$$\theta = -\Big[\xi^{ij} - \frac{1}{2}\eta^i\eta^j\Big]^{-1}\eta, \quad \omega = \Big[\xi^{ij} - \frac{1}{2}\eta^i\eta^j\Big]^{-1}.$$

The image of the Hessian domain $(\mathbf{R}^n \times \Omega, D, g = Dd\varphi)$ by the gradient mapping is a real Siegel domain (cf. Definition 10.4)

$$\Big\{ (\eta, \xi) \in \mathbf{R}^n \times \mathbf{S}_n \mid \xi - \frac{1}{2}\eta \, {}^t\eta > 0 \Big\}.$$

The Legendre transform φ' of φ is given by

$$\varphi'(\eta, \xi) = -\frac{1}{2}\log \det \Big(\xi - \frac{1}{2}\eta \, {}^t\eta\Big) - \frac{n}{2}.$$

Example 6.7. Let $\rho : \mathbf{R}^+ \longrightarrow \mathbf{S}_n$ be a mapping defined by $\rho(\omega) = \omega I_n$ where I_n is the unit matrix of degree n. By Proposition 6.1, the potential function for the family of probability distributions induced by ρ is given by

$$\varphi(\theta, \omega) = \frac{1}{2}\Big(\frac{1}{\omega}\,{}^t\theta\theta - n \log \omega\Big).$$

Put

$$\eta^i = -\frac{\partial \varphi}{\partial \theta_i} = -\frac{\theta_i}{\omega}, \quad \xi = -\frac{\partial \varphi}{\partial \omega} = \frac{1}{2}\Big(\frac{{}^t\theta\theta}{\omega^2} + \frac{n}{\omega}\Big).$$

Then

$$\theta = -\frac{n}{2}\eta\Big(\xi - \frac{1}{2}\,{}^t\eta\eta\Big)^{-1}, \quad \omega = \frac{n}{2}\Big(\xi - \frac{1}{2}\,{}^t\eta\eta\Big)^{-1}.$$

This means that the image of the Hessian domain $(\mathbf{R}^n \times \mathbf{R}^+, D, g = Dd\varphi)$ by the gradient mapping is given by a domain over an elliptic paraboloid,

$$\Big\{ (\eta, \xi) \in \mathbf{R}^n \times \mathbf{R} \mid \xi - \frac{1}{2}\,{}^t\eta\eta > 0 \Big\}.$$

The Legendre transform φ' of φ is expressed by

$$\varphi'(\eta, \xi) = -\frac{n}{2} \log \left(\xi - \frac{1}{2} \, {}^t\eta\eta \right) + \frac{n}{2} \log \frac{n}{2} - \frac{n}{2}.$$

By Proposition 3.8 the Hessian sectional curvature of the dual Hessian structure $(D', g = D'd\varphi')$ is a constant with value $\frac{2}{n}$.

Suppose that there exist a Lie subgroup G of $GL(V)$ acting transitively on Ω, and a matrix representation $\boldsymbol{f}$ of G satisfying

(**C**$_2$) $\rho(s\omega) = \boldsymbol{f}(s)\rho(\omega)\, {}^t\boldsymbol{f}(s)$ for $s \in G$ and $\omega \in V$.

Then G acts on $\mathbf{R}^n \times \Omega$ by $s(\theta, \omega) = (\boldsymbol{f}(s)\theta, s\omega)$. Since $\varphi(\boldsymbol{f}(s)\theta, s\omega) - \varphi(\theta, \omega)$ is a constant, the Hessian metric $g = Dd\varphi$ is invariant under G.

The following theorem is due to [Rothaus (1960)].

Theorem 6.1. *Let Ω be a regular convex cone admitting a transitive Lie group G and let ρ be a mapping satisfying the conditions* (**C**$_1$) *and* (**C**$_2$). *Then*

$$\left\{ (\xi, \theta, \omega) \in \mathbf{R} \times \mathbf{R}^n \times \Omega \mid \xi - {}^t\theta\rho(\omega)^{-1}\theta > 0 \right\}$$

is a homogeneous regular convex cone. Conversely, any homogeneous regular convex cone is obtained in this way from a lower-dimensional homogeneous regular convex cone Ω and ρ satisfying the conditions (**C**$_1$) *and* (**C**$_2$).

For a classification of ρ for self-dual homogeneous regular convex cones satisfying the conditions (**C**$_1$) and (**C**$_2$), the interested reader may refer to [Satake (1972)].

Example 6.8. Let $M(n, \mathbf{R})$ be the set of all real matrices of degree n and let $End(M(n, \mathbf{R}))$ be the set of all endomorphisms of $M(n, \mathbf{R})$. Define $\rho : \mathbf{S}_n \longrightarrow End(M(n, \mathbf{R}))$ by

$$\rho(\omega)x = \omega x + x\omega \quad for \ \omega \in \mathbf{S}_n \ and \ x \in M(n, \mathbf{R}).$$

Then $\rho(\omega)$ is symmetric with respect to an inner product $\langle x, y \rangle = \mathrm{Tr}\, {}^txy$ and is positive definite for $\omega \in \mathbf{S}_n^+$. Define a matrix representation $\boldsymbol{f}$ of the orthogonal group $O(n)$ on the space $M(n, \mathbf{R})$ by

$$\boldsymbol{f}(s)x = sx^ts.$$

Then

$$ {}^t\boldsymbol{f}(s)x = {}^tsxs, \quad \rho(\boldsymbol{f}(s)\omega) = \boldsymbol{f}(s)\rho(\omega)^t\boldsymbol{f}(s).$$

Introducing $\mathbf{A}_n = \{x \in M(n, \mathbf{R}) \mid {}^t x = -x\}$, we have

$$M(n, \mathbf{R}) = \mathbf{S}_n + \mathbf{A}_n,$$

$$\rho(\omega)\mathbf{S}_n \subset \mathbf{S}_n, \qquad \rho(\omega)\mathbf{A}_n \subset \mathbf{A}_n.$$

Thus ρ induces equivariant linear mappings ρ^+ and ρ^- from $\mathbf{S}_n$ into $End(\mathbf{S}_n)$ and $End(\mathbf{A}_n)$ respectively. The Hessian structure on $\mathbf{A}_n \times \mathbf{S}_n^+$ induced by ρ^- is deeply related to the theory of stable feedback systems [Ohara and Amari (1994)].

Example 6.9. Let Ω be a homogeneous self-dual regular convex cone with vertex 0 in a real vector space V. Since the linear automorpism group G of Ω is self-dual by Theorem 4.7, it is completely reducible and so reductive. Thus G admits a discrete subgroup Γ such that the quotient space $\Gamma \backslash \Omega$ is a compact flat manifold. By the study of [Satake (1972)] there exists a linear mapping ρ from V into $\mathbf{S}_n$ satisfying the conditions $(\mathbf{C}_1)$ and $(\mathbf{C}_2)$. Let $E(\Gamma \backslash \Omega, \rho)$ be the vector bundle over $\Gamma \backslash \Omega$ associated with the universal covering $\pi : \Omega \longrightarrow \Gamma \backslash \Omega$ and ρ. Since the Hessian structure $(D, g = Dd\varphi)$ on $\mathbf{R}^n \times \Omega$ is Γ-invariant, it induces a Hessian structure on $E(\Gamma \backslash \Omega, \rho)$.

Chapter 7

Cohomology on flat manifolds

In this chapter we study cohomology on flat manifolds. Using a flat connection we define a certain cohomology similar to the Dolbeault cohomology. The cohomology theory plays an important role in the study of flat manifolds. In section **7.1** we prove fundamental identities of the exterior product operator, the interior product operator and the star operator on the space $\mathcal{A}^{p,q}$ consisting of all (p,q)-forms. Using a flat connection, in section **7.2** we construct a cochain complex $\{\sum_p \mathcal{A}^{p,q}, \partial\}$ similar to the Dolbeault complex. We define a Laplacian with respect to the coboundary operator ∂, and prove a duality theorem which corresponds to Kodaira-Serre's duality theorem. In section **7.3** we generalize Koszul's vanishing theorem, and also demonstrate as an application that an affine Hopf manifold does not admit any Hessian metric. We proceed to prove in section **7.4** a vanishing theorem similar to Kodaira-Nakano's vanishing theorem. In section **7.5**, taking an appropriate flat line bundle L over a Hessian manifold, we obtain basic identities on the space of L-valued (p,q)-forms, which are similar to the Kählerian identities. In section **7.6** we define affine Chern classes for a flat vector bundle over a flat manifold. The first affine Chern class for a Hessian manifold is represented by the second Koszul form.

7.1 (p,q)-forms on flat manifolds

Throughout sections 7.1-7.5 we will always assume that a flat manifold (M, D) is compact and oriented. Let T^* be the cotangent bundle over M. We denote by $(\overset{p}{\wedge} T^*)\otimes(\overset{q}{\wedge} T^*)$ the tensor product of vector bundles $\overset{p}{\wedge} T^*$ and $\overset{q}{\wedge} T^*$, and by $\mathcal{A}^{p,q}$ the space of all smooth sections of $(\overset{p}{\wedge} T^*)\otimes(\overset{q}{\wedge} T^*)$. An element in $\mathcal{A}^{p,q}$ is called a (p,q)-form. Using an affine coordinate system a (p,q)-form ω is expressed by

$$\omega = \sum \omega_{i_1 \cdots i_p \bar{j}_1 \cdots \bar{j}_q}(dx^{i_1} \wedge \cdots \wedge dx^{i_p}) \otimes (dx^{\bar{j}_1} \wedge \cdots \wedge dx^{\bar{j}_q})$$
$$= \sum \omega_{I_p, \bar{J}_q} dx^{I_p} \otimes dx^{\bar{J}_q},$$

where

$$\omega_{I_p, \bar{J}_q} = \omega_{i_1 \cdots i_p \bar{j}_1 \cdots \bar{j}_q}, \quad dx^{I_p} = dx^{i_1} \wedge \cdots \wedge dx^{i_p}, \quad dx^{\bar{J}_q} = dx^{\bar{j}_1} \wedge \cdots \wedge dx^{\bar{j}_q}.$$

Let $\mathfrak{F}$ be the set of all smooth functions on M and let $\mathfrak{X}$ be the set of all smooth vector fields on M. A (p,q)-form ω is identified with a $\mathfrak{F}$-multilinear mapping

$$\omega : \overbrace{\mathfrak{X} \times \cdots \times \mathfrak{X}}^{p \ terms} \times \overbrace{\mathfrak{X} \times \cdots \times \mathfrak{X}}^{q \ terms} \longrightarrow \mathfrak{F},$$

such that $\omega(X_1, \cdots, X_p; Y_1, \cdots, Y_q)$ is skew symmetric with respect to $\{X_1, \cdots, X_p\}$ and $\{Y_1, \cdots, Y_q\}$ respectively.

Definition 7.1. For $\omega \in \mathcal{A}^{p,q}$ and $\eta \in \mathcal{A}^{r,s}$ we define the exterior product $\omega \wedge \eta \in \mathcal{A}^{p+r,q+s}$ by

$$(\omega \wedge \eta)(X_1, \cdots, X_{p+r}; Y_1, \cdots, Y_{q+s})$$
$$= \frac{1}{p!r!q!s!} \sum_{\sigma,\tau} \epsilon_\sigma \epsilon_\tau \omega(X_{\sigma(1)}, \cdots, X_{\sigma(p)}; Y_{\tau(1)}, \cdots, Y_{\tau(q)})$$
$$\times \eta(X_{\sigma(p+1)}, \cdots, X_{\sigma(p+r)}; Y_{\tau(q+1)}, \cdots, Y_{\tau(q+s)}),$$

where σ (resp. τ) is a permutation on $p+r$ (resp. $q+s$) letters, and ϵ_σ (resp. ϵ_τ) is the sign of σ (resp. τ).

For $\omega = \sum \omega_{I_p, \bar{J}_q} dx^{I_p} \otimes dx^{\bar{J}_q}$ and $\eta = \sum \eta_{K_r, \bar{L}_s} dx^{K_r} \otimes dx^{\bar{L}_s}$ we have

$$\omega \wedge \eta = \sum \omega_{I_p, \bar{J}_q} \eta_{K_r, \bar{L}_s}(dx^{I_p} \wedge dx^{K_r}) \otimes (dx^{\bar{J}_q} \wedge dx^{\bar{L}_s}),$$

where the symbols $\wedge$ of the right-hand side represent the ordinary exterior product.

Definition 7.2. For $\omega \in \mathcal{A}^{r,s}$ we define an exterior product operator by

$$e(\omega) : \omega \in \mathcal{A}^{p,q} \longrightarrow \omega \wedge \eta \in \mathcal{A}^{p+r,q+s}.$$

Definition 7.3. For $X \in \mathfrak{X}$ we define interior product operators by

$$i(X) : \mathcal{A}^{p,q} \longrightarrow \mathcal{A}^{p-1,q}, \quad i(X)\omega = \omega(X, \cdots; \cdots),$$
$$\bar{i}(X) : \mathcal{A}^{p,q} \longrightarrow \mathcal{A}^{p,q-1}, \quad \bar{i}(X)\omega = \omega(\cdots; X, \cdots).$$

As for the case of the ordinary interior product operator, it is straightforward to see that

$$i(X)(\omega \wedge \eta) = i(X)\omega \wedge \eta + (-1)^p \omega \wedge i(X)\eta,$$
$$\bar{i}(X)(\omega \wedge \eta) = \bar{i}(X)\omega \wedge \eta + (-1)^q \omega \wedge \bar{i}(X)\eta, \quad for \ \omega \in \mathcal{A}^{p,q}. \quad (7.1)$$

We select a Riemannian metric g on M. The volume element v determined by g is expressed by

$$v = (\det[g_{ij}])^{\frac{1}{2}} dx^1 \wedge \cdots \wedge dx^n.$$

We identify v with $v \otimes 1 \in \mathcal{A}^{n,0}$ and set $\bar{v} = 1 \otimes v \in \mathcal{A}^{0,n}$.

Definition 7.4. We define the star operator $\star : \mathcal{A}^{p,q} \longrightarrow \mathcal{A}^{n-p,n-q}$ by

$$(\star\omega)(X_1, \cdots, X_{n-p}; Y_1, \cdots, Y_{n-q})v \wedge \bar{v}$$
$$= \omega \wedge \bar{i}(X_1)g \wedge \cdots \wedge \bar{i}(X_{n-p})g \wedge i(Y_1)g \wedge \cdots \wedge i(Y_{n-q})g.$$

Then

$$\star(v \wedge \bar{v}) = 1, \qquad \star 1 = v \wedge \bar{v}, \qquad\qquad\qquad (7.2)$$
$$(\star\omega)(X_1, \cdots, X_{n-p}; Y_1, \cdots, Y_{n-q}) \qquad\qquad\qquad (7.3)$$
$$= \star(\omega \wedge \bar{i}(X_1)g \wedge \cdots \wedge \bar{i}(X_{n-p})g \wedge i(Y_1)g \wedge \cdots \wedge i(Y_{n-q})g),$$

for $\omega \in \mathcal{A}^{p,q}$.

Lemma 7.1. *The following identities hold on the space* $\mathcal{A}^{p,q}$.

(1) $i(X)\star = (-1)^p \star e(\bar{i}(X)g)$, $\qquad \bar{i}(X)\star = (-1)^q \star e(i(X)g)$.
(2) $\star i(X) = (-1)^{p+1} e(\bar{i}(X)g)\star$, $\qquad \star \bar{i}(X) = (-1)^{q+1} e(i(X)g)\star$.

Proof. From equation (7.3) we obtain

$$(i(X) \star \omega)(X_1, \cdots, X_{n-p-1}; Y_1, \cdots, Y_{n-q})$$
$$= (\star\omega)(X, X_1, \cdots, X_{n-p-1}; Y_1, \cdots, Y_{n-q})$$
$$= \star\Big(\omega \wedge \bar{i}(X)g \wedge \bar{i}(X_1)g \wedge \cdots \wedge \bar{i}(X_{n-p-1})g \wedge i(Y_1)g \wedge \cdots \wedge i(Y_{n-q})g\Big)$$
$$= (-1)^p \star \Big(e(\bar{i}(X)g)\omega \wedge \bar{i}(X_1)g \wedge \cdots \wedge \bar{i}(X_{n-p-1})g \wedge i(Y_1)g$$
$$\wedge \cdots \wedge i(Y_{n-q})g\Big)$$
$$= \Big((-1)^p \star e(\bar{i}(X)g)\omega\Big)(X_1, \cdots, X_{n-p-1}; Y_1, \cdots, Y_{n-q}),$$

$$(\star i(X)\omega)(X_1,\cdots,X_{n-p+1};Y_1,\cdots,Y_{n-q})$$

$$= \star\Big(i(X)\omega \wedge \bar{i}(X_1)g \wedge \cdots \wedge \bar{i}(X_{n-p+1})g \wedge i(Y_1)g \wedge \cdots \wedge i(Y_{n-q})g\Big)$$

$$= \star\Big\{ \sum_{i=1}^{n-p+1} (-1)^{p+i}\omega \wedge \bar{i}(X_1)g \wedge \cdots \wedge i(X)\bar{i}(X_i)g \wedge \cdots \wedge \bar{i}(X_{n-p+1})g$$

$$\wedge i(Y_1) \wedge \cdots \wedge i(Y_{n-q})g\Big\}$$

$$= (-1)^p \sum_{i=1}^{n-p+1} (-1)^i \Big(i(X)g\Big)(X_i)\, \star \Big(\omega \wedge \bar{i}(X_1)g \wedge \cdots \wedge \bar{i}(\hat{X}_i)g \wedge \cdots$$

$$\wedge \bar{i}(X_{n-p+1})g \wedge i(Y_1)g \wedge \cdots \wedge i(Y_{n-q})g\Big)$$

$$= (-1)^p \sum_{i=1}^{n-p+1} (-1)^i \Big(\bar{i}(X)g\Big)(X_i)$$

$$\times (\star\omega)(X_1,\cdots,\hat{X}_i,\cdots,X_{n-p+1};Y_1,\cdots,Y_{n-q})$$

$$= (-1)^{p+1}\Big(\bar{i}(X)g \wedge \star\omega\Big)(X_1,\cdots,X_{n-p+1};Y_1,\cdots,Y_{n-q}),$$

where the symbol $\hat{\ }$ indicates that this term is omitted. $\qquad\square$

Lemma 7.2. *For $\omega \in \mathcal{A}^{p,q}$ we have*

$$\star\star\omega = (-1)^{(p+q)(n+1)}\omega.$$

Proof. By (7.2), (7.3) and Lemma 7.1 we obtain

$$(\star\star\omega)(X_1,\cdots,X_p;Y_1,\cdots,Y_q)$$

$$= \star\Big(\star\omega \wedge \bar{i}(X_1)g \wedge \cdots \wedge \bar{i}(X_p) \wedge i(Y_1)g \wedge \cdots \wedge i(Y_q)g\Big)$$

$$= (-1)^{(p+q)(n+1)} \star \Big(\star\bar{i}(Y_q)\cdots\bar{i}(Y_1)i(X_p)\cdots i(X_1)\omega\Big)$$

$$= (-1)^{(p+q)(n+1)}\omega(X_1,\cdots,X_p;Y_1,\cdots,Y_q). \qquad\square$$

Let $\{E_1,\cdots,E_n\}$ be an orthonormal frame field with respect to the Riemannian metric g; $g(E_i,E_j) = \delta_{ij}$. Considering g as an element in $\mathcal{A}^{1,1}$ we set

$$\theta^j = \bar{i}(E_j)g \in \mathcal{A}^{1,0}, \qquad \bar{\theta}^j = i(E_j)g \in \mathcal{A}^{0,1}.$$

It follows from Lemma 7.1 that the following identities hold on $\mathcal{A}^{p,q}$,

$$i(E_j) = (-1)^{p+1} \star^{-1} e(\theta^j)\star, \qquad \bar{i}(E_j) = (-1)^{q+1} \star^{-1} e(\bar{\theta}^j)\star,$$

$$e(\theta_j) = (-1)^p \star^{-1} i(E_j)\star, \qquad e(\bar{\theta}_j) = (-1)^q \star^{-1} \bar{i}(E_j)\star. \qquad (7.4)$$

For simplicity we use the following notation,

$$I_p = (i_1, \cdots, i_p), \qquad i_1 < \cdots < i_p,$$
$$I_{n-p} = (i_{p+1}, \cdots, i_n), \qquad i_{p+1} < \cdots < i_n,$$

where $(I_p, I_{n-p}) = (i_1, \cdots, i_p, i_{p+1}, \cdots, i_n)$ is a permutation of $(1, \cdots, n)$. Using the orthonormal frame $\omega \in \mathcal{A}^{p,q}$ is expressed as

$$\omega = \sum \omega_{i_1, \cdots, i_p j_1, \cdots, j_q} \theta^{i_1} \wedge \cdots \wedge \theta^{i_p} \wedge \bar{\theta}^{j_1} \wedge \cdots \wedge \bar{\theta}^{j_q}$$
$$= \sum_{I_p, J_q} \omega_{I_p, J_q} \theta^{I_p} \wedge \bar{\theta}^{J_q},$$

where $I_p = (i_1, \cdots, i_p)$, $J_q = (j_1, \cdots, j_q)$ and $\theta^{I_p} = \theta^{i_1} \wedge \cdots \wedge \theta^{i_p}$, $\bar{\theta}^{J_q} = \bar{\theta}^{j_1} \wedge \cdots \wedge \bar{\theta}^{j_q}$.

We then have

$$\star\omega = \sum_{I_p, J_q} \omega_{I_p, J_q} \epsilon(I_p, I_{n-p}) \epsilon(J_q, J_{n-q}) \theta^{I_{n-p}} \wedge \bar{\theta}^{J_{n-q}}. \tag{7.5}$$

For $\omega = \sum_{I_p, J_q} \omega_{I_p, J_q} \theta^{I_p} \wedge \bar{\theta}^{J_q}$ and $\eta = \sum_{I_p, J_q} \eta_{I_p, J_q} \theta^{I_p} \wedge \bar{\theta}^{J_q} \in \mathcal{A}^{p,q}$ we put

$$\langle \omega, \eta \rangle = \sum_{I_p, J_q} \omega_{I_p, J_q} \eta_{I_p, J_q},$$

and define an inner product (ω, η) on $\mathcal{A}^{p,q}$ by

$$(\omega, \eta) = \int_M \langle \omega, \eta \rangle v.$$

From equation (7.5) we obtain the following lemma.

Lemma 7.3. *For ω and $\eta \in \mathcal{A}^{p,q}$ we have*

$$\omega \wedge \star\eta = \eta \wedge \star\omega = \langle \omega, \eta \rangle v \wedge \bar{v}.$$

Let K and K^* be line bundles over M defined by

$$K = \overset{n}{\wedge} T^*, \quad K^* = \overset{n}{\wedge} T,$$

where T (resp. T^*) is the tangent (resp. cotangent) bundle over M. We denote by $\mathcal{A}^{p,q}(K^*)$ the space of all smooth sections of $(\overset{p}{\wedge} T^*) \otimes (\overset{q}{\wedge} T^*) \otimes K^*$, and define isomorphisms $\kappa : \mathcal{A}^{p,q} \longrightarrow \mathcal{A}^{p,q}(K^*)$ and $C : \mathcal{A}^{p,n}(K^*) \longrightarrow \mathcal{A}^p$ by

$$\kappa(\omega) = \omega \otimes v^*, \qquad C((\eta \otimes \bar{v}) \otimes v^*) = \eta,$$

where v^* is the dual section of v,

$$v^* = (\det[g_{ij}])^{-\frac{1}{2}} \left(\frac{\partial}{\partial x^1} \wedge \cdots \wedge \frac{\partial}{\partial x^n} \right).$$

Definition 7.5. We denote by $\tilde{\star}$ the composition of the mappings κ and $\star$

$$\tilde{\star} = \kappa\star : \mathcal{A}^{p,q} \longrightarrow \mathcal{A}^{n-p,n-q}(K^*).$$

Lemma 7.4. *We have*

$$(\omega,\ \eta) = \int_M C(\omega \wedge \tilde{\star}\eta).$$

Proof. By Lemma 7.3 we obtain

$$C(\omega \wedge \tilde{\star}\eta) = C(\omega \wedge (\star\eta \otimes v^*)) = C((\omega \wedge \star\eta) \otimes v^*)$$
$$= C(\langle\omega,\eta\rangle v \otimes \bar{v} \otimes v^*) = \langle\omega,\eta\rangle v. \qquad \square$$

Lemma 7.5. *Let $\rho \in \mathcal{A}^{1,1}$. The adjoint operator $i(\rho)$ of $e(\rho)$ is given by*

$$i(\rho) = (-1)^{p+q} \star^{-1} e(\rho) \star \quad on \ \mathcal{A}^{p,q}.$$

Proof. It follows from Lemmata 7.2 and 7.3 that

$$\omega \wedge \star e(\rho)\eta = e(\rho)\eta \wedge \star\omega = (-1)^{p+q}\eta \wedge \star \star^{-1} e(\rho) \star \omega. \qquad \square$$

We denote by $\Pi_{p,q}$ the projection from $\sum_{r,s} \mathcal{A}^{r,s}$ to $\mathcal{A}^{p,q}$, and define an operator

$$\Pi = \sum_{p,q}(n - p - q)\Pi_{p,q}.$$

Proposition 7.1. *The following equations hold.*

(1) $e(g) = \sum_j e(\theta^j)e(\bar{\theta}^j)$.

(2) $i(g) = \sum_j i(E_j)\bar{i}(E_j)$.

(3) $[\Pi, e(g)] = -2e(g), \quad [\Pi, i(g)] = 2i(g), \quad [i(g), e(g)] = \Pi.$

Proof. Assertion (1) follows from $g = \sum_j \theta^j \wedge \bar{\theta}^j$. By Lemma 7.5 and (7.4) we obtain

$$i(g) = (-1)^{p+q} \star^{-1} e(g) \star$$
$$= \sum_j \{(-1)^{p+1} \star^{-1} e(\theta^j)\star\}\{(-1)^{q+1} \star^{-1} e(\bar{\theta}^j)\star\}$$
$$= \sum_j i(E_j)\bar{i}(E_j).$$

Since $\sum_j e(\theta^j)i(E_j) = p$ and $\sum_j e(\bar{\theta}^j)\bar{i}(E_j) = q$ on $\mathcal{A}^{p,q}$, we have

$$i(g)e(g) = \sum_{j,k} i(E_j)e(\theta^k)\bar{i}(E_j)e(\bar{\theta}^k)$$
$$= \sum_{j,k}\{\delta_j^k - e(\theta^k)i(E_j)\}\{\delta_j^k - e(\bar{\theta}^k)\bar{i}(E_j)\}$$
$$= n - p - q + e(g)i(g).$$

This implies $[i(g), e(g)] = \Pi$. $\qquad \square$

7.2 Laplacians on flat manifolds

In this section we construct a certain cochain complex on a flat manifold (M, D) analogous to the Dolbeault complex on a complex manifold, and prove a duality theorem similar to the Kodaira-Serre's duality theorem on a complex manifold. We first define two coboundary operators ∂ and $\bar{\partial}$ by using the flat connection.

Definition 7.6. We define $\partial : \mathcal{A}^{p,q} \longrightarrow \mathcal{A}^{p+1,q}$ and $\bar{\partial} : \mathcal{A}^{p,q} \longrightarrow \mathcal{A}^{p,q+1}$ by

$$\partial = \sum_i e(dx^i) D_{\partial/\partial x^i}, \quad \bar{\partial} = \sum_i e(\bar{dx}^i) D_{\partial/\partial x^i},$$

where dx^i is identified with $dx^i \otimes 1 \in \mathcal{A}^{1,0}$, and $\bar{dx}^i = 1 \otimes dx^i \in \mathcal{A}^{0,1}$.

Then it is easy to see that

$$\partial\partial = 0, \quad \bar{\partial}\bar{\partial} = 0.$$

Thus the pairs $\left\{ \sum_p \mathcal{A}^{p,q}, \partial \right\}$ and $\left\{ \sum_q \mathcal{A}^{p,q}, \bar{\partial} \right\}$ are cochain complexes with coboundary operators ∂ and $\bar{\partial}$ respectively. These complexes correspond to the Dolbeault complex on a complex manifold [Kobayashi (1987)][Kodaira (1986)][Morrow and Kodaira (1971)] [Wells (1979)].

Let F be a flat vector bundle over M and let $\mathcal{A}^{p,q}(F)$ be the space of all smooth sections of $(\overset{p}{\wedge} T^*) \otimes (\overset{q}{\wedge} T^*) \otimes F$. We extend the coboundary operators ∂ and $\bar{\partial}$ to the space $\mathcal{A}^{p,q}(F)$ as follows. Since the vector bundle F is flat, we choose local frame fields such that the transition functions are constants. Let $s = \{s_1, \cdots, s_r\}$ be such a local frame field of F. For $\sum_i \omega^i \otimes s_i \in \mathcal{A}^{p,q}(F)$ where $\omega^i \in \mathcal{A}^{p,q}$ we define ∂ and $\bar{\partial}$ by

$$\partial(\sum_i \omega^i \otimes s_i) = \sum_i (\partial\omega^i) \otimes s_i \in \mathcal{A}^{p+1,q}(F),$$

$$\bar{\partial}(\sum_i \omega^i \otimes s_i) = \sum_i (\bar{\partial}\omega^i) \otimes s_i \in \mathcal{A}^{p,q+1}(F).$$

Then the pairs $\left\{ \sum_p \mathcal{A}^{p,q}(F), \partial \right\}$ and $\left\{ \sum_q \mathcal{A}^{p,q}(F), \bar{\partial} \right\}$ are cochain complexes.

Definition 7.7. Let F^* be the dual bundle of F. For a local frame field $\{s_1, \cdots, s_r\}$ as above we denote by $\{s^{*1}, \cdots, s^{*r}\}$ the dual frame field of $\{s_1, \cdots, s_r\}$. Choosing a fiber metric a on F, we define an isomorphism $\tilde{\star}_F : \mathcal{A}^{p,q}(F) \longrightarrow \mathcal{A}^{n-p,n-q}(F^* \otimes K^*)$ by

$$\tilde{\star}_F \left(\sum_i \omega^i \otimes s_i \right) = \sum_{i,j} \left\{ \left(G^{-\frac{1}{2}} a_{ij} \right)(\star\omega^j) \otimes s^{*i} \right\} \otimes \left(\frac{\partial}{\partial x^1} \wedge \cdots \wedge \frac{\partial}{\partial x^n} \right)$$

where $a_{ij} = a(s_i, s_j)$ and $G = \det[g_{ij}]$.

For $\omega = \sum_i \omega^i \otimes s_i \in \mathcal{A}^{p,q}(F)$ and $\eta = \sum_i \eta_i \otimes s^{*i} \in \mathcal{A}^{r,s}(K^* \otimes F^*)$ we set

$$\omega \wedge \eta = \sum_i \omega^i \wedge \eta_i \in \mathcal{A}^{p+r,q+s}(K^*).$$

Definition 7.8. We define an inner product on $\mathcal{A}^{p,q}(F)$ by

$$(\omega,\ \eta) = \int_M C(\omega \wedge \tilde{\star}_F \eta).$$

Proposition 7.2. *Let δ_F and $\bar{\delta}_F$ be the adjoint operators of ∂ and $\bar{\partial}$ with respect to the above inner product respectively. On the space $\mathcal{A}^{p,q}(F)$ we have*

$$\delta_F = (-1)^p \tilde{\star}_F^{-1} \partial\, \tilde{\star}_F, \qquad \bar{\delta}_F = (-1)^q \tilde{\star}_F^{-1} \bar{\partial}\, \tilde{\star}_F.$$

Proof. Since $C\partial = dC$ on $\mathcal{A}^{p,n}(K^*)$, we have for $\omega \in \mathcal{A}^{p,q}(F)$ and $\eta \in \mathcal{A}^{p-1,q}(F)$

$$(\delta_F \omega,\ \eta) = \int_M C\Big(\sum_i \partial \eta^i \wedge (\tilde{\star}_F\, \omega)_i\Big)$$

$$= \int_M dC\Big(\sum_i \eta^i \wedge (\tilde{\star}_F\, \omega)_i\Big) + (-1)^p \int_M C\Big(\eta \wedge \tilde{\star}_F\, \tilde{\star}_F^{-1} \partial \tilde{\star}_F\, \omega\Big)$$

$$= \big((-1)^p \tilde{\star}_F^{-1} \partial \tilde{\star}_F\, \omega,\ \eta\big). \qquad \square$$

Definition 7.9. We define the Laplacians $\square_F$ and $\bar{\square}_F$ with respect to ∂ and $\bar{\partial}$ by

$$\square_F = \partial \delta_F + \delta_F \partial, \qquad \bar{\square}_F = \bar{\partial}\bar{\delta}_F + \bar{\delta}_F \bar{\partial}.$$

For a trivial line bundle the Laplacians will be denoted by $\square$ and $\bar{\square}$.

Let us choose a fiber metric b on $F^* \otimes K^*$ by

$$b(s^{*i} \otimes v^*, s^{*j} \otimes v^*) = a^{ij}$$

where a^{ij} is the (i,j)-component of the inverse matrix of $[a_{ij}]$. Identifying $(F^* \otimes K^*)^* \otimes K^*$ with F and using the fiber metric b we obtain

$$\tilde{\star}_{F^* \otimes K^*}\Big\{\sum_i \eta_i \otimes \Big(s^{*i} \otimes \Big(\frac{\partial}{\partial x^1} \wedge \cdots \wedge \frac{\partial}{\partial x^n}\Big)\Big)\Big\}$$

$$= \sum_i \Big(\sum_j (\det[g_{kl}])^{\frac{1}{2}} a^{ij} \star \eta_j\Big) \otimes s_i,$$

and so

$$\tilde{\star}_{F^* \otimes K^*}\, \tilde{\star}_F = (-1)^{(p+q)(n+1)} \quad \text{on } \mathcal{A}^{p,q}(F),$$

$$\tilde{\star}_F\, \tilde{\star}_{F^* \otimes K^*} = (-1)^{(p+q)(n+1)} \quad \text{on } \mathcal{A}^{p,q}(F^* \otimes K^*).$$

Theorem 7.1.

(1) $\Box_{F^* \otimes K^*} \tilde{\star}_F = \tilde{\star}_F \Box_F$.

(2) $\Box_F \tilde{\star}_{F^* \otimes K^*} = \tilde{\star}_{F^* \otimes K^*} \Box_{F^* \otimes K^*}$.

Proof. By Proposition 7.2 we have

$$\partial \tilde{\star}_F = (-1)^p \tilde{\star}_F \delta_F \quad \text{on} \quad \mathcal{A}^{p,q}(F).$$

Again by Proposition 7.2, we obtain

$$\begin{aligned}
\delta_{F^* \otimes K^*} \tilde{\star}_F &= (-1)^{n+p} \tilde{\star}_{F^* \otimes K^*}^{-1} \partial \tilde{\star}_{F^* \otimes K^*} \tilde{\star}_F \\
&= (-1)^{p+1} \tilde{\star}_F \partial \quad \text{on} \quad \mathcal{A}^{p,q}(F).
\end{aligned}$$

Hence it follows that

$$\begin{aligned}
\Box_{F^* \otimes K^*} \tilde{\star}_F &= \left(\partial \, \delta_{F^* \otimes K^*} + \delta_{F^* \otimes K^*} \, \partial \right) \tilde{\star}_F \\
&= (-1)^{p+1} \partial \tilde{\star}_F \partial + (-1)^p \delta_{F^* \otimes K^*} \tilde{\star}_F \delta_F \\
&= \tilde{\star}_F \delta_F \partial + \tilde{\star}_F \partial \delta_F \\
&= \tilde{\star}_F \Box_F.
\end{aligned}$$

Assertion (2) is a consequence of substituting $F^* \otimes K^*$ for F in (1). □

Definition 7.10. A form $\omega \in \mathcal{A}(F)^{p,q}$ is said to be $\Box_F$-harmonic if

$$\Box_F \omega = 0.$$

We denote by $H_{\Box_F}^{p,q}$ the set of all $\Box_F$-harmonic forms in $\mathcal{A}(F)^{p,q}$.

By Theorem 7.1 we obtain the following duality theorem. This is similar to the Kodaira-Serre's duality theorem for a complex manifold [Kobayashi (1997, 1998)][Morrow and Kodaira (1971)][Wells (1979)].

Theorem 7.2 (Duality Theorem). *The mapping $\tilde{\star}_F$ induces an isomorphism from $H_{\Box_F}^{p,q}$ to $H_{\Box_{F^* \otimes K^*}}^{n-p,n-q}$.*

Let us denote by $H_{\partial}^{p,q}(F)$ the p-th cohomology group of the cochain complex $\left\{ \sum_p \mathcal{A}^{p,q}(F), \partial \right\}$. A form $\omega \in \mathcal{A}^{p,q}$ is called D-parallel if $D\omega = 0$. Let $\mathsf{P}_D^q(F)$ be a sheaf of germs of F-valued D-parallel q-forms, and let $H^p(\mathsf{P}_D^q(F))$ be the p-th cohomology group with coefficients in $\mathsf{P}_D^q(F)$.

Theorem 7.3. *We have the following canonical isomorphisms.*

$$H_{\Box_F}^{p,q} \cong H_{\partial}^{p,q}(F) \cong H^p(\mathsf{P}_D^q(F)).$$

Proof.   It follows from harmonic theory [Kobayashi (1997, 1998)][Kodaira (1986)] that

$$H^{p,q}_{\square_F} \cong H^{p,q}_{\bar\partial}(F).$$

Let $\mathsf{A}^{p,q}(F)$ be a sheaf of germs of smooth sections of $(\overset{p}{\wedge} T^*) \otimes (\overset{q}{\wedge} T^*) \otimes F$. Then

$$0 \longrightarrow \mathsf{P}^q(F) \longrightarrow \mathsf{A}^{0,q}(F) \overset{\partial}{\longrightarrow} \mathsf{A}^{1,q}(F) \overset{\partial}{\longrightarrow} \mathsf{A}^{2,q}(F) \overset{\partial}{\longrightarrow} \cdots$$

is a fine resolution of $\mathsf{P}^q(F)$ [Kobayashi (1997, 1998)][Kodaira (1986)]. Hence

$$H^{p,q}_{\bar\partial}(F) \cong H^p(\mathsf{P}^q_D(F)).$$

$\square$

7.3   Koszul's vanishing theorem

J.L. Koszul proved a vanishing theorem for a hyperbolic flat manifold, and applied the result to study deformations of hyperbolic flat connections [Koszul (1968a)][Koszul (1968b)]. In this section we state Koszul's vanishing theorem in a generalized form and prove, as an application, that an affine Hopf manifold cannot admit any Hessian metric.

Theorem 7.4. *Let (M, D) be a compact flat manifold and let F be a flat vector bundle over M. Suppose that M admits a Riemannian metric g, a Killing vector field H and a fiber metric a on F satisfying the following conditions.*

(1) *$D_X H = X$ for $X \in \mathfrak{X}D$*

(2) *Choosing frame fields on F whose transition functions are all constants, there exists a constant $c \neq 0$ such that*

$$Ha(s_i, s_j) = ca(s_i, s_j),$$

for each such frame field $\{s_1, \cdots, s_r\}$.

Then we have

$$H^{p,0}_{\bar\partial}(F) = \{0\} \quad for \ \ p \geq 1.$$

Proof.   We denote by $\mathcal{L}_H$ Lie differentiation with respect to H. We claim

$$\mathcal{L}_H \sigma = D_H \sigma + p\sigma \quad \text{for} \ \ \sigma \in \mathcal{A}^{p,0}. \tag{7.6}$$

Since $[H, X_i] = D_H X_i - D_{X_i} H = D_H X_i - X_i$, we obtain

$$(\mathcal{L}_H \sigma)(X_1, \cdots, X_p)$$
$$= H(\sigma(X_1, \cdots, X_p)) - \sum_i \sigma(X_1, \cdots, [H, X_i], \cdots, X_p)$$
$$= H(\sigma(X_1, \cdots, X_p)) - \sum_i \sigma(X_1, \cdots, D_H X_i, \cdots, X_p) + p\sigma(X_1, \cdots, X_p)$$
$$= (D_H \sigma)(X_1, \cdots, X_p) + p\sigma(X_1, \cdots, X_p).$$

Let $\omega = \sum_i \omega^i s_i$ and $\eta = \sum_i \eta^i s_i \in \mathcal{A}^{p,0}(F)$. Then

$$(\omega, \eta) = \int_M C(\omega \wedge \tilde{\ast}_F \eta) = \int_M \langle \omega, \eta \rangle v,$$

where v is the volume element of g, and $\langle \omega, \eta \rangle = \sum_{i,j} g(\omega^i, \eta^j) a(s_i, s_j)$. It follows from the above condition (2), (7.6) and $\mathcal{L}_H g = 0$ that

$$\mathcal{L}_H \langle \omega, \eta \rangle$$
$$= \sum_{i,j} (\mathcal{L}_H g(\omega^i, \eta^j)) a(s_i, s_j) + \sum_{i,j} g(\omega^i, \eta^j)(\mathcal{L}_H a(s_i, s_j))$$
$$= \sum_{i,j} \{g(\mathcal{L}_H \omega^i, \eta^j) + g(\omega^i, \mathcal{L}_H \eta^j)\} a(s_i, s_j) + c \sum_{i,j} g(\omega^i, \eta^j) a(s_i, s_j)$$
$$= \langle D_H \omega, \eta \rangle + \langle \omega, D_H \eta \rangle + (2p + c)\langle \omega, \eta \rangle.$$

By Stokes' theorem and $\mathcal{L}_H v = 0$ we have

$$\int_M (\mathcal{L}_H \langle \omega, \eta \rangle) v = \int_M \mathcal{L}_H (\langle \omega, \eta \rangle v)$$
$$= \int_M (di(H) + i(H)d)(\langle \omega, \eta \rangle v)$$
$$= \int_M di(H)(\langle \omega, \eta \rangle v) = 0.$$

With these expressions we obtain

$$(D_H \omega, \eta) + (\omega, D_H \eta) + (2p + c)(\omega, \eta) = 0. \tag{7.7}$$

Since $\partial i(H) + i(H)\partial = \mathcal{L}_H$ on $A^{p,0}$, it follows from (7.6) that

$$(\partial i(H) + i(H)\partial)\omega = D_H \omega + p\omega, \tag{7.8}$$

so

$$\partial i(H)\partial \omega = \partial(\partial i(H) + i(H)\partial)\omega = \partial D_H \omega + p\partial \omega,$$
$$\partial i(H)\partial \omega = (\partial i(H) + i(H)\partial)\partial \omega = D_H \partial \omega + (p + 1)\partial \omega.$$

Hence

$$(\partial D_H - D_H \partial)\omega = \partial\omega. \tag{7.9}$$

Note that ω is $\Box_F$-harmonic if and only if

$$\partial\omega = 0, \qquad \delta_F\omega = 0.$$

Let $\omega \in H^{p,0}_{\Box_F}$. By (7.9) we have

$$\partial D_H\omega = (D_H\partial + \partial)\omega = 0.$$

It follows from equations (7.7) and (7.9) that, for $\eta \in A^{p-1,0}(F)$,

$$
\begin{aligned}
(\delta_F D_H\omega, \ \eta) &= (D_H\omega, \ \partial\eta) \\
&= -(\omega, \ D_H\partial\eta) - (2p+c)(\omega, \ \partial\eta) \\
&= -(\omega, \ \partial D_H\eta) + (\omega, \ \partial\eta) - (2p+c)(\omega, \ \partial\eta) \\
&= (\delta_F\omega, \ -D_H\eta - (2p+c-1)\eta) = 0,
\end{aligned}
$$

and so

$$\delta_F D_H\omega = 0.$$

Hence

$$D_H\omega \in H^{p,0}_{\Box_F}.$$

Since $\partial i(H)\omega = (\partial i(H) + i(H)\partial)\omega = D_H\omega + p\omega$ by equation (7.8), we have

$$\partial i(H)\omega \in H^{p,0}_{\Box_F}.$$

This implies

$$(\partial i(H)\omega, \ \partial i(H)\omega) = (\delta_F \partial i(H)\omega, \ i(H)\omega) = 0.$$

Therefore

$$D_H\omega = -p\omega. \tag{7.10}$$

If we put $\eta = \omega$, from equation (7.10) the equality (7.7) may be reduced to

$$c(\omega, \ \omega) = 0.$$

Hence $\omega = 0$ and

$$H^{p,0}_{\Box_F} = \{0\}.$$

This result, together with Theorem 7.3, implies

$$H^{p,0}_{\partial}(F) = \{0\}.$$

$\Box$

Remark 7.1. A vector field H on a flat manifold (M, D) is called a **radiant vector fiels** if it satisfies the condition (1) of Theorem 7.4. A flat manifold admitting such an H is said to be **radiant** [Fried, Goldman and Hirsch (1981)].

Suppose we have a Riemannian metric g and a vector field H satisfying condition (1) of Theorem 7.4. Then

$$
\begin{aligned}
Hg_{ij} &= Hg\left(\frac{\partial}{\partial x^i}, \frac{\partial}{\partial x^j}\right) \\
&= g\left(\left[H, \frac{\partial}{\partial x^i}\right], \frac{\partial}{\partial x^j}\right) + g\left(\frac{\partial}{\partial x^i}, \left[H, \frac{\partial}{\partial x^j}\right]\right) \\
&= g\left(D_H \frac{\partial}{\partial x^i} - \frac{\partial}{\partial x^i}, \frac{\partial}{\partial x^j}\right) + g\left(\frac{\partial}{\partial x^i}, D_H \frac{\partial}{\partial x^j} - \frac{\partial}{\partial x^j}\right) \\
&= -2g_{ij}.
\end{aligned}
$$

Hence

$$
Hg_{ij} = -2g_{ij}, \qquad Hg^{ij} = 2g^{ij}. \tag{7.11}
$$

Let $\mathbf{T}_s^r = (\overset{r}{\otimes} T) \otimes (\overset{s}{\otimes} T^*)$ be a tensor bundle over M of contravariant degree r and covariant degree s. Then $\mathbf{T}_s^r$ is a flat vector bundle and admits a fiber metric induced by g,

$$
g\left(s_{i_1 \cdots i_r}^{j_1 \cdots j_s}, s_{k_1 \cdots k_r}^{l_1 \cdots l_s}\right) = g_{i_1 k_1} \cdots g_{i_r k_r} g^{j_1 l_1} \cdots g^{j_s l_s},
$$

where $\left\{s_{i_1 \cdots i_r}^{j_1 \cdots j_s} = \dfrac{\partial}{\partial x^{i_1}} \otimes \cdots \otimes \dfrac{\partial}{\partial x^{i_r}} \otimes dx^{j_1} \otimes \cdots \otimes dx^{j_s}\right\}$ is a local frame field of $\mathbf{T}_s^r$. It follows from relations (7.11) that

$$
Hg\left(s_{i_1 \cdots i_r}^{j_1 \cdots j_s}, s_{k_1 \cdots k_r}^{l_1 \cdots l_s}\right) = 2(s - r)g\left(s_{i_1 \cdots i_r}^{j_1 \cdots j_s}, s_{k_1 \cdots k_r}^{l_1 \cdots l_s}\right).
$$

Corollary 7.1. *Let (M, D) be a compact flat manifold admitting a Riemannian metric g and a vector field H satisfying the condition (1) of Theorem 7.4. Then we have*

$$
H_{\bar{\partial}}^{p,0}(\mathbf{T}_s^r) = \{0\} \quad \text{for } r \neq s \text{ and } p \geq 1.
$$

Let Ω be a regular convex cone in $\mathbf{R}^n$ with vertex 0 and let $(\tilde{D}, \tilde{g})$ be the canonical Hessian structure on Ω. A 1-parameter transformation group $x \longrightarrow e^t x$ of Ω induces a vector field $\tilde{H}$;

$$
\tilde{H} = \sum_i \tilde{x}^i \frac{\partial}{\partial \tilde{x}^i},
$$

where $\{\tilde{x}^1, \cdots, \tilde{x}^n\}$ is an affine local coordinate system on Ω. Hence

$$
\tilde{D}_{\tilde{X}} \tilde{H} = \tilde{X}.
$$

Since the 1-parameter transformation group leaves $\tilde{g}$ invariant, $\tilde{H}$ is a Killing vector field with respect to $\tilde{g}$. Suppose further that Ω admits a discrete subgroup Γ of the linear automorphism group of Ω acting properly discontinuously and freely on Ω, and that the quotient space $\Gamma \backslash \Omega$ is compact. Then $\tilde{D}$ induces a flat connection D on $\Gamma \backslash \Omega$. Because $\tilde{H}$ is invariant by the linear automorphism group of Ω, there exists a vector field H on $\Gamma \backslash \Omega$ such that

$$H_{\pi(\tilde{x})} = \pi_{*\tilde{x}}(\tilde{H}_{\tilde{x}}),$$

where π is the projection from Ω onto $\Gamma \backslash \Omega$. Since $\tilde{g}$ is Γ-invariant, there exists a Riemannian metric g on $\Gamma \backslash \Omega$ such that

$$\tilde{g} = \pi^* g.$$

Then H is a Killing vector field with respect to g and satisfies

$$D_X H = X.$$

Corollary 7.2. *Let Ω be a regular convex cone in $\mathbf{R}^n$ with vertex 0. Suppose that Ω admits a discrete subgroup Γ of the linear automorphism group of Ω acting properly discontinuously and freely on Ω, and that the quotient space $\Gamma \backslash \Omega$ is compact. For a tensor bundle $\mathbf{T}_s^r$ on $\Gamma \backslash \Omega$, we have*

$$H_{\partial}^{p,0}(\mathbf{T}_s^r) = \{0\} \quad \text{for } r \neq s \text{ and } p \geq 1.$$

We set $\mathbf{R}^{n*} = \mathbf{R}^n - \{0\}$. Let Γ^* be the group generated by $\mathbf{R}^{n*} \ni x \longrightarrow 2x \in \mathbf{R}^{n*}$. Then Γ^* acts on $\mathbf{R}^{n*}$ properly discontinuously and freely, and the quotient space $\Gamma^* \backslash \mathbf{R}^{n*}$ is a compact flat manifold and is called an **affine Hopf manifold**. Let $\tilde{H} = \sum_i \tilde{x}^i \dfrac{\partial}{\partial \tilde{x}^i}$. Then, as above, a flat connection D and a vector field H on $\Gamma^* \backslash \mathbf{R}^{n*}$ are induced by the flat connection $\tilde{D}$ and the vector field $\tilde{H}$ on $\mathbf{R}^n$, and satisfy

$$D_X H = X.$$

Let g^* be a Riemannian metric on $\mathbf{R}^{n*}$ defined by

$$g^* = \frac{1}{2\sum_i (\tilde{x}^i)^2} \sum_i (d\tilde{x}^i)^2.$$

Then

$$\mathcal{L}_{\tilde{H}} g^* = 0.$$

In fact, putting $f^* = \dfrac{1}{2\sum_i (\tilde{x}^i)^2}$, it follows from $\tilde{H}f^* + 2f^* = 0$ that

$$(\mathcal{L}_{\tilde{H}} g^*)\left(\frac{\partial}{\partial \tilde{x}^i}, \frac{\partial}{\partial \tilde{x}^j}\right)$$

$$= \tilde{H}g^*\left(\frac{\partial}{\partial \tilde{x}^i}, \frac{\partial}{\partial \tilde{x}^j}\right) - g^*\left(\left[\tilde{H}, \frac{\partial}{\partial \tilde{x}^i}\right], \frac{\partial}{\partial \tilde{x}^j}\right) - g^*\left(\frac{\partial}{\partial \tilde{x}^i}, \left[\tilde{H}, \frac{\partial}{\partial \tilde{x}^j}\right]\right)$$

$$= \tilde{H}g_{ij}^* + 2g_{ij}^* = (\tilde{H}f^* + 2f^*)\delta_{ij} = 0.$$

This implies, in particular, that g^* is Γ^*-invariant. Hence $\Gamma^* \backslash \mathbf{R}^{n*}$ admits a Riemannian metric g such that the pullback on $\mathbf{R}^{n*}$ of g coincides with g^* and the vector field H is a Killing vector field with respect to g. Hence we have the following corollary.

Corollary 7.3. *Let* $\mathbf{T}_s^r$ *be a tensor bundle over an affine Hopf manifold. Then we have*

$$H_{\bar{\partial}}^{p,0}(\mathbf{T}_s^r) = \{0\} \quad for\ r \neq s\ and\ p \geq 1.$$

Corollary 7.4. *An affine Hopf manifold cannot admit any Hessian metric.*

Proof. Suppose that an affine Hopf manifold $\Gamma^* \backslash \mathbf{R}^{n*}$ admits a Hessian metric g. Considering g as an element in $\mathcal{A}^{1,0}(\mathbf{T}_1) = \mathcal{A}^{1,1}$, we know $\partial g = 0$ by Lemma 7.6. Since $H_{\bar{\partial}}^{1,0}(\mathbf{T}_1) = \{0\}$ by Corollary 7.3, there exists $\omega = \sum \omega_{\bar{j}} dx^{\bar{j}} \in \mathcal{A}^{0,0}(\mathbf{T}_1) = \mathcal{A}^{0,1}$ such that $g = \partial \omega$. Then ω is a closed 1-form because g is a covariant symmetric tensor of degree 2. The relation $g = \partial \omega$ is equivalent to $g = D\omega$. Hence, by Corollary 8.3 and Theorem 8.3, the universal covering manifold is a regular convex cone. On the other hand, $\mathbf{R}^{n*}$ contains straight lines and is not a convex set. This is a contradiction.

$\square$

Corollary 7.4 also follows immediately from Theorem 8.2.

7.4 Laplacians on Hessian manifolds

In this section we consider the Laplacian determined by a Hessian structure, and prove a certain vanishing theorem for a Hessian manifold similar to that of Kodaira-Nakano's theorem for a Kählerian manifold.

Let (M, D) be a flat manifold and let $\mathbf{T}$ and $\bar{\mathbf{T}}$ be tensor bundles over M. We denote by $\Gamma(\mathbf{T})C\Gamma(\bar{\mathbf{T}})$ and $\Gamma(\mathbf{T} \otimes \bar{\mathbf{T}})$ the spaces of all smooth sections of $\mathbf{T}C\bar{\mathbf{T}}$ and $\mathbf{T} \otimes \bar{\mathbf{T}}$ respectively. Let g be a Riemannian metric on M and let ∇ be the Levi-Civita connection for g. Using D and ∇, we define two connections on $\mathbf{T} \otimes \bar{\mathbf{T}}$.

Definition 7.11. We define two connections $\mathfrak{D}$ and $\bar{\mathfrak{D}}$ on $\mathbf{T} \otimes \bar{\mathbf{T}}$ by

$$\mathfrak{D}_X \sigma = (2\gamma_X \otimes \bar{I} + D_X)\sigma,$$
$$\bar{\mathfrak{D}}_X \sigma = (I \otimes 2\gamma_X + D_X)\sigma \quad \text{for } \sigma \in \Gamma(\mathbf{T} \otimes \bar{\mathbf{T}}),$$

where I and $\bar{I}$ are the identity mappings on $\Gamma(\mathbf{T})$ and $\Gamma(\bar{\mathbf{T}})$ respectively, and $\gamma_X = \nabla_X - D_X$.

Lemma 7.6. *Let us consider g as an element in $\mathcal{A}^{1,1} = \Gamma(T^* \otimes T^*)$. Then the following conditions are equivalent.*

(1) *g is a Hessian metric.*
(2) *$\partial g = 0$ ($\bar{\partial} g = 0$).*
(3) *$\mathfrak{D}_X g = 0$ ($\bar{\mathfrak{D}}_X g = 0$).*

The Hessian structure (D, g) is of Koszul type (cf. Definition 2.2) if and only if g is exact with respect to ∂.

Proof. The equivalence of (1)-(3) follows from Proposition 2.1 and the following equalities

$$\partial g = \partial\Big(\sum_{i,j} g_{ij} dx^i \otimes dx^j\Big) = \sum_{k,i,j} e(dx^k) \frac{\partial g_{ij}}{\partial x^k} dx^i \otimes dx^j$$

$$= \sum_j \Big(\sum_{k<i} \Big(\frac{\partial g_{ij}}{\partial x^k} - \frac{\partial g_{kj}}{\partial x^i}\Big)(dx^k \wedge dx^i)\Big) \otimes dx^j,$$

$$(\mathfrak{D}_X g)(Y; Z) = X g(Y; Z) - g(2\gamma_X Y + D_X Y; Z) - g(Y; D_X Z)$$
$$= g(\nabla_X Y; Z) + g(Y; \nabla_X Z) - g(2\gamma_X Y + D_X Y; Z) - g(Y; D_X Z)$$
$$= -g(\gamma_X Y; Z) + g(Y; \gamma_X Z). \qquad \square$$

Henceforth in this section we assume that g is a Hessian metric.

Proposition 7.3. *For the connections $\mathfrak{D}$ and $\bar{\mathfrak{D}}$ we have*

(1) *$\nabla_X = \dfrac{1}{2}(\mathfrak{D}_X + \bar{\mathfrak{D}}_X)$.*
(2) *The curvature tensors $R_\mathfrak{D}$ of $\mathfrak{D}$ and $R_{\bar{\mathfrak{D}}}$ of $\bar{\mathfrak{D}}$ vanish,*

$$R_\mathfrak{D}(X, Y) = \mathfrak{D}_X \mathfrak{D}_Y - \mathfrak{D}_Y \mathfrak{D}_X - \mathfrak{D}_{[X,Y]} = 0,$$
$$R_{\bar{\mathfrak{D}}}(X, Y) = \bar{\mathfrak{D}}_X \bar{\mathfrak{D}}_Y - \bar{\mathfrak{D}}_Y \bar{\mathfrak{D}}_X - \bar{\mathfrak{D}}_{[X,Y]} = 0.$$

(3) *Let v be the volume element of g. We identify v with $v \otimes 1 \in \mathcal{A}^{n,0}$ and set $\bar{v} = 1 \otimes v \in \mathcal{A}^{0,n}$. Then $v \wedge \bar{v} = v \otimes v \in \mathcal{A}^{n,n}$, and*

$$\mathfrak{D}_X(v \wedge \bar{v}) = 0, \qquad \bar{\mathfrak{D}}_X(v \wedge \bar{v}) = 0.$$

Proof. By the definition of $\mathfrak{D}$ and $\bar{\mathfrak{D}}$, and by Theorem 2.2, we have

$$\mathfrak{D}_X(fu \otimes \bar{w}) = f\{(D'_X u) \otimes \bar{w} + u \otimes (D_X \bar{w})\} + (Xf)u \otimes \bar{w}, \quad (7.12)$$
$$\bar{\mathfrak{D}}_X(fu \otimes \bar{w}) = f\{(D_X u) \otimes \bar{w} + u \otimes (D'_X \bar{w})\} + (Xf)u \otimes \bar{w},$$

for $u \in \Gamma(\mathbf{T})$ and $\bar{w} \in \Gamma(\bar{\mathbf{T}})$ and a function f on M. This proves assertion (1). Using this assertion, together with the fact that the curvature tensors R_D of D and $R_{D'}$ of D' vanish, we obtain

$$R_{\mathfrak{D}}(X,Y)(fu \otimes \bar{w}) = f(R_{D'}(X,Y)u) \otimes \bar{w} + fu \otimes (R_D(X,Y)\bar{w}) = 0,$$
$$R_{\bar{\mathfrak{D}}}(X,Y)(fu \otimes \bar{w}) = f(R_D(X,Y)u) \otimes \bar{w} + fu \otimes (R_{D'}(X,Y)\bar{w}) = 0.$$

By the definition of α we have

$$\mathfrak{D}_X(v \wedge \bar{v}) = (2\gamma_X \otimes \bar{I} + D_X)(v \otimes v)$$
$$= (2\gamma_X v) \otimes v + (D_X v) \otimes v + v \otimes D_X v$$
$$= (-2\alpha(X)v) \otimes v + \alpha(X)v \otimes v + v \otimes \alpha(X)v = 0. \qquad \square$$

We note here that for $\omega \in \mathcal{A}^{p,q} = \Gamma((\overset{p}{\wedge} T^*) \otimes (\overset{q}{\wedge} T^*))$ we have

$$(\mathfrak{D}_X \omega)(Y_1, \cdots, Y_p; Z_1, \cdots, Z_q) \qquad (7.13)$$
$$= X\omega(Y_1, \cdots, Y_p; Z_1, \cdots, Z_q)$$
$$- \sum_{i=1}^{p} \omega(Y_1, \cdots, D'_X Y_i, \cdots, Y_p; Z_1, \cdots, Z_q)$$
$$- \sum_{j=1}^{q} \omega(Y_1, \cdots, Y_p; Z_1, \cdots, D_X Z_j, .., Z_q),$$

$$(\bar{\mathfrak{D}}_X \omega)(Y_1, \cdots, Y_p; Z_1, \cdots, Z_q) \qquad (7.14)$$
$$= X\omega(Y_1, \cdots, Y_p; Z_1, \cdots, Z_q)$$
$$- \sum_{i=1}^{p} \omega(Y_1, \cdots, D_X Y_i, \cdots, Y_p; Z_1, \cdots, Z_q)$$
$$- \sum_{j=1}^{q} \omega(Y_1, \cdots, Y_p; Z_1, \cdots, D'_X Z_j, .., Z_q).$$

Lemma 7.7. *Considering g as an element in $\mathcal{A}^{1,1}$ we obtain*

(1) $\mathfrak{D}_X i(Y) g = i(D'_X Y) g, \quad \mathfrak{D}_X \bar{i}(Y) g = \bar{i}(D_X Y) g.$
(2) $\bar{\mathfrak{D}}_X i(Y) g = i(D_X Y) g, \quad \bar{\mathfrak{D}}_X \bar{i}(Y) g = \bar{i}(D'_X Y) g.$

Proof. We have

$$(\mathfrak{D}_X i(Y) g)(Z) = X((i(Y) g)(Z)) - (i(Y) g)(D_X Z)$$
$$= X g(Y; Z) - g(Y; D_X Z) = g(D'_X Y; Z)$$
$$= (i(D'_X Y) g)(Z).$$

The same approach also proves the other equalities. □

Let us express the coboundary operators and the dual operators using $\mathfrak{D}$ and $\bar{\mathfrak{D}}$. Let $\{E_1, \cdots, E_n\}$ be an orthonormal frame field with respect to g, and set $\theta^j = \bar{i}(E_j) g \in \mathcal{A}^{1,0}$ and $\bar{\theta}^j = i(E_j) g \in \mathcal{A}^{0,1}$.

Proposition 7.4. *The coboundary operators ∂ and $\bar{\partial}$ are expressed by*

$$\partial = \sum_j e(\theta^j) \mathfrak{D}_{E_j}, \qquad \bar{\partial} = \sum_j e(\bar{\theta}^j) \bar{\mathfrak{D}}_{E_j}.$$

Proof. For a p-form ω we have

$$\sum \theta^j \wedge D_{E_j} \omega = \sum dx^j \wedge D_{\partial / \partial x^j} \omega = d\omega = \sum \theta^j \wedge \nabla_{E_j} \omega,$$

and so

$$\sum \theta^j \wedge D_{E_j} \omega = \sum \theta^j \wedge D'_{E_j} \omega.$$

Put $\partial_{\mathfrak{D}} = \sum_j e(\theta^j) \mathfrak{D}_{E_j}$. Let f be a function and let u and ω be be a p-form and a q-form respectively. With expressions (7.12) we obtain

$$\partial_{\mathfrak{D}} (f u \otimes w) = \sum e(\theta^j) \mathfrak{D}_{E_j} (f u \otimes w)$$
$$= f \sum e(\theta^j)(D'_{E_j} u) \otimes w + f \sum e(\theta^j)(u \otimes D_{E_j} w) + \sum e(\theta^j)(E_j f)(u \otimes w)$$
$$= f(\sum e(\theta^j) D_{E_j} u) \otimes w + f \sum e(\theta^j)(u \otimes D_{E_j} w) + \sum e(\theta^j)(E_j f)(u \otimes w)$$
$$= \sum e(\theta^j) \{(f D_{E_j} u) \otimes w + f u \otimes D_{E_j} w + (E_j f) u \otimes w\}$$
$$= \sum e(\theta^j) D_{E_j} (f u \otimes w) = \sum e(dx^j) D_{\partial / \partial x^j} (f u \otimes w)$$
$$= \partial(f u \otimes w).$$

Thus the first assertion is proved. The second assertion may be proved by using the same method. □

Proposition 7.5. *We have*

$$\mathfrak{D}_X \star = \star \bar{\mathfrak{D}}_X.$$

Proof. Let $\omega \in \mathcal{A}^{p,q}$. By equation (7.3) and Lemma 7.7 we have

$$\bar{\mathfrak{D}}_X(\omega \wedge \bar{i}(X_1)g \wedge \cdots \wedge \bar{i}(X_{n-p})g \wedge i(Y_1)g \wedge \cdots \wedge i(Y_{n-q})g)$$
$$= \bar{\mathfrak{D}}_X\omega \wedge \bar{i}(X_1)g \wedge \cdots \wedge \bar{i}(X_{n-p})g \wedge i(Y_1)g \wedge \cdots \wedge i(Y_{n-q})g$$
$$+ \sum_j \omega \wedge \bar{i}(X_1)g \wedge \cdots \wedge \bar{i}(D'_X X_j)g \wedge \cdots \wedge \bar{i}(X_{n-p})g \wedge i(Y_1)g \wedge \cdots \wedge i(Y_{n-q})g$$
$$+ \sum_k \omega \wedge \bar{i}(X_1)g \wedge \cdots \wedge \bar{i}(X_{n-p})g \wedge i(Y_1)g \wedge \cdots \wedge i(D_X Y_k)g \wedge \cdots \wedge i(Y_{n-q})g$$
$$= (\star\bar{\mathfrak{D}}_X\omega)(X_1, \cdots, X_{n-p}; Y_1, \cdots, Y_{n-q})v \wedge \bar{v}$$
$$+ \sum_j (\star\omega)(X_1, \cdots, D'_X X_j, \cdots, X_{n-p}; Y_1, \cdots, Y_{n-q})v \wedge \bar{v}$$
$$+ \sum_k (\star\omega)(X_1, \cdots, X_{n-p}; Y_1, \cdots, D_X Y_k, \cdots, Y_{n-q})v \wedge \bar{v}.$$

Using the fact $\mathfrak{D}_X = \bar{\mathfrak{D}}_X$ on $\mathcal{A}^{n,n}$, Lemma 7.6 (3), and equations (7.3), (7.13), we obtain

$$\bar{\mathfrak{D}}_X(\omega \wedge \bar{i}(X_1)g \wedge \cdots \wedge \bar{i}(X_{n-p})g \wedge i(Y_1)g \wedge \cdots \wedge i(Y_{n-q})g)$$
$$= \mathfrak{D}_X(\omega \wedge \bar{i}(X_1)g \wedge \cdots \wedge \bar{i}(X_{n-p})g \wedge i(Y_1)g \wedge \cdots \wedge i(Y_{n-q})g)$$
$$= \mathfrak{D}_X\{(\star\omega)(X_1, \cdots, X_{n-p}; Y_1, \cdots, Y_{n-q})v \wedge \bar{v}\}$$
$$= \mathfrak{D}_X\{(\star\omega)(X_1, \cdots, X_{n-p}; Y_1, \cdots, Y_{n-q})\}v \wedge \bar{v}$$
$$= (\mathfrak{D}_X \star \omega)(X_1, \cdots, X_{n-p}; Y_1, \cdots, Y_{n-q})v \wedge \bar{v}$$
$$+ \sum_j (\star\omega)(X_1, \cdots, D'_X X_j, \cdots, X_{n-p}; Y_1, \cdots, Y_{n-q})v \wedge \bar{v}$$
$$+ \sum_k (\star\omega)(X_1, \cdots, X_{n-p}; Y_1, \cdots, D_X Y_k, \cdots, Y_{n-q})v \wedge \bar{v}.$$

The above equations imply $\star\bar{\mathfrak{D}}_X\omega = \mathfrak{D}_X \star \omega$. $\square$

Proposition 7.6. *The dual operator δ of ∂ is expressed by*

$$\delta = -\sum_j i(E_j)\bar{\mathfrak{D}}_{E_j} + i(X_\alpha),$$

where X_α is a vector field determined by $\alpha(Y) = g(X_\alpha, Y)$ for the first Koszul form α.

Proof. Let $\omega \in \mathcal{A}^{p,q}$. It follows from (7.4) and Propositions 7.2, 7.4 and 7.5 that

$$\delta\omega = (-1)^p \tilde{\star}^{-1} \partial \tilde{\star}\omega, = (-1)^p \star^{-1} \sqrt{G}\partial\Big(\frac{1}{\sqrt{G}} \star \omega\Big)$$

$$= (-1)^p \star^{-1}\Big\{\partial \star \omega + \sqrt{G}\partial\Big(\frac{1}{\sqrt{G}}\Big) \wedge \star\omega\Big\}$$

$$= (-1)^p\Big\{\sum_j \star^{-1}e(\theta^j)\mathfrak{D}_{E_j} \star \omega - \star^{-1}\alpha \wedge \star\omega\Big\}$$

$$= (-1)^p\Big\{\sum_j \star^{-1}e(\theta^j) \star \bar{\mathfrak{D}}_{E_j}\omega - \star^{-1}e(\alpha) \star \omega\Big\}$$

$$= \Big\{-\sum_j i(E_j)\bar{\mathfrak{D}}_{E_j} + i(X_\alpha)\Big\}\omega.$$

$\square$

Corollary 7.5. *The following relations hold.*

(1) $\delta g = 1 \otimes \alpha$.

(2) $\square g = \beta$.

Proof. It follows from Proposition 7.6 and Lemma 7.6 that

$$\delta g = i(X_\alpha)g = g(X_\alpha;\ ) = 1 \otimes \alpha.$$

Hence

$$\square g = (\partial\delta + \delta\partial)g = \partial\delta g = \partial(1 \otimes \alpha) = \beta.$$

$\square$

Lemma 7.8. *We have*

$$\delta e(g) + e(g)\delta = -\bar{\partial} + e(1 \otimes \alpha).$$

Proof. By Propositions 7.4 and 7.6 we have

$$\delta e(g) = -\sum_j i(E_j)\bar{\mathfrak{D}}_{E_j}e(g) + i(X_\alpha)e(g)$$

$$= -\sum_j i(E_j)e(g)\bar{\mathfrak{D}}_{E_j} + e(1 \otimes \alpha) - e(g)i(X_\alpha)$$

$$= -\sum_j(-e(g)i(E_j) + e(\bar{\theta}^j))\bar{\mathfrak{D}}_{E_j} + e(1 \otimes \alpha) - e(g)i(X_\alpha)$$

$$= -e(g)\delta - \bar{\partial} + e(1 \otimes \alpha).$$

$\square$

Let F be a flat line bundle over M with a fiber metric a. Choosing frame fields of F such that the transition functions are constant, we define a closed 1-form A and a symmetric bilinear form B by

$$A = -d\log a(s, s),$$
$$B = -Dd\log a(s, s),$$

where s is such a frame of F. The forms A and B are called **the first Koszul form** and **the second Koszul form** with respect to the fiber metric a, respectively.

Let $K = \overset{n}{\bigwedge} T^*$ be the canonical line bundle over M. For an affine coordinate system $\{x_\lambda^1, \cdots, x_\lambda^n\}$ on M we set

$$s_\lambda = dx_\lambda^1 \wedge \cdots \wedge dx_\lambda^n.$$

Then

$$s_\mu = J_{\lambda\mu} s_\lambda,$$

where $J_{\lambda\mu}$ are non-zero constants. Thus $K = \overset{n}{\bigwedge} T^*$ is a flat line bundle. Suppose that M admits a Hessian structure (D, g). Expressing the volume element v of g by $v = v_\lambda s_\lambda$, we have

$$g(s_\lambda, s_\lambda) = \frac{1}{v_\lambda^2} g(v, v).$$

Since $g(v, v)$ is a positive constant, the Koszul forms with respect to the fiber metric g on K are given by

$$A = 2\alpha, \qquad B = 2\beta.$$

Definition 7.12. A flat line bundle F is said to be positive (resp. negative) if it admits a fiber metric such that the corresponding second Koszul form B is positive definite (resp. negative definite).

Lemma 7.9. *The adjoint operator δ_F of ∂ is expressed as*

$$\delta_F = \delta + i(X_A),$$

where X_A is a vector field given by $A(Y) = g(X_A, Y)$.

Proof. It follows from Propostion 7.2 and relations (7.4) that on the space $\mathcal{A}^{p,q}(F)$ we have

$$\begin{aligned}
\delta_F &= (-1)^p \tilde{\star}_F^{-1} \partial \tilde{\star}_F = (-1)^p a^{-1} \tilde{\star}^{-1} \partial (a\tilde{\star}) \\
&= (-1)^p \star^{-1} e(-A) \star + (-1)^p \tilde{\star}^{-1} \partial \tilde{\star} \\
&= i(X_A) + \delta.
\end{aligned}$$

$\square$

Proposition 7.7. *We have*

(1) $[\Box_F, e(g)] = e(B + \beta)$.
(2) $[\Box_F, i(g)] = -i(B + \beta)$.

Proof. It follows from Lemmata 7.8 and 7.9 that

$$\begin{aligned}
\Box_F e(g) &= \partial(\delta + i(X_A))e(g) + (\delta + i(X_A))\partial e(g) \\
&= \partial\delta e(g) - \delta e(g)\partial + \partial i(X_A)e(g) + i(X_A)\partial e(g) \\
&= \partial(-e(g)\delta - \bar\partial + e(1\otimes\alpha)) + (e(g)\delta + \bar\partial - e(1\otimes\alpha))\partial \\
&\quad + \partial i(X_A)e(g) + i(X_A)\partial e(g) \\
&= e(g)(\partial\delta + \delta\partial) - \partial\bar\partial + \bar\partial\partial + \partial e(1\otimes\alpha) - e(1\otimes\alpha)\partial \\
&\quad + \partial i(X_A)e(g) - i(X_A)e(g)\partial.
\end{aligned}$$

Using the following relations,

$$\partial\bar\partial - \bar\partial\partial = 0, \qquad \partial e(1\otimes\alpha) - e(1\otimes\alpha)\partial = e(\beta),$$

$$\begin{aligned}
\partial i(X_A)&e(g) - i(X_A)e(g)\partial \\
&= \partial(e(1\otimes A) - e(g)i(X_A)) - (e(1\otimes A) - e(g)i(X_A))\partial \\
&= e(B) + e(g)(\partial i(X_A) + i(X_A)\partial),
\end{aligned}$$

we obtain

$$\Box_F e(g) = e(g)\Box_F + e(B + \beta).$$

Assertion (2) follows from (1) by taking the adjoint operators. $\square$

Proposition 7.8. *For $\omega \in H_{\Box_F}^{p,q}$ we have*

(1) $(e(B + \beta)i(g)\omega, \omega) \leq 0$.
(2) $(i(g)e(B + \beta)\omega, \omega) \geq 0$.
(3) $([i(g), e(B + \beta)]\omega, \omega) \geq 0$.

Proof. By Proposition 7.7 we have

$$\Box_F e(g)\omega = e(B + \beta)\omega, \quad \Box_F i(g)\omega = -i(B + \beta)\omega,$$

hence

$$\begin{aligned}
0 &\leq (\Box_F e(g)\omega, e(g)\omega) = (e(B + \beta)\omega, e(g)\omega) \\
&= (i(g)e(B + \beta)\omega, \omega), \\
0 &\leq (\Box_F i(g)\omega, i(g)\omega) = (-i(B + \beta)\omega, i(g)\omega) \\
&= -(\omega, e(B + \beta)i(g)\omega).
\end{aligned}$$

Assertion (3) follows from assertions (1) and (2). $\square$

The vanishing theorem due to Kodaira-Nakano plays an important role in the theory of complex manifolds. For example, it is used to prove that a Hodge manifold is algebraic [Kobayashi (1987)][Morrow and Kodaira (1971)][Wells (1979)]. The following is a vanishing theorem for a flat manifold similar to that of Kodaira-Nakano.

Theorem 7.5 (Vanishing Theorem). *Let (M, D) be an oriented compact flat manifold and let $K = \overset{n}{\wedge} T^*$ be the canonical line bundle over M. For a line bundle F over M we have*

(1) *If the line bundle $2F + K$ is positive, then $H^{p,q}_{\Box_F} = 0$ for $p + q > n$.*
(2) *If the line bundle $2F + K$ is negative, then $H^{p,q}_{\Box_F} = 0$ for $p + q < n$.*

Proof. Suppose that $2F + K$ is negative. Then $2F + K$ admits a fiber metric $\tilde{a}$ such that the corresponding second Koszul form $\tilde{B}$ is negative definite. Put $g = -\tilde{B}$. Then (D, g) is a Hessian structure. Denoting by β the second Koszul form of (D, g), we can choose a fiber metric a on F satisfying $2(B + \beta) = \tilde{B}$, where B is the second Koszul form for a. It follows from Proposition 7.8 that for $\omega \in H^{p,q}_{\Box_F}$

$$([i(g), e(g)]\omega, \, \omega) = ([i(g), -2e(B + \beta)]\omega, \, \omega) \leq 0.$$

Hence by Proposition 7.1 we have

$$(n - p - q)(\omega, \, \omega) \leq 0.$$

This shows that if $n - p - q > 0$, then $H^{p,q}_{\Box_F} = 0$. Thus the proof of assertion (2) is complete. Assertion (1) follows from (2) and Theorem 7.2. $\qquad \Box$

Expressions in affine coordinate systems

Using an affine coordinate system $\{x^1, \cdots, x^n\}$ we give here local expressions of covariant differentials of $\mathfrak{D}$, $\bar{\mathfrak{D}}$ and the coboundary operator ∂C the dual operator δ and the Laplacian $\Box$.

Let $\omega \in \mathcal{A}^{p,q} = \Gamma((\overset{p}{\wedge} T^*) \otimes (\overset{q}{\wedge} T^*))$ and let the local expression be given by

$$\omega = \sum \omega_{i_1 \cdots i_p \bar{j}_1 \cdots \bar{j}_q} (dx^{i_1} \wedge \cdots \wedge dx^{i_p}) \otimes (dx^{\bar{j}_1} \wedge \cdots \wedge dx^{\bar{j}_q}).$$

Put

$$\mathfrak{D}_k \omega_{i_1 \cdots i_p \bar{j}_1 \cdots \bar{j}_q} = \left(\mathfrak{D}_{\partial / \partial x^k} \omega \right)_{i_1 \cdots i_p \bar{j}_1 \cdots \bar{j}_q},$$

$$\bar{\mathfrak{D}}_{\bar{k}} \omega_{i_1 \cdots i_p \bar{j}_1 \cdots \bar{j}_q} = (\bar{\mathfrak{D}}_{\partial / \partial x^{\bar{k}}} \omega)_{i_1 \cdots i_p \bar{j}_1 \cdots \bar{j}_q}.$$

Then

$$\mathfrak{D}_k \omega_{i_1 \cdots i_p \bar{j}_1 \cdots \bar{j}_q} = \frac{\partial \omega_{i_1 \cdots i_p \bar{j}_1 \cdots \bar{j}_q}}{\partial x^k} - 2 \sum_\lambda \Gamma^l_{k\, i_\lambda} \omega_{i_1 \cdots (l)_\lambda \cdots i_p \bar{j}_1 \cdots \bar{j}_q},$$

$$\bar{\mathfrak{D}}_{\bar{k}} \omega_{i_1 \cdots i_p \bar{j}_1 \cdots \bar{j}_q} = \frac{\partial \omega_{i_1 \cdots i_p \bar{j}_1 \cdots \bar{j}_q}}{\partial x^k} - 2 \sum_\mu \Gamma^{\bar{l}}_{k\, \bar{j}_\mu} \omega_{i_1 \cdots i_p \bar{j}_1 \cdots (\bar{l})_\mu \cdots \bar{j}_q}.$$

$$\mathfrak{D}_k g_{i\bar{j}} = 0, \quad \mathfrak{D}_k g^{i\bar{j}} = 0.$$

$$\bar{\mathfrak{D}}_{\bar{k}} g_{i\bar{j}} = 0, \quad \bar{\mathfrak{D}}_{\bar{k}} g^{i\bar{j}} = 0.$$

$$\mathfrak{D}_i \alpha_{\bar{j}} = \beta_{i\bar{j}}, \quad \mathfrak{D}_i \alpha^j = \beta^j_i.$$

$$\bar{\mathfrak{D}}_{\bar{i}} \alpha_j = \beta_{j\bar{i}}, \quad \bar{\mathfrak{D}}_{\bar{i}} \alpha^{\bar{j}} = \beta^{\bar{j}}_{\bar{i}}.$$

$$[\mathfrak{D}_k, \bar{\mathfrak{D}}_{\bar{l}}] \omega_{i_1 \cdots i_p \bar{j}_1 \cdots \bar{j}_q}$$
$$= 2 \sum_\lambda Q^m{}_{k\bar{l}i_\lambda} \omega_{i_1 \cdots (m)_\lambda \cdots i_p \bar{j}_1 \cdots \bar{j}_q} - 2 \sum_\mu Q^{\bar{m}}{}_{\bar{l}k\bar{j}_\mu} \omega_{i_1 \cdots i_p \bar{j}_1 \cdots (\bar{m})_\mu \cdots \bar{j}_q}.$$

$$(\partial \omega)_{i_1 \cdots i_{p+1} \bar{j}_1 \cdots \bar{j}_q} = \sum_\lambda (-1)^{\lambda-1} \mathfrak{D}_{i_\lambda} \omega_{i_1 \cdots \hat{i}_\lambda \cdots i_{p+1} \bar{j}_1 \cdots \bar{j}_q}.$$

$$(\delta \omega)_{i_1 \cdots i_{p-1} \bar{j}_1 \cdots \bar{j}_q} = -g^{s\bar{r}} \bar{\mathfrak{D}}_{\bar{r}} \omega_{si_1 \cdots i_{p-1} \bar{j}_1 \cdots \bar{j}_q} + \alpha^s \omega_{si_1 \cdots i_{p-1} \bar{j}_1 \cdots \bar{j}_q}.$$

From the above expressions we may further obtain

$$(\Box \omega)_{i_1 \cdots i_p \bar{j}_1 \cdots \bar{j}_q}$$
$$= -g^{s\bar{r}} \bar{\mathfrak{D}}_{\bar{r}} \mathfrak{D}_s \omega_{i_1 \cdots i_p \bar{j}_1 \cdots \bar{j}_q} + \alpha^s \mathfrak{D}_s \omega_{i_1 \cdots i_p \bar{j}_1 \cdots \bar{j}_q}$$
$$\quad - g^{s\bar{r}} \sum_\lambda [\mathfrak{D}_{i_\lambda}, \bar{\mathfrak{D}}_{\bar{r}}] \omega_{i_1 \cdots (s)_\lambda \cdots i_p \bar{j}_1 \cdots \bar{j}_q} + \sum_\lambda \beta^s_{i_\lambda} \omega_{i_1 \cdots (s)_\lambda \cdots i_p \bar{j}_1 \cdots \bar{j}_q}$$
$$= -g^{s\bar{r}} \bar{\mathfrak{D}}_{\bar{r}} \mathfrak{D}_s \omega_{i_1 \cdots i_p \bar{j}_1 \cdots \bar{j}_q} + \alpha^s \mathfrak{D}_s \omega_{i_1 \cdots i_p \bar{j}_1 \cdots \bar{j}_q}$$
$$\quad - \sum_\lambda \beta^s_{i_\lambda} \omega_{i_1 \cdots (s)_\lambda \cdots i_p \bar{j}_1 \cdots \bar{j}_q} + 2 \sum_{\lambda,\mu} Q^{\bar{t}s}{}_{i_\lambda \bar{j}_\mu} \omega_{i_1 \cdots (s)_\lambda \cdots i_p \bar{j}_1 \cdots (\bar{t})_\mu \cdots \bar{j}_q}.$$

7.5   Laplacian $\Box_L$

In the theory of Kählerian manifolds, it is essential that the Kählerian forms are harmonic and the exterior product operators by the Kählerian forms are commutative with the Laplacians [Kobayashi (1997, 1998)][Morrow and Kodaira (1971)][Wells (1979)]. However, for a Hessian manifold (M, D, g), we know by Corollary 7.5 and Proposition 7.7 that

$$\square g = \beta, \qquad [\square, e(g)] = e(\beta).$$

Hence the Hessian metric g is not always harmonic with respect to the Laplacian $\square$Cand the exterior operator $e(g)$ is not necessarily commutative with $\square$. We cannot therefore expect simple identities on Hessian manifolds in the manner of the Kählerian identities.

Let (M, D, g) be an oriented Hessian manifold and let v be the volume element of g. For each affine coordinate system $\{x_\lambda^1, \cdots, x_\lambda^n\}$ we define a positive function G_λ by

$$v = (G_\lambda)^{\frac{1}{2}} dx_\lambda^1 \wedge \cdots \wedge dx_\lambda^n.$$

We introduce

$$K_{\lambda\mu} = \left(\frac{G_\lambda}{G_\mu}\right)^{\frac{1}{2}}.$$

Then $\{K_{\lambda\mu}\}$ form transition functions on the flat line bundle $K = \overset{n}{\wedge} T^*$. Let L be a flat line bundle with transition functions $\{(K_{\lambda\mu})^{-\frac{1}{2}}\}$ and a fiber metric $\{(G_\lambda)^{\frac{1}{2}}\}$. We shall show that g is $\square_L$-harmonic and $e(g)$ is commutative with $\square_L$, and prove simple identities similar to the Kählerian identities.

Proposition 7.9. *The adjoint operators of ∂ and $\bar\partial$ are given by*

(1) $\delta_L = -\sum_j i(E_j)\bar{\mathfrak{D}}_{E_j}$.
(2) $\bar\delta_L = -\sum_j \bar{i}(E_j)\mathfrak{D}_{E_j}$.

Proof. By Propositions 7.2, 7.4, 7.5 and relations (7.4) we obtain

$$\delta_L = (-1)^p \tilde{\star}_L^{-1} \partial \tilde{\star}_L = \sum_j (-1)^p \tilde{\star}_L^{-1} e(\theta^j) \tilde{\star}_L \tilde{\star}_L^{-1} \mathfrak{D}_{E_j} \tilde{\star}_L$$

$$= -\sum_j i(E_j)\bar{\mathfrak{D}}_{E_j}.$$

$\square$

By Proposition 7.9 and Lemma 7.6 we have the following corollary.

Corollary 7.6. *The Hessian metric g is harmonic with respect to $\square_L$ and $\bar\square_L$, that is,*

(1) $\square_L g = 0$,
(2) $\bar\square_L g = 0$.

Proposition 7.10.

(1) $\delta_L e(g) + e(g)\delta_L = -\bar{\partial}.$
(2) $\bar{\delta}_L e(g) + e(g)\bar{\delta}_L = -\partial.$

Proof. It follows from Lemma 7.6 and Propositions 7.4 and 7.9 that

$$\delta_L e(g) = -\sum_j i(E_j)\bar{\mathfrak{D}}_{E_j} e(g) = -\sum_j i(E_j)e(g)\bar{\mathfrak{D}}_{E_j}$$

$$= -\sum_j \{i(E_j)g \wedge \bar{\mathfrak{D}}_{E_j} - g \wedge i(E_j)\bar{\mathfrak{D}}_{E_j}\}$$

$$= -\sum_j e(\bar{\theta}^j)\bar{\mathfrak{D}}_{E_j} + e(g)\sum_j i(E_j)\bar{\mathfrak{D}}_{E_j}$$

$$= -\bar{\partial} - e(g)\delta_L.$$

Thus assertion (1) is proved. Assertion (2) may be proved similarly. □

Lemma 7.10.

(1) $\partial\bar{\delta}_L = \bar{\delta}_L\partial.$
(2) $\bar{\partial}\delta_L = \delta_L\bar{\partial}.$

Proof. By Proposition 7.10 we obtain

$$\partial\bar{\delta}_L = -(\bar{\delta}_L e(g) + e(g)\bar{\delta}_L)\bar{\delta}_L = -\bar{\delta}_L e(g)\bar{\delta}_L$$

$$= \bar{\delta}_L(-\bar{\delta}_L e(g) - e(g)\bar{\delta}_L) = \bar{\delta}_L\partial.$$ □

Theorem 7.6. *We have the following identities for the Laplacians.*

(1) $\Box_L = \bar{\Box}_L.$
(2) $\Box_L$ *is commutative with the following operators*
$$\tilde{\ast}_L, \quad e(g), \quad i(g), \quad \partial, \quad \bar{\partial}, \quad \delta_L, \quad \bar{\delta}_L.$$

Proof. Let K^* and L^* be the dual line bundles of K and L respectively. Identifying $L^* \otimes K^*$ with L, by Theorem 7.1 we obtain

$$\Box_L \tilde{\ast}_L = \tilde{\ast}_L \Box_L.$$

It follows form Proposition 7.10 that

$$\Box_L = \partial\delta_L + \delta_L\partial$$

$$= -(\bar{\delta}_L e(g) + e(g)\bar{\delta}_L)\delta_L - \delta_L(\bar{\delta}_L e(g) + e(g)\bar{\delta}_L)$$

$$= \bar{\delta}_L(\delta_L e(g) + \bar{\partial}) + (e(g)\delta_L + \bar{\partial})\bar{\delta}_L - e(g)\bar{\delta}_L\delta_L - \delta_L\bar{\delta}_L e(g)$$

$$= \bar{\partial}\bar{\delta}_L + \bar{\delta}_L\bar{\partial} + (\bar{\delta}_L\delta_L - \delta_L\bar{\delta}_L)e(g) + e(g)(\delta_L\bar{\delta}_L - \bar{\delta}_L\delta_L)$$

$$= \bar{\Box}_L.$$

Again by Proposition 7.10 we have

$$
\begin{aligned}
\Box_L e(g) &= (\partial \delta_L + \delta_L \partial) e(g) = \partial(-\bar{\partial} - e(g)\delta_L) - \delta_L e(g)\partial \\
&= -\partial\bar{\partial} + e(g)\partial\delta_L + (e(g)\delta_L + \bar{\partial})\partial = e(g)(\partial\delta_L + \delta_L\partial) \\
&= e(g)\Box_L.
\end{aligned}
$$

By Lemma 7.10 we obtain

$$
\Box_L \bar{\partial} = (\partial\delta_L + \delta_L\partial)\bar{\partial} = \partial\bar{\partial}\delta_L + \delta_L\bar{\partial}\partial = \bar{\partial}\partial\delta_L + \bar{\partial}\delta_L\partial = \bar{\partial}\Box_L.
$$

The other cases are straightforward. $\qquad\Box$

By Theorem 7.2 we also have the following corollary.

Corollary 7.7 (Duality Theorem for Hessian Manifolds).

(1) *The mapping $\tilde{*}_L$ induces a linear isomorphism from $H^{p,q}_{\Box_L}$ to $H^{n-p,n-q}_{\Box_L}$.*
(2) *Let $h^{p,q}_L = \dim H^{p,q}_{\partial}(L)$. Then*

$$
h^{p,q}_L = h^{q,p}_L = h^{n-p,n-q}_L.
$$

It is known that the space of harmonic forms on a compact Kählerian manifold admits the **Lefschetz decomposition** [Wells (1979)]. We state here without proof that similar results hold for Hessian manifolds.

Theorem 7.7. *The space $H^{p,q}_{\Box_L}$ is decomposed into the direct sum*

$$
H^{p,q}_{\Box_L} = \sum_s e(g)^s H^{p-s,q-s}_{\Box_L,0},
$$

where

$$
H^{p,q}_{\Box_L,0} = \{\omega \in H^{p,q}_{\Box_L} \mid i(g)\omega = 0\}.
$$

7.6 Affine Chern classes of flat manifolds

Using an analogous method of [Bott and Chern (1965)] we define affine Chern classes of flat vector bundles. This notion is similar to that of Chern classes of complex manifolds [Kobayashi and Nomizu (1963, 1969)][Kobayashi (1997, 1998)][Wells (1979)].

Let F be a flat vector bundle of rank m over a flat manifold (M, D) and let F^* be the dual bundle of F. An element $\varphi \in \mathcal{A}^{p,q}(F \otimes F^*)$ is identified with a matrix $[\varphi^i_j]$ of degree m whose (i,j) component φ^i_j is an element in $\mathcal{A}^{p,q}$. For $\varphi = [\varphi^i_j] \in \mathcal{A}^{p,q}(F \otimes F^*)$ and $\psi = [\psi^k_l] \in \mathcal{A}^{r,s}(F \otimes F^*)$ we define $[\varphi, \psi] \in \mathcal{A}^{p+r,q+s}(F \otimes F^*)$ by

$$
[\varphi, \psi] = \varphi \wedge \psi - (-1)^{pr+qs}\psi \wedge \varphi,
$$

where the (i, j) component of $[\varphi, \psi]$ is given by

$$[\varphi, \psi]^i_j = \sum_k \varphi^i_k \wedge \psi^k_j - (-1)^{pr+qs} \sum_k \psi^i_k \wedge \varphi^k_j.$$

Lemma 7.11. *Let* $f(X_1, \cdots, X_k)$ *be a* $GL(m, \mathbf{R})$-*invariant symmetric multilinear form on* $\mathfrak{gl}(m, \mathbf{R})$. *Let* $\varphi_i \in \mathcal{A}^{p_i, q_i}(F \otimes F^*)$ $(1 \leq i \leq k)$ *and* $\psi \in \mathcal{A}^{p,q}(F \otimes F^*)$. *Then*

$$\sum_i (-1)^{(p_1 + \cdots + p_i)p + (q_1 + \cdots + q_i)q} f(\varphi_1, \cdots, [\varphi_i, \psi], \cdots, \varphi_k) = 0.$$

Proof. It is enough to show the equality for $\varphi_i = \omega_i X_i$ and $\psi = \omega Y$ where $\omega_i \in \mathcal{A}^{p_i, q_i}$, $\omega \in \mathcal{A}^{p,q}$ and $X_i, Y \in \mathfrak{gl}(m, \mathbf{R})$. Since

$$[\omega_i X_i, \omega Y] = (\omega_i \wedge \omega) X_i Y - (-1)^{p_i p + q_i q}(\omega \wedge \omega_i) Y X_i,$$

we obtain

$$\sum_i (-1)^{(p_1 + \cdots + p_i)p + (q_1 + \cdots + q_i)q} f(\omega_1 X_1, \cdots, [\omega_i X_i, \omega Y], \cdots, \omega_k X_k)$$

$$= \sum_i (-1)^{(p_1 + \cdots + p_i)p + (q_1 + \cdots + q_i)q}$$

$$\times (\omega_1 \wedge \cdots \wedge (\omega_i \wedge \omega) \wedge \cdots \wedge \omega_k) f(X_1, \cdots, X_i Y, \cdots, X_k)$$

$$- \sum_i (-1)^{(p_1 + \cdots + p_{i-1})p + (q_1 + \cdots + q_{i-1})q}$$

$$\times (\omega_1 \wedge \cdots \wedge (\omega \wedge \omega_i) \wedge \cdots \wedge \omega_k) f(X_1, \cdots, Y X_i, \cdots, X_k)$$

$$= (\omega \wedge \omega_1 \wedge \cdots \wedge \omega_k) \sum_i \Big\{ f(X_1, \cdots, X_i Y, \cdots, X_k)$$

$$- f(X_1, \cdots, Y X_i, \cdots, X_k) \Big\}$$

$$= (\omega \wedge \omega_1 \wedge \cdots \wedge \omega_k) \sum_i f(X_1, \cdots, [X_i, Y], \cdots, X_k)$$

$$= 0.$$

$\square$

Let h be a fiber metric on F. Choosing local frames of F so that the transition functions are constants, we define a matrix $H(s) = [h(s_i, s_j)]$ for each such frame $s = \{s_1, \cdots, s_m\}$. Then $H(s)^{-1}\partial H(s)$ is an element in $\mathcal{A}^{1,0}(F \otimes F^*)$ independent of the choice of $\{s\}$, and is denoted by A_hG

$$A_h = H(s)^{-1}\partial H(s).$$

We define an element $B_h \in \mathcal{A}^{1,1}(F \otimes F^*)$ by

$$B_h = \bar{\partial} A_h.$$

Lemma 7.12. *We have*

(1) $\partial A_h = -A_h \wedge A_h$.
(2) $\partial B_h = -[B_h, A_h]$.

Proof. Since $\partial H(s)^{-1} = -H(s)^{-1}(\partial H(s))H(s)^{-1}$, we have

$$\partial A_h = \partial H(s)^{-1} \wedge \partial H(s) = -H(s)^{-1}(\partial H(s))H(s)^{-1} \wedge \partial H(s)$$
$$= -H(s)^{-1}\partial H(s) \wedge H(s)^{-1}\partial H(s) = -A_h \wedge A_h,$$

whereupon we may obtain

$$\partial B_h = \bar{\partial}\partial A_h = \bar{\partial}(-A_h \wedge A_h) = -(\bar{\partial} A_h \wedge A_h + A_h \wedge \bar{\partial} A_h)$$
$$= -(B_h \wedge A_h + A_h \wedge B_h) = -[B_h, A_h]. \qquad \square$$

Let $f(X_1, \cdots, X_k)$ be a $GL(m, \mathbf{R})$-invariant symmetric multilinear form on $\mathfrak{gl}(m, \mathbf{R})$. We define $f(B_h) \in \mathcal{A}^{k,k}$ by

$$f(B_h) = f(B_h, \cdots, B_h).$$

We then have the following lemma.

Lemma 7.13. *We have*

$$\partial f(B_h) = 0, \qquad \bar{\partial} f(B_h) = 0.$$

Proof. Denoting by X_j^i the (i,j)-component of $X \in \mathfrak{gl}(m, \mathbf{R})$, we express f by

$$f(X_1, \cdots, X_k) = \sum c_{i_1 \cdots i_k}^{j_1 \cdots j_k} (X_1)_{j_1}^{i_1} \cdots (X_k)_{j_k}^{i_k}.$$

Then

$$f(B_h) = \sum c_{i_1 \cdots i_k}^{j_1 \cdots j_k} (B_h)_{j_1}^{i_1} \wedge \cdots \wedge (B_h)_{j_k}^{i_k}.$$

Since $\bar{\partial} B_h = \bar{\partial}(\bar{\partial} A_h) = 0$ we have

$$\bar{\partial} f(B_h) = \sum c_{i_1 \cdots i_k}^{j_1 \cdots j_k} \sum_r (-1)^r (B_h)_{j_1}^{i_1} \wedge \cdots \wedge \bar{\partial}(B_h)_{j_r}^{i_r} \wedge \cdots \wedge (B_h)_{j_k}^{i_k} = 0.$$

By Lemmata 7.11 and 7.12 we obtain

$$\partial f(B_k) = \sum c_{i_1 \cdots i_k}^{j_1 \cdots j_k} \sum_r (-1)^r (B_h)_{j_1}^{i_1} \wedge \cdots \wedge \partial(B_h)_{j_r}^{i_r} \wedge \cdots \wedge (B_h)_{j_k}^{i_k}$$

$$= \sum_r (-1)^r \sum c_{i_1 \cdots i_k}^{j_1 \cdots j_k} (B_h)_{j_1}^{i_1} \wedge \cdots \wedge (-[B_h, A_h]_{j_r}^{i_r}) \wedge \cdots \wedge (B_h)_{j_k}^{i_k}$$

$$= -\sum_r (-1)^r f(B_h, \cdots, \overset{r}{[B_h, A_h]}, \cdots, B_h)$$

$$= 0. \qquad \square$$

Let $\{h_t\}$ be a family of fiber metrics on F parametrized by t. For the sake of clarity we set

$$A_t = A_{h_t}, \qquad B_t = B_{h_t}.$$

Putting $H_t(s) = [h_t(s_i, s_j)]$, we define $L_t \in F \otimes F^*$ by

$$L_t = H_t(s)^{-1}\frac{d}{dt}H_t(s),$$

and $f^*(B_t; L_t) \in \mathcal{A}^{k-1,k-1}$ by

$$f^*(B_t; L_t) = \sum_i f(B_t, \cdots, \overset{i}{L_t}, \cdots, B_t).$$

By Lemma 7.11 and $\partial B_t = -[B_t, A_t]$ we obtain

$$\partial f^*(B_t; L_t)$$
$$= \sum_i \sum_{j<i}(-1)^{j-1} f(\cdots, \partial \overset{j}{B_t}, \cdots \overset{i}{L_t}, \cdots)$$
$$+ \sum_i (-1)^{i-1} f(\cdots, \partial \overset{i}{L_t}, \cdots) + \sum_i \sum_{j>i}(-1)^{j-2} f(\cdots, \overset{i}{L_t}, \cdots, \partial \overset{j}{B_t}, \cdots)$$
$$= \sum_i \Big\{ \sum_{j<i}(-1)^j f(\cdots, [\overset{j}{B_t}, A_t], \cdots, \overset{i}{L_t}, \cdots)$$
$$+ (-1)^{i-1} f(\cdots, [\overset{i}{L_t}, A_t], \cdots) + \sum_{j>i}(-1)^{j-1} f(\cdots, \overset{i}{L_t}, \cdots, [\overset{j}{B_t}, A_t], \cdots) \Big\}$$
$$- \sum_i (-1)^{i-1} f(\cdots, [\overset{i}{L_t}, A_t], \cdots) + \sum_i (-1)^{i-1} f(\cdots, \partial \overset{i}{L_t}, \cdots)$$
$$= \sum_i (-1)^{i-1} f(B_t, \cdots, \partial L_t - \overset{i}{[L_t}, A_t], \cdots, B_t).$$

We have

$$\partial L_t = \partial \Big(H_t(s)^{-1}\frac{d}{dt}H_t(s) \Big)$$
$$= -H_t(s)^{-1}(\partial H_t(s))H_t(s)^{-1}\frac{d}{dt}H_t(s) + H_t(s)^{-1}\partial\frac{d}{dt}H_t(s)$$
$$= -A_t L_t + \frac{d}{dt}\Big(H_t(s)^{-1}\partial H_t(s) \Big) + H_t(s)^{-1}\Big(\frac{d}{dt}H_t(s) \Big)H_t(s)^{-1}\partial H_t(s)$$
$$= -A_t L_t + \frac{d}{dt}A_t + L_t A_t$$
$$= [L_t, A_t] + \frac{d}{dt}A_t.$$

Hence

$$\partial f^*(B_t; L_t) = \sum_i (-1)^{i-1} f(B_t, \cdots, \overset{i}{\frac{d}{dt}A_t}, \cdots, B_t).$$

Since $\bar{\partial}B_t = 0$, we obtain

$$\bar{\partial}\partial f^*(B_t; L_t)$$

$$= \sum_i (-1)^{i-1} \sum_{j<i} (-1)^{j-1} f(\cdots, \overset{j}{\bar{\partial}B_t}, \cdots, \overset{i}{\frac{d}{dt}A_t}, \cdots) + \sum_i f(\cdots, \overset{i}{\bar{\partial}\frac{d}{dt}A_t}, \cdots)$$

$$+ \sum_i (-1)^{i-1} \sum_{j>i} (-1)^{j-2} f(\cdots, \overset{i}{\frac{d}{dt}A_t}, \cdots, \overset{j}{\bar{\partial}B_t}, \cdots)$$

$$= \sum_i f(B_t, \cdots, \overset{i}{\frac{d}{dt}B_t}, \cdots, B_t)$$

$$= \frac{d}{dt} f(B_t, \cdots, B_t).$$

Hence we have the following lemma.

Lemma 7.14.

$$f(B_1) - f(B_0) = \bar{\partial}\partial \int_0^1 f^*(B_t; L_t) dt.$$

Definition 7.13. We define a cohomology group $\hat{H}^k(M)$ of M by

$$\hat{H}^k(M) = \{\omega \in \mathcal{A}^{k,k} \mid \partial\omega = 0, \ \bar{\partial}\omega = 0\}/\partial\bar{\partial}\mathcal{A}^{k-1,k-1}.$$

Let $f_k(X)$ be a $GL(m, \mathbf{R})$-invariant homogeneous polynomial of degree k on $\mathfrak{gl}(m, \mathbf{R})$ determined by

$$\det(I - tX) = \sum_{k=0}^m t^k f_k(X),$$

where I is the unit matrix of degree m. We define $c_k(F, h) \in \mathcal{A}^{k,k}$ by

$$c_k(F, h) = f_k(B_h).$$

Then, by Lemma 7.13,

$$\partial c_k(F, h) = 0, \qquad \bar{\partial}c_k(F, h) = 0,$$

and so, by Lemma 7.14, we have the following theorem.

Theorem 7.8. *The element in $\hat{H}^k(M)$ represented by $c_k(F, h)$ is independent of the choice of a fiber metric h.*

Definition 7.14. We denote by $\hat{c}_k(F)$ the element in $\hat{H}^k(M)$ represented by $c_k(F, h)$, and call it the k-th **affine Chern class** of F. The k-th affine Chern class of the tangent bundle T of M is said to be the k-th affine Chern class of M and is denoted by $\hat{c}_k(M)$.

Proposition 7.11. *The first affine Chern class $\hat{c}_1(M)$ of a Hessian manifold (M, D, g) is represented by -2β where β is the second Koszul form of (D, g).*

Proof. By Proposition 2.2 and Definition 3.1 we have

$$A_g = 2 \sum \left(\gamma^i_{\ jl} dx^l \right) \otimes \left(\frac{\partial}{\partial x^i} \otimes dx^j \right),$$

$$B_g = 2 \sum \left(\frac{\partial \gamma^i_{\ jl}}{\partial x^{\bar{k}}} dx^l \otimes dx^{\bar{k}} \right) \otimes \left(\frac{\partial}{\partial x^i} \otimes dx^j \right)$$

$$= 2 \sum \left(Q^i_{\ j\bar{k}l} dx^l \otimes dx^{\bar{k}} \right) \otimes \left(\frac{\partial}{\partial x^i} \otimes dx^j \right)$$

$$= 2Q,$$

where Q is the Hessian curvature tensor (cf. Definition 3.1). Hence

$$c_k(T, g) = f_k(2Q) = \frac{(-2)^k}{k!} \sum \delta^{j_1 \cdots j_k}_{i_1 \cdots i_k} Q^{i_1}_{j_1} \wedge \cdots \wedge Q^{i_k}_{j_k},$$

where $Q^i_j = \sum Q^i_{\ j\bar{k}l} dx^l \otimes dx^{\bar{k}}$, and

$$c_1(T, g) = -2 \sum Q^i_{\ i\bar{k}l} dx^l \otimes dx^{\bar{k}} = -2\beta. \qquad \square$$

Proposition 7.12. *Let (M, D, g) be an n-dimensional Hessian manifold. For the sake of clarity, we set $c_k = c_k(T, g)$ and $g^{n-2} = \overbrace{g \wedge \cdots \wedge g}^{n-2 \text{ terms}}$. Then we have*

(1) $c_2 = 2\left(\beta \wedge \beta - Q^i_j \wedge Q^j_i \right) C$

(2) $(c_1 \wedge c_1) \wedge g^{n-2} = 4(n-2)! \left\{ \left(\operatorname{Tr} \hat{\beta} \right)^2 - \operatorname{Tr} \hat{\beta}^2 \right\} v \otimes v C$

(3) $c_2 \wedge g^{n-2} = 2(n-2)! \left\{ \left(\operatorname{Tr} \hat{\beta} \right)^2 - 2\operatorname{Tr} \hat{\beta}^2 + \operatorname{Tr} \hat{Q}^2 \right\} v \otimes v,$

where $\hat{\beta}$ is a tensor field of type $(1,1)$ defined by $\beta^i_j = g^{ik}\beta_{kj}$ and $\hat{Q}$ is a linear mapping given in Definition 3.4.

Proof. From the proof of Proposition 7.11 and Proposition 3.4 (3) we have

$$c_2 = 2\left(Q^i_i \wedge Q^j_j - Q^i_j \wedge Q^j_i \right) = 2\left(\beta \wedge \beta - Q^i_j \wedge Q^j_i \right).$$

Thus the proof of assertion (1) is completed. Let $\theta^1, \cdots, \theta^n$ be an orthonormal frame of T^*M; $g = \sum_i \theta^i \otimes \theta^i$, and let $\beta = \sum_{i,j} \beta_{ij} \theta^i \otimes \theta^j$. Then we have

$$(\beta \wedge \beta) \wedge g^{n-2}$$
$$= \sum \beta_{i_1 j_1} \beta_{i_2 j_2} (\theta^{i_1} \wedge \theta^{i_2} \wedge \theta^{i_3} \wedge \cdots \wedge \theta^{i_n}) \otimes (\theta^{j_1} \wedge \theta^{j_2} \wedge \theta^{i_3} \wedge \cdots \wedge \theta^{i_n})$$
$$= \sum (\beta_{i_1 i_1} \beta_{i_2 i_2} - \beta_{i_1 i_2} \beta_{i_2 i_1})(\theta^{i_1} \wedge \cdots \wedge \theta^{i_n}) \otimes (\theta^{i_1} \wedge \cdots \wedge \theta^{i_n})$$
$$= (n-2)! \left\{ \sum_{i \neq j} \beta_{ii} \beta_{jj} - \sum_{i \neq j} \beta_{ij}^2 \right\} v \otimes v$$
$$= (n-2)! \left\{ (\operatorname{Tr} \hat{\beta})^2 - \operatorname{Tr} \hat{\beta}^2 \right\} v \otimes v.$$

The above expression, together with the relation $c_1 = -2\beta$, imply (2). We have

$$Q_j^i \wedge Q_i^j \wedge g^{n-2}$$
$$= \sum Q^i{}_{jp_1 q_1} Q^j{}_{ip_2 q_2} (\theta^{q_1} \wedge \theta^{q_2} \wedge \theta^{p_3} \wedge \cdots \wedge \theta^{p_n}) \otimes (\theta^{p_1} \wedge \theta^{p_2} \wedge \theta^{p_3} \wedge \cdots \wedge \theta^{p_n})$$
$$= \sum (Q^i{}_{jp_1 p_1} Q^j{}_{ip_2 p_2} - Q^i{}_{jp_1 p_2} Q^j{}_{ip_2 p_1})(\theta^{p_1} \wedge \cdots \wedge \theta^{p_n}) \otimes (\theta^{p_1} \wedge \cdots \wedge \theta^{p_n})$$
$$= (n-2)! \sum_{p \neq q} \{ Q^i{}_{jpp} Q^j{}_{iqq} - Q^i{}_{jpq} Q^j{}_{jqp} \} v \otimes v$$
$$= (n-2)! \{ \beta^i{}_j \beta^j{}_i - Q^i{}_j{}^k{}_l Q^j{}_i{}^l{}_k \} v \otimes v$$
$$= (n-2)! \{ \operatorname{Tr} \hat{\beta}^2 - \operatorname{Tr} \hat{Q}^2 \} v \otimes v.$$

From the above expression, together with assertions (1) and (2) we obtain assertion (3). □

The notion of Einstein-Hessian metrics was first produced by [Cheng and Yau (1982)]. The following proposition is an anlogy of the Miyaoka-Yau inequality for Einstein-Kähler manifolds, which was first proved in [Miyaoka (1977)] for the case $n = 2$, and the general case was obtained by combining the results of [Chen and Ogiue (1975)] and [Yau (1977)].

Proposition 7.13. *Let (M, D, g) be an n-dimensional Hessian manifold. An element $\omega = f(v \otimes v) \in \mathcal{A}^{n,n}$ is said to be non-negative $\omega \geq 0$, if f is non-negative $f \geq 0$. Suppose that the Hessian structure (D, g) is Einstein-Hessian. Then*

$$\{ -n(c_1 \wedge c_1) + 2(n+1)c_2 \} \wedge g^{n-2} \geq 0.$$

The equality holds if and only if the Hessian sectional curvature of (D, g) is a constant.

Proof.   Since (D, g) is Einstein-Hessian it follows that

$$\beta_j^i = \frac{1}{n}\delta_j^i \operatorname{Tr} \hat{\beta}.$$

and so

$$\operatorname{Tr} \hat{\beta}^2 = \beta_j^i \, \beta_i^j = \frac{1}{n}(\operatorname{Tr} \hat{\beta})^2.$$

By Proposition 7.12 we have

$$(c_1 \wedge c_1) \wedge g^{n-2} = 4(n-2)! \left\{ \frac{n-1}{n}(\operatorname{Tr} \hat{\beta})^2 \right\} v \otimes v,$$

$$c_2 \wedge g^{n-2} = 2(n-2)! \left\{ \frac{n-2}{n}(\operatorname{Tr} \hat{\beta})^2 + \operatorname{Tr} \hat{Q}^2 \right\} v \otimes v.$$

From the above equations we further have

$$\left\{ -n(c_1 \wedge c_1) + 2(n+1)c_2 \right\} \wedge g^{n-2}$$

$$= 4(n+1)((n-2)!) \left\{ \operatorname{Tr} \hat{Q}^2 - \frac{2}{n(n+1)}(\operatorname{Tr} \hat{\beta})^2 \right\} v \otimes v.$$

Applying Theorem 3.3 to the above formula we obtain the desired result.

$\square$

Chapter 8

Compact Hessian manifolds

In section **8.1** we survey affine developments, exponential mappings and universal coverings for flat manifolds by using Koszul's method. Applying these results, we prove in section **8.2** the convexity and the hyperbolicity of Hessian manifolds. In section **8.3** we restate a theorem due to [Cheng and Yau (1986)] [Pogorelov (1978)] and Calabi's theorem [Calabi (1958)] as follows: If the first Koszul form α on a Hessian manifold (M, D, g) vanishes, then the Levi-Civita connection ∇ for g coincides with D. For a compact Hessian manifold we show an integral formula of the second Koszul form β, and prove that β cannot be negative definite and that $\beta = 0$ if and only if $\nabla = D$. [Delanoe (1989)] and [Cheng and Yau (1982)] proved a certain problem, which is analogous to the Calabi conjecture for Kählerian manifolds and concerned with the first affine Chern class $\hat{c}_1(M)$.

8.1 Affine developments and exponential mappings for flat manifolds

Following Koszul's method we define the affine development of a flat manifold, and study the relation between the affine development and the exponential mapping [Koszul (1965)].

Let (M, D) be a connected simply connected flat manifold. Recall that (M, D) is flat means that D has both zero curvature and zero torsion. We select a point $o \in M$. Since the curvature of D is vanishes and M is simply connected, for each $u \in T_o M$ there exists a unique D-parallel vector field P_u on M such that the value of P_u at o coincides with u. We define a $T_o M$-valued 1-form ω on M by

$$\omega(P_u) = u \quad \text{for } u \in T_o M.$$

Since $(D_X\omega)(P_u) = X\omega(P_u) - \omega(D_X P_u) = 0$, ω is D-parallel and so, since D is torsion-free, $d\omega = 0$. Because $\dot{\sigma}(t)(\omega(\dot{\sigma}(t))) = \omega(D_{\dot{\sigma}(t)}\dot{\sigma}(t))$, a smooth curve $\sigma(t)$ on M is a geodesic if and only if $\omega(\dot{\sigma}(t))$ is a constant vector

Let C be a curve on M from o to a. Since M is simply connected and $d\omega = 0$, by Stokes's theorem, the value $\int_C \omega$ does not depend on the choice of curve C joining o and a, but depends only on a. We can therefore denote the value by

$$F(a) = \int_o^a \omega.$$

We shall show that the mapping $F : M \ni a \longrightarrow F(a) \in T_o M$ is an affine mapping. Let $\sigma(t)$ be a geodesic on (M, D). Then

$$F(\sigma(t)) = \int_o^{\sigma(0)} \omega + \int_{\sigma(o)}^{\sigma(t)} \omega = \int_o^{\sigma(0)} \omega + \int_0^t \omega(\dot{\sigma}(t))dt$$

$$= \int_o^{\sigma(0)} \omega + t\omega(\dot{\sigma}(t)),$$

and $\omega(\dot{\sigma}(t))$ is a constant vector. Thus $F(\sigma(t))$ is a line on $T_o M$. Hence $F : M \longrightarrow T_o M$ is an affine mapping. Since $dF = \omega$, the rank of F is equal to $\dim M$ at any point of M. Thus F is an affine immersion from M into $T_o M$.

Definition 8.1. The affine immersion $F : M \longrightarrow T_o M$ is called the **affine development** of (M, D).

We denote by G the group of all affine transformations of (M, D). Let $s \in G$ and $u \in T_o M$. Since $s_* P_u$ is D-parallel, there exists a unique element $\boldsymbol{f}(s)u \in T_o M$ such that

$$s_* P_u = P_{\boldsymbol{f}(s)u}.$$

Then $u \longrightarrow \boldsymbol{f}(s)u$ is a linear transformation of $T_o M$, and $s \longrightarrow \boldsymbol{f}(s)$ is a linear representation of G on $T_o M$.

Let $s \in G$ and $a \in M$. We denote by C_a and C_{so} curves from o to a and from o to so respectively. Since a curve $C_{so} + sC_a$ joining C_{so} and sC_a is a curve from o to sa, and given also that $s^*\omega = \boldsymbol{f}(s)\omega$, we have

$$F(sa) = \int_{C_{so}+sC_a} \omega = \int_{C_{so}} \omega + \int_{sC_a} \omega$$

$$= F(so) + \int_{C_a} s^*\omega$$

$$= F(so) + \boldsymbol{f}(s)F(a).$$

We define a mapping $q : G \longrightarrow T_o M$ by $q(s) = F(so)$. Then

$$F(sa) = f(s)F(a) + q(s),$$
$$q(rs) = f(r)q(s) + q(r).$$

This means that $s \longrightarrow a(s) = (f(s), q(s))$ is an affine representation of G on $T_o M$ (cf. section **1.1**) and

$$F(sa) = a(s)F(a).$$

Therefore (F, a) is an equivariant affine immersion from (M, G) into $(T_o M, A(T_o M))$ where $A(T_o M)$ is the group of all affine transformations of $T_o M$.

Definition 8.2. We denote by $\exp_o^D$ the exponential mapping at $o \in M$, and by $\mathcal{E}_o$ the domain of definition for $\exp_o^D$.

Theorem 8.1. *Let (M, D) be a connected simply connected flat manifold. Then*

(1) $\exp_o^D : \mathcal{E}_o \longrightarrow M$ *is an affine mapping, and $F \circ \exp_o^D$ is the identity mapping on $\mathcal{E}_o$. In particular, $\exp_o^D$ is injective.*

(2) *If $\mathcal{E}_o$ is convex, then $\exp_o^D$ is an affine isomorphism.*

Proof. Let $\sigma(t) = \exp_o^D tu$ where $u \in \mathcal{E}_o$. Then

$$F(\exp_o^D u) = \int_o^{\sigma(1)} \omega = \int_0^1 \omega(\dot{\sigma}(t))dt = \int_0^1 u dt = u.$$

Hence $F \circ \exp_o^D$ is the identity mapping on $\mathcal{E}_o$, and $\exp_o^D$ is injective. Let $\tau(t) = u + tv$ be an arbitrary geodesic on $\mathcal{E}_o$. Differentiating both sides of $\tau(t) = (F \circ \exp_o^D)(\tau(t))$ by t, we have $v = \dot{\tau}(t) = \omega((\exp_o^D)_*(\dot{\tau}(t)))$. Thus $\exp_o^D \tau(t)$ is a geodesic on M. This shows that $\exp_o^D : \mathcal{E}_o \longrightarrow M$ is an affine mapping. Suppose that $\mathcal{E}_o$ is convex. We shall show that $\exp_o^D : \mathcal{E}_o \longrightarrow M$ is surjective. Let a be an arbitrary point of the closure of $\exp_o^D \mathcal{E}_o$ and let $u = F(a)$. Then since $F \circ \exp_o^D$ is the identity mapping on $\mathcal{E}_o$, u is an element of the closure of $\mathcal{E}_o$. Since $\mathcal{E}_o$ is convex, we know $tu \in \mathcal{E}_o$ for all $0 \le t < 1$. Since the rank of F at a is equal to $\dim M$, we can choose a sufficiently small neighourhood W of u and a mapping $h : W \longrightarrow M$ such that $F \circ h$ is the identity mapping on W and $h(u) = a$. Because the point a is contained in the closure of $\exp_o^D \mathcal{E}_o$, the set $h(W) \cap \exp_o^D \mathcal{E}_o$ is a non-empty open set and $F(h(W) \cap \exp_o^D \mathcal{E}_o) \subset W \cap \mathcal{E}_o$. For any element $v \in F(h(W) \cap \exp_o^D \mathcal{E}_o)$, we have $\exp_o^D v \in h(W)$ and $F(\exp_o^D v) = v = F(h(v))$. Since F is injective on $h(W)$, we obtain $h(v) = \exp_o^D v$, and so h and $\exp_o^D$

coincide on $F(h(W) \cap \exp_o^D \mathcal{E}_o)$. On the other hand, h and $\exp_o^D$ are both affine mappings from $W \cap \mathcal{E}_o$ to M, and coincide on $F(h(W) \cap \exp_o^D \mathcal{E}_o)$. They therefore also coincide on $W \cap \mathcal{E}_o$. Hence, if $t \longrightarrow 1$ then $\exp_o^D tu \longrightarrow h(u) = a$. This implies $u \in \mathcal{E}_o$ and $a = h(u) = \exp_o^D u \in \exp_o^D \mathcal{E}_o$, and so $\exp_o^D \mathcal{E}_o$ is a closed set. Therefore $\exp_o^D \mathcal{E}_o$ is a non-empty open and closed subset of M, and so $M = \exp_o^D \mathcal{E}_o$. $\square$

Corollary 8.1. *Let (M, D) be a connected flat manifold. Then*

(1) $\exp_o^D : \mathcal{E}_o \longrightarrow M$ *is an affine mapping and the rank is equal to* $\dim M$ *at each point.*

(2) *If $\mathcal{E}_o$ is convex, then $\exp_o^D : \mathcal{E}_o \longrightarrow M$ is the universal covering of M.*

Proof. Assertion (1) follows from Theorem 8.1 (1). Suppose that $\mathcal{E}_0$ is convex. Let $\pi : \tilde{M} \longrightarrow M$ be the universal covering of M, and let $\tilde{D}$ be the flat connection on $\tilde{M}$ induced by D. Choosing a point $\tilde{o}$ such that $\pi(\tilde{o}) = o$ we denote by $\exp_{\tilde{o}}^{\tilde{D}}$ the exponential mapping at $\tilde{o}$, and by $\tilde{\mathcal{E}}_{\tilde{o}}$ the domain of definition for $\exp_{\tilde{o}}^{\tilde{D}}$. Since $\pi_{*\tilde{o}} : T_{\tilde{o}} \tilde{M} \longrightarrow T_o M$ is a linear isomorphism and $\pi_{*\tilde{o}}(\tilde{\mathcal{E}}_{\tilde{o}}) = \mathcal{E}_o$, it follows that $\tilde{\mathcal{E}}_{\tilde{o}}$ is convex. Moreover, by Theorem 8.1 (2), $\exp_{\tilde{o}}^{\tilde{D}} : \tilde{\mathcal{E}}_{\tilde{o}} \longrightarrow \tilde{M}$ is is an affine isomorphism. Since $\exp_o^D \circ \pi_{*\tilde{o}} = \pi \circ \exp_{\tilde{o}}^{\tilde{D}}$, we know that $\exp_o^D \circ \pi_{*\tilde{o}} : \tilde{\mathcal{E}}_{\tilde{o}} \longrightarrow M$ is the universal covering of M, and so $\exp_o^D : \mathcal{E}_o \longrightarrow M$ is also the universal covering of M. This proves (2). $\square$

Proof of Proposition 1.1. The assertion (2) is obvious. We shall prove the assertion (1). Let D be a flat connection on a manifold M. By Corollary 8.1 (1), for any point $o \in M$ there exists a neighbourhood U of $0 \in \mathcal{E}_o$ such that the restriction $\exp_o^D |_U$ of $\exp_o^D$ to U is a diffeomorphism from U to $\exp_o^D(U)$. For an affine coordinate system $\{y^1, \cdots, y^n\}$ on $T_o M$ we set

$$x^i = y^i \circ \exp_o^D |_U^{-1} .$$

Then $\{x^1, \cdots, x^n\}$ is a local coordinate system around o satisfying $D_{\partial/\partial x^i} \partial/\partial x^j = 0$. It is easy to see that the changes between such local coordinate systems are affine transformations. Thus the asserton (1) is proved. $\square$

8.2 Convexity of Hessian manifolds

In this section we prove that the universal covering of a quasi-compact Hessian manifold is isomorphic to a convex domain in $\mathbf{R}^n$.

A diffeomorphism s of a manifold M is called an automorphism of a flat manifold (M, D) if it leaves the flat connection D invariant. The set of all automorphisms of (M, D) forms a Lie group (cf. [Kobayashi (1972)]) and is denoted by $Aut(M, D)$. A diffeomorphism s of M is said to be an automorphism of a Hessian manifold (M, D, g) if it preserves both D and g. The set of all automorphisms of (M, D, g) forms a Lie subgroup of $Aut(M, D)$ and is denoted by $Aut(M, D, g)$.

Definition 8.3. Let G be a subgroup of $Aut(M, D)$. If there exists a compact subset C of M such that $M = GC$, then $G \backslash M$ is said to be **quasi-compact**.

Theorem 8.2. *Let (M, D, g) be an n-dimensional connected Hessian manifold. Suppose that $Aut(M, D, g)$ contains a subgroup G such that $G \backslash M$ is quasi-compact. Then the universal covering of M is isomorphic to a convex domain in $\mathbf{R}^n$.*

Corollary 8.2. *Suppose that an n-dimensional Hessian manifold (M, D, g) is compact, or M admits a transitive subgroup of $Aut(M, D, g)$. Then the universal covering of M is isomorphic to a convex domain in $\mathbf{R}^n$.*

In order to prove Theorem 8.2, we shall first prepare some lemmata.

Lemma 8.1. *Let $(\Omega, D, g = Dd\varphi)$ be a Hessian domain in $\mathbf{R}^n$. Suppose that the potential φ satisfies*

$$\lim_{t \to 1^-} \varphi(a + tb) = \infty$$

for all a and $b \in \mathbf{R}^n$ such that $a + tb \in \Omega$ for $0 \le t < 1$ and $a + b \notin \Omega$. Then Ω is isomorphic to a convex domain.

Proof. For points $a_1, \cdots, a_p$ in $\mathbf{R}^n$ we denote by $env(a_1, \cdots, a_p)$ the minimum convex subset of $\mathbf{R}^n$ containing $a_1, \cdots, a_p$. Since any two points in Ω are joined by continuous segments of lines, to prove the convexity of Ω it is sufficient to prove: If $env(a, b) \subset \Omega$ and $env(a, c) \subset \Omega$, then $env(a, b, c) \subset \Omega$. Without loss of generality we may assume that a is the origin 0. Put $I = \{t \in \mathbf{R} \mid env(0, b, tc) \subset \Omega\}$. Then I is an open interval containing 0. Let $\tau \in I$ and $e \in env(0, b, \tau c)$. Since φ is a convex function, we know

$$\varphi(e) \le \max(\varphi(0), \varphi(b), \varphi(\tau c)).$$

If $0 \le \tau \le 1$, then $\tau c \in env(0, c)$, so $\varphi(\tau c) \le \max(\varphi(0), \varphi(c))$, and hence

$$\varphi(e) \le \max(\varphi(0), \varphi(b), \varphi(c)).$$

Put $\nu = \sup I$ and suppose $\nu \leq 1$. Then there exists a point d contained in $env(0, b, \nu c)$ but not in Ω. Let $0 \leq \theta < 1$. Then $\theta\nu < 1$, $\theta\nu \in I$ and $\theta d \in env(0, b, \theta\nu c) \subset \Omega$, and we have

$$\varphi(\theta d) \leq \max(\varphi(0), \varphi(b), \varphi(c)).$$

Alternatively, it follows from the condition for φ that

$$\lim_{\theta \to 1^-} \varphi(\theta d) = \infty,$$

which is a contradiction. Hence $\nu = \sup I > 1$. Thus we have $1 \in I$ and $env(0, b, c) \subset \Omega$. $\qquad\square$

Let TM be the tangent bundle over a Hessian manifold (M, D, g) with projection $\pi : TM \longrightarrow M$. We denote by $\exp^D$ the exponential mapping given by D, and by $\mathcal{E}$ the domain of definition for $\exp^D$. For $y \in TM$ we denote by $|y|$ the length of the tangent vector, and set

$$\lambda(y) = \sup\{t \in \mathbf{R} \mid ty \in \mathcal{E}\},$$
$$I(y) = \{t \in \mathbf{R} \mid -\lambda(-y) < t < \lambda(y)\}.$$

There then exists a parametrized family of functions $\{\varphi_{(y,t)} \mid t \in I(y)\}$ such that

(i) $\varphi_{(y,t)}$ is a potential defined on a neighbourhood of a point $c_y(t)$.
(ii) If t and t' are sufficiently near, then $\varphi_{(y,t)}$ and $\varphi_{(y,t')}$ coincide on a small neighbourhood of $c_y(t)$.

Such a family $\{\varphi_{(y,t)} \mid t \in I(y)\}$ is called the **family of potentials along the geodesic** $c_y(t) = \exp^D ty$. We introduce

$$h_y(t) = \varphi_{(y,t)}(c_y(t)) - \varphi_{(y,0)}(c_y(0)) - (y\varphi_{(y,0)})t.$$

Then $h_y(t)$ is a function of t depending only on y not on the choice of potentials along c_y. In fact, choosing another family of potentials $\{\tilde{\varphi}_{(y,t)} \mid t \in I(y)\}$ along a geodesic c_y we put

$$\tilde{h}_y(t) = \tilde{\varphi}_{(y,t)}(c_y(t)) - \tilde{\varphi}_{(y,0)}(c_y(0)) - (y\tilde{\varphi}_{(y,0)})t.$$

Then $\tilde{h}_y(t) - h_y(t)$ is a polynomial in t of degree at most 1, and takes the value 0 in a neighbourhood of $t = 0$. Hence $\tilde{h}_y(t)$ coincides with $h_y(t)$. The function $h_y(t)$ restricted to $(0, \lambda(y))$ satisfies the following equalities,

$$\frac{d^2}{dt^2} h_y(t) = g(\dot{c}_y(t), \dot{c}(t)) > 0,$$
$$\frac{d}{dt} h_y(t) = \dot{c}_y(t)\varphi_{(y,t)} - \dot{c}(0)\varphi_{(y,0)} > 0,$$
$$h_y(t) > 0.$$

Let $s \in Aut(M, D, g)$. Since $sc_y = c_{s_*y}$, if $\{\varphi_{(y,t)} \mid t \in I(y)\}$ is a family of potentials along a geodesic c_y, then $\{\varphi_{(y,t)} \circ s^{-1} \mid t \in I(y)\}$ is a family of potentials along $sc_y = c_{s_*y}$. Hence

$$
\begin{aligned}
h_{s_*y}(t) &= (\varphi_{(y,t)} \circ s^{-1})(c_{s_*y}(t)) - (\varphi_{(y,0)} \circ s^{-1})(c_{s_*y}(0)) \\
&\quad -((s_*y)(\varphi_{(y,0)} \circ s^{-1}))t \\
&= \varphi_{(y,t)}(c_y(t)) - \varphi_{(y,0)}(c_y(0)) - (y\varphi_{(y,0)})t \\
&= h_y(t).
\end{aligned}
$$

Since $G\backslash M$ is quasi-compact, there exists a compact subset C of M such that $M = GC$. Then we can choose a positive number ϵ so that if $y \in TM$ is in $\pi^{-1}C$ and the length $|y|$ of y satisfies $|y| \leq \epsilon$, so $y \in \mathcal{E}$. Given also that $M = GC$, if $y \in TM$ and $|y| \leq \epsilon$ then $y \in \mathcal{E}$. We set $A = \{y \in TM \mid |y| = \epsilon\}$. Denoting by A_C the set of all elements in A such that the origins are contained in C, we have $A = GA_C$. A function defined by $\mathcal{E} \ni y \longrightarrow h_y(1) \in \mathbf{R}^+$ is G-invariant. Since A_C is compact, $h_y(1)$ attains a positive minimum r on A_C. Furthermore, given that $h_y(1)$ is G-invariant and $A = GA_C$, for all $y \in A$ we have

$$
h_y(1) \geq r > 0.
$$

Lemma 8.2. $\displaystyle\lim_{t \to \lambda(y)} h_y(t) = \infty.$

Proof. Let $t_0 \in (0, \lambda(y))$. Then

$$
c_{\mu \dot{c}_y(t_0)}(t) = c_y(\mu t + t_0),
$$

where μ is sufficiently near 0. Put $z = c_y(t_0)$. Then $\dfrac{\epsilon}{|z|}z \in A \subset \mathcal{E}$ and $c_z(t) = c_y(t_0 + t)$. Hence $t_0 + \dfrac{\epsilon}{|z|} \in I(y)$. Let $\{\varphi_{(y,t)}\}$ be a family of potentials along c_y, then $\{\varphi_{(z,t)} = \varphi_{(y,t_0+t)}\}$ is a family of potentials along

c_z. Hence

$$h_y\left(t_0 + \frac{\epsilon}{|z|}\right)$$

$$= \varphi_{(y,t_0+\frac{\epsilon}{|z|})}\left(c_y\left(t_0 + \frac{\epsilon}{|z|}\right)\right) - \varphi_{(y,0)}(c_y(0)) - (y\varphi_{(y,0)})\left(t_0 + \frac{\epsilon}{|z|}\right)$$

$$= \varphi_{(z,\frac{\epsilon}{|z|})}\left(c_z\left(\frac{\epsilon}{|z|}\right)\right) - \varphi_{(z,0)}(c_z(0)) - (z\varphi_{(z,0)})\frac{\epsilon}{|z|}$$

$$+ \varphi_{(y,t_0)}(c_y(t_0)) - \varphi_{(y,0)}(c_y(0)) - (y\varphi_{(y,0)})t_0$$

$$+ (\dot{c}_y(t_0)\varphi_{(y,t_0)} - \dot{c}_y(0)\varphi_{(y,0)})\frac{\epsilon}{|z|}$$

$$= h_z\left(\frac{\epsilon}{|z|}\right) + h_y(t_0) + \left(\frac{d}{dt}\bigg|_{t=t_0} h_y(t)\right)\frac{\epsilon}{|z|}$$

$$= h_y(t_0) + h_{(\frac{\epsilon}{|z|})z}(1) + \left(\frac{d}{dt}\bigg|_{t=t_0} h_y(t)\right)\frac{\epsilon}{|z|}.$$

Therefore

$$h_y\left(t_0 + \frac{\epsilon}{|z|}\right) > h_y(t_0) + r,$$

which implies

$$\lim_{t \to \lambda(y)} h_y(t) = \infty. \qquad \square$$

Since $\exp_o^D : \mathcal{E}_o \longrightarrow M$ is an affine mapping by Corollary 8.1, the induced metric

$$\tilde{g} = (\exp_o^D)^* g$$

is a Hessian metric on $\mathcal{E}_o$. By the same argument as in the proof of Proposition 2.1, we obtain the following lemma.

Lemma 8.3. *There exists a potential $\tilde{\varphi}$ of $\tilde{g}$ on $\mathcal{E}_o$.*

Lemma 8.4. *Let $y \in T_oM$. If $\lambda(y) < \infty$, then*

$$\lim_{t \to \lambda(y)} \tilde{\varphi}(ty) = \infty.$$

Proof. Let $\{\varphi_{(y,t)}\}$ be a family of potentials of g along the geodesic $c_y(t)$. Put $\tilde{\varphi}_{(y,t)} = \varphi_{(y,t)} \circ \exp_o^D$. Then $\{\tilde{\varphi}_{(y,t)}\}$ is a potential of $\tilde{g}$ along the geodesic ty $(0 \le t < \lambda(y))$, and $a(t) = \tilde{\varphi}(ty) - \tilde{\varphi}_{(y,t)}(ty)$ is a polynomial of t with degree 1. By Lemma 8.2 we have

$$\lim_{t \to \lambda(y)} \tilde{\varphi}(ty) = \lim_{t \to \lambda(y)} (\varphi_{(y,t)}(c_y(t)) + a(t))$$

$$= \lim_{t \to \lambda(y)} \{h_y(t) + \varphi_{(y,0)}(c_y(0)) + (y\varphi_{(y,0)})t + a(t)\}$$

$$= \infty. \qquad \square$$

Proof of Theorem 8.2. By Lemmata 8.1 and 8.4 the domain $\mathcal{E}_o$ is convex and by Corollary 8.1 $\exp_o^D : \mathcal{E}_o \longrightarrow M$ is the universal covering of M. □

Inspired by the work [Kaup (1968)], J. Vey introduced the notion of hyperbolicity for flat manifolds [Vey (1968)]. For flat manifolds M and N, we denote by $\mathit{Aff}(N, M)$ the space of all affine mappings from N to M with compact open topology.

Definition 8.4. A flat manifold M is said to be **hyperbolic** if, for any flat manifold N, the natural mapping given by

$$(f, x) \in \mathit{Aff}(N, M) \times N \longrightarrow (f(x), x) \in M \times N$$

is proper.

The following Theorem 8.3 is due to [Vey (1968)].

Theorem 8.3. *A flat manifold M is hyperbolic if and only if the universal covering of M is isomorphic to a regular convex domain.*

Theorem 8.4. *A hyperbolic flat manifold (M, D) admits a Hessian metric $g = D\omega$ of Koszul type (cf. Definition 2.2).*

Proof. By Theorem 8.3, the universal covering Ω of a hyperbolic flat manifold (M, D) is a regular convex domain in $\mathbf{R}^n$. Hence, by Proposition 4.2, there exists an affine coordinate system $\{y^1, \cdots, y^n\}$ such that $y^i > 0$ on Ω for all i, and the tube domain $T\Omega = \mathbf{R}^n + \sqrt{-1}\Omega$ over Ω is holomorphically isomorphic to a bounded domain in $\mathbf{C}^n$. Therefore $T\Omega$ admits the Bergmann volume element v^T, and the Bergmann metric g^T,

$$v^T = (\sqrt{-1})^{n^2} K dz^1 \wedge \cdots \wedge dz^n \wedge d\bar{z}^1 \wedge \cdots \wedge d\bar{z}^n,$$

$$g^T = \sum_{i,j} \frac{\partial^2 \log K}{\partial z^i \partial \bar{z}^j} dz^i d\bar{z}^j,$$

where $z^i = x^i + \sqrt{-1}y^i$ and $\{x^1, \cdots, x^n\}$ is an affine coordinate system on $\mathbf{R}^n$. Since v^T is invariant under parallel translations $z^i \longrightarrow z^i + a^i$ ($a^i \in \mathbf{R}$), the function K depends only on variables $\{y^1, \cdots, y^n\}$. Since M is a quotient space $\Gamma \backslash \Omega$ of Ω by a discrete subgroup Γ of the affine automorphism group of Ω, each $s \in \Gamma$ induces a holomorphic automorphism of $T\Omega$ which leaves v^T invariant. Hence a volume element on Ω defined by

$$v = \sqrt{K} dy^1 \wedge \cdots \wedge dy^n$$

is s invariant. The first Koszul form α and the second Koszul form β with respect to v are therefore given by

$$\alpha = \frac{1}{2} \sum_i \frac{\partial \log K}{\partial y^i} dy^i,$$

$$\beta = \frac{1}{2} \sum_{i,j} \frac{\partial^2 \log K}{\partial y^i \partial y^j} dy^i dy^j.$$

Thus β coincides with the restriction of $2g^T$ on $\sqrt{-1}\Omega$, and hence β is positive definite on Ω. Since α is Γ-invariant closed 1-form on Ω, M carries a closed 1-form ω such that the pull back by the projection π from Ω to $M = \Gamma \backslash \Omega$ coincides with α, that is, $\pi^* \omega = \alpha$. Since $\pi^*(D\omega) = \beta$, we know that $g = D\omega$ is positive definite, and so is a Hessian metric of Koszul type.

$\square$

Theorem 8.5. *Let (M, D) be a connected flat manifold with a Lie subgroup G of $\mathrm{Aut}(M, D)$ such that $G \backslash M$ is quasi-compact. Suppose that M admits a Hessian metric $g = D\omega$ of Koszul type where ω is G-invariant. Then (M, D) is hyperbolic.*

Corollary 8.3. *A compact connected flat manifold (M, D) is hyperbolic if and only if (M, D) admits a Hessian metric $g = D\omega$ of Koszul type.*

For the proof of Theorem 8.5, we require the following Lemma.

Lemma 8.5. *Under the same assumption of Theorem 8.5, if $\omega(y) \geq 0$ for $y \neq 0 \in TM$, then $\lambda(y) < \infty$.*

Proof. There exists a family of potentials $\{\varphi_{(y,t)} \mid t \in I(y)\}$ of $g = D\omega$ along a geodesic $c_y(t) = \exp^D ty$ such that

$$\omega = d\varphi_{(y,t)}.$$

Since $\dfrac{d}{dt} h_y(t) = \omega(\dot{c}_y(t)) - \omega(\dot{c}(0)) > 0$ for $t \in (0, \lambda(y))$, it follows that

$$\omega(\dot{c}_y(1)) > \omega(y), \qquad y \in \mathcal{E}.$$

We set $B = \{y \in TM \mid |y| = \epsilon, \omega(y) \geq 0\}$ and denote by B_C the set of all elements in B whose origins are contained in C. Then $B = GB_C \subset \mathcal{E}$. Let L and N be the maximum values of G-invariant functions $\omega(y)$ and $\omega(y)\omega(\dot{c}_y(1))^{-1}$ on a compact set B_C. Then

$$\omega(y) \leq L,$$

$$\omega(y)\omega(\dot{c}_y(1))^{-1} \leq N < 1, \qquad y \in B.$$

We define a sequence of numbers $t_0 = 0 < t_1 < \cdots < t_i < t_{i+1} < \cdots < \lambda(y)$ by induction as follows. Supposing that t_j is defined for all $j \le i$ we define t_{i+1} by the condition $(t_{i+1} - t_i)|\dot{c}_y(t_i)| = \epsilon$. Put $z_i = (t_{i+1} - t_i)\dot{c}_y(t_i)$. Then $z_i \in B \subset \mathcal{E}$ and $\exp^D z_i = c_{(t_{i+1}-t_i)\dot{c}_y(t_i)}(1) = c_y(t_{i+1}) = \exp^D t_{i+1}y$, hence $t_{i+1} < \lambda(y)$D Since $c_{z_i}(t) = c_y((t_{i+1} - t_i)t + t_i)$, we have $\dot{c}_{z_i}(t) = (t_{i+1} - t_i)\dot{c}_y((t_{i+1} - t_i)t + t_i)$ and so $\dot{c}_{z_i}(1) = (t_{i+1} - t_i)\dot{c}_y(t_{i+1})$. Using this and $\omega(z_i) \le N\omega(\dot{c}_{z_i}(1))$ we obtain

$$\omega(\dot{c}_y(t_i)) \le N\omega(\dot{c}_y(t_{i+1})).$$

Since $\omega(\dot{c}_y(t))$ is monotonically increasing on $(0, \lambda(y))$ we have

$$(t_{i+1} - t_i)\omega(\dot{c}_y(t_i)) \le \int_{t_i}^{t_{i+1}} \omega(\dot{c}_y(t))dt = \int_0^1 \omega(\dot{c}_{z_i}(t))dt.$$

By $h_y(t) = \int_0^t \left(\frac{d}{ds}h_y(s)\right)ds = \int_0^t \omega(\dot{c}_y(s))ds - \omega(y)t$ we obtain

$$\int_0^1 \omega(\dot{c}_{z_i}(t))dt = h_{z_i}(1) + \omega(z_i).$$

Putting $a_i = \omega(\dot{c}_y(t_i))$ and denoting by R maximum of a function $z \longrightarrow h_z(1)$ on B we obtain

$$(t_{i+1} - t_i)a_i < R + L,$$

$$a_i \le Na_{i+1}.$$

This implies

$$t_{i+1} - t_1 < \frac{1}{a_1}(R + L)(1 - N)^{-1}.$$

By $t_1|y| = \epsilon$ and by $t_1 a_1 = t_1\omega(\dot{c}_y(t_1)) \ge t_1 \int_0^1 \omega(\dot{c}_y(t_1 t))dt = \int_0^1 \omega(\dot{c}_{t_1 y}(t))dt = h_{t_1 y}(1) + \omega(t_1 y) \ge r$ we have

$$t_{i+1} \le \left\{1 + \frac{1}{r}(R + L)(1 - N)^{-1}\right\}\frac{\epsilon}{|y|}.$$

Alternatively, since

$$h_y(t_{i+1}) - h_y(t_i) = \int_{t_i}^{t_{i+1}} \omega(\dot{c}_y(t))dt - (t_{i+1} - t_i)\omega(y)$$

$$= \int_0^1 \omega(\dot{c}_{z_i}(t))dt - (t_{i+1} - t_i)\omega(y)$$

$$= h_{z_i}(1) + \omega(z_i) - (t_{i+1} - t_i)\omega(y)$$

$$= h_{z_i}(1) + (t_{i+1} - t_i)(\omega(\dot{c}_y(t_i)) - \omega(\dot{c}_y(0)))$$

$$\ge r$$

it follows that

$$\lim_{i \to \infty} h_y(t_i) = \infty.$$

Hence $\lim_{i \to \infty} t_i = \lambda(y)$ and

$$\lambda(y) \leq \left\{ 1 + \frac{1}{r}(R + L)(1 - N)^{-1} \right\} \frac{\epsilon}{|y|}.$$

$\square$

Proof of Theorem 8.5.   Suppose that there exists a Hessian metric $g = D\omega$ of Koszul type such that ω is G-invariant. By the proof of Theorem 8.2, we know that $\exp_o^D : \mathcal{E}_o \longrightarrow M$ is the universal covering of M, and $\mathcal{E}_o$ is a convex domain. It follows from Lemma 8.5 that M does not admit a geodesic $\varphi_y(t)$ such that $y \neq 0$ and $I(y) = (-\infty, \infty)$. Thus $\mathcal{E}_0$ is a regular convex domain, and so the flat manifold (M, D) is hyperbolic. $\square$

8.3   Koszul forms on Hessian manifolds

In this section we shall describe an important role of the Koszul forms on Hessian manifolds. Let β be the the second Koszul form of a Hessian manifold (M, D, g). Then the first affine Chern class $\hat{c}_1(M)$ of (M, D, g) is represented by -2β (cf. Proposition 7.11). Cheng-Yau and Delanoë proved a certain problem related to the first affine Chern class $\hat{c}_1(M)$ and the second Koszul form β, which is similar to the Calabi conjecture for Kählerian manifolds [Delanoe (1989)][Cheng and Yau (1982)].

We note here that a theorem due to Cheng-Yau and Pogorelov may be restated in terms of the first Koszul form α as follows.

Theorem 8.6. *Let* $(\mathbf{R}^n, D, g = Dd\varphi)$ *be a Hessian domain. If the first Koszul form* α *vanishes identically, then* φ *is a polynomial of degree 2 and the Levi-Civita connection* ∇ *of* g *coincides with* D.

This theorem was first proved by [Jörgens (1954)] in the case $n = 2$ using complex anlytic techniques. In the cases $n = 3, 4, 5$ it was proved in [Calabi (1958)] using affine differential techniques. The general case was proved independently by [Cheng and Yau (1986)] and [Pogorelov (1978)].

A theorem of [Calabi (1958)] is restated as follows

Theorem 8.7. *Let* $(\Omega, D, g = Dd\varphi)$ *be a Hessian domain. Suppose that* g *is complete and the first Koszul form* α *vanishes identically. Then the Levi-Civita connection* ∇ *of* g *coincides with* D.

The following Theorem 8.8 and Corollary 8.4 suggest that the second Koszul form β plays an important role for the theory of compact Hessian manifolds.

Theorem 8.8. *Let (M, D, g) be a compact oriented Hessian manifold and let α and β be the first and the second Koszul forms respectively. Then we obtain*

(1) $\displaystyle\int_M \beta^i{}_i\, v = \int_M \alpha_i \alpha^i\, v \geq 0,$
    where v is the volume element determined by g.

(2) *If $\displaystyle\int_M \beta^i{}_i\, v = 0$, then the Levi-Civita connection ∇ of g coincides with D.*

Corollary 8.4. *Let (M, D, g) be a compact oriented Hessian manifold. Then we have*

(1) *β cannot be negative definite.*
(2) *If β is negative semi-definite, then the Levi-Civita connection ∇ of g coincides with D.*

Proof of Theorem 8.8. By Proposition 3.4, α_i is given by

$$\alpha_i = \gamma^r{}_{ri}. \tag{8.1}$$

Let $\alpha_{i;j}$ be the i-th component of $\nabla_{\partial/\partial x^j}\alpha$. Then

$$\alpha_{i;j} = \beta_{ij} - \alpha_r \gamma^r{}_{ij}, \qquad \alpha^i{}_{;i} = \beta^i{}_i - \alpha_r \alpha^r.$$

Applying Green's theorem it follows that

$$\int_M (\beta^i{}_i - \alpha_r \alpha^r)\, v = \int_M \alpha^i{}_{;i}\, v = 0,$$

which implies (1).

Suppose $\displaystyle\int_M \beta^i{}_i v = 0$, then by (1) we have

$$\alpha = 0. \tag{8.2}$$

Let R_{jk} be the component of the Ricci tensor of g. By Proposition 2.3 and (8.2) we obtain

$$R_{jk} = R^s{}_{jsk} = \gamma^r{}_{js}\gamma^s{}_{kr} - \alpha_r \gamma^r{}_{jk} = \gamma^r{}_{js}\gamma^s{}_{kr}. \tag{8.3}$$

Hence the scalar curvature R is given by

$$R = R^k{}_k = \gamma_{rst}\gamma^{rst}. \tag{8.4}$$

Let us compute the Laplacian ΔR of the scalar curvature R. We denote by $\gamma_{ijk;l}$ the component of $\nabla_{\partial/\partial x^l} \gamma$. Then by Proposition 2.2 we have

$$\gamma_{ijk;l} = \frac{\partial \gamma_{ijk}}{\partial x^l} - \gamma_{rjk}\gamma^r_{il} - \gamma_{irk}\gamma^r_{jl} - \gamma_{ijr}\gamma^r_{kl}$$

$$= \frac{1}{2}\frac{\partial^2 g_{ij}}{\partial x^k \partial x^l} - g^{rs}(\gamma_{rjk}\gamma_{sil} + \gamma_{irk}\gamma_{sjl} + \gamma_{ijr}\gamma_{skl}).$$

Hence $\gamma_{ijk;l}$ is symmetric with respect to i, j, k, l,

$$\gamma_{ijk;l} = \gamma_{ljk;i}. \tag{8.5}$$

By Proposition 2.3, relations (8.1), (8.2) and (8.5), together with the Ricci formula we have

$$g^{rs}\gamma_{ijk;r;s} = g^{rs}\gamma_{rjk;i;s} = g^{rs}(\gamma_{rjk;i;s} - \gamma_{rjk;s;i}) + g^{rs}\gamma_{rjk;s;i} \tag{8.6}$$

$$= -g^{rs}(\gamma_{pjk}R^p_{ris} + \gamma_{rpk}R^p_{jis} + \gamma_{rjp}R^p_{kis}) + g^{rs}\gamma_{rsk;j;i}$$

$$= \gamma_{pjk}R^p_{\ i} - \gamma^s_{pk}(\gamma^p_{qi}\gamma^q_{js} - \gamma^p_{qs}\gamma^q_{ji})$$

$$- \gamma^s_{jp}(\gamma^p_{qi}\gamma^q_{ks} - \gamma^p_{qs}\gamma^q_{ki}) + \alpha_{k;j;i}$$

$$= \gamma^{pqs}(\gamma_{qsi}\gamma_{pjk} + \gamma_{spj}\gamma_{qki} + \gamma_{spk}\gamma_{qji})$$

$$- \gamma^p_{qi}\gamma^q_{sj}\gamma^s_{pk} - \gamma^p_{qi}\gamma^q_{sk}\gamma^s_{pj}.$$

It follows from Proposition 2.3 and relations (8.3), (8.4) and (8.6) that

$$\frac{1}{2}\Delta R = \frac{1}{2}g^{ij}R_{;i;j} \tag{8.7}$$

$$= \gamma^{ijk}g^{rs}\gamma_{ijk;r;s} + \gamma^{ijk;l}\gamma_{ijk;l}$$

$$= 3\gamma^{ijk}\gamma^{pqr}\gamma_{qri}\gamma_{pjk} - 2\gamma^{ijk}\gamma^p_{qi}\gamma^q_{rj}\gamma^r_{pk} + \gamma^{ijk;l}\gamma_{ijk;l}$$

$$= R_{ij}R^{ij} + R_{ijkl}R^{ijkl} + \gamma_{ijk;l}\gamma^{ijk;l}$$

$$\geq 0.$$

Thus by the E. Hopf's lemma [Kobayashi and Nomizu (1963, 1969)] R is a constant and

$$\Delta R = 0.$$

Therefore by the relations (8.4) and (8.7) we have

$$R_{ij} = 0, \qquad \gamma_{ijk} = 0.$$

This implies $\nabla = D$. $\qquad\qquad\qquad\qquad\qquad\qquad\qquad\qquad\quad$ $\square$

In the section 7.6 we defined a cohomology group $\hat{H}^k(M)$ of a flat manifold (M, D) by

$$\hat{H}^k(M) = \{\omega \in \mathcal{A}^{k,k} \mid \partial\omega = 0, \bar{\partial}\omega = 0\}/\partial\bar{\partial}\mathcal{A}^{k-1,k-1}.$$

For a volume element ω on M, as in Definition 3.2, we define a closed 1-form α_ω, and a symmetric bilinear form β_ω by

$$D_X \omega = \alpha_\omega(X)\omega,$$

$$\beta_\omega = D\alpha_\omega,$$

and call these the first and the second Koszul forms with respect to the volume element ω respectively.

Consider β_ω as an element in $\mathcal{A}^{1,1}$. Then

$$\partial \beta_\omega = 0, \quad \bar{\partial} \beta_\omega = 0.$$

We denote by $[\beta_\omega]$ the element in $\hat{H}^1(M)$ represented by β_ω. Let ω' be another volume element, then there exists a function f on M such that

$$\omega' = f\omega.$$

Since

$$\alpha_{\omega'} = \alpha_\omega + d\log|f|$$

it follows that

$$\beta_{\omega'} = \beta_\omega + \partial\bar{\partial}\log|f|.$$

Therefore $[\beta_{\omega'}] = [\beta_\omega]$, that is, $[\beta_\omega]$ is independent of the choice of volume element ω.

For compact Kählerian manifolds the following problem was prorosed by [Calabi (1954)][Calabi (1955)].

Calabi Conjecture Let M be a compact Kählerian manifold with Kählerian metric g and let $c_1(M)$ be the first Chern class of M. For an arbitrary closed $(1,1)$ form ρ representing $c_1(M)$, does there exist a unique Kählerian metric $\tilde{g}$ such that the Ricci form of $\tilde{g}$ coincides with ρ, and the Kählerian forms of $\tilde{g}$ and g are cohomologous in the Dolbeault cohomology classes ?

The complete solution of Calabi Conjecture was presented by [Yau (1977)]

We have seen in Proposition 7.11 that the first affine Chern class $\hat{c}_1(M)$ of a Hessian manifold (M, D, g) is represented by $[-2\beta]$. Cheng-Yau and Delanoë proved independently the following result analogous to the Calabi conjecture.

Theorem 8.9. *Let (M, D, g) be a compact oriented Hessian manifold. For an arbitrary representative $-2\beta_{\tilde{v}}$ of the first affine Chern class $\hat{c}_1(M)$ there exists a Hessian metric $\tilde{g}$ such that the second Koszul form of $\tilde{g}$ coincides with $\beta_{\tilde{v}}$, and $[\tilde{g}] = [g]$ as elements in $\hat{H}^1(M)$.*

Proof.   For the proof the interested reader may refer to [Cheng and Yau (1982)][Delanoe (1989)].   □

By Corollary 8.4 and Theorem 8.9 we have the following corollaries.

Corollary 8.5. *Let M be a compact oriented Hessian manifold. The second Koszul form β_ω for any volume element ω cannot be negative definite.*

Corollary 8.6. *Let (M, D, g) be a compact oriented Hessian manifold. Suppose that M admits a D-parallel volume element. Then there exists a Hessian metric $\tilde{g}$ on (M, D) such that the Levi-Civita connection of $\tilde{g}$ coincides with D.*

The following theorem was also proved in [Cheng and Yau (1982)].

Theorem 8.10. *Let (M, D, g) be a compact Hessian manifold. Suppose that M admits a volume element ω such that β_ω is positive definite. Then there exists an Einstein-Hessian metric $\tilde{g}$ whose second Koszul form $\beta_{\tilde{g}}$ coincides with $\tilde{g}$.*

Chapter 9

Symmetric spaces with invariant Hessian structures

In this chapter we study symmetric homogeneous spaces with invariant Hessian structures. Following Koszul's approach, in section **9.1** we relate invariant flat affine connections to affine representations of Lie algebras [Koszul (1961)]. In the section **9.2** we characterize invariant Hessian metrics by affine representations of Lie algebras. Applying these results, we show that a homogeneous space of a semisimple Lie group does not admit any invariant Hessian structure. In section **9.3** we give a correspondence between symmetric homogeneous spaces with invariant Hessian structures and certain commutative algebras by using affine representations of Lie algebras. Investigating the structure of the commutative algebra, we prove that a simply connected symmetric homogeneous space with invariant Hessian structure is a direct product of a Euclidean space and a homogeneous self-dual regular convex cone.

9.1 Invariant flat connections and affine representations

Let G be a connected Lie group and let G/K be a homogeneous space on which G acts effectively. In this section we give a bijective correspondence between the set of G-invariant flat connections on G/K and the set of a certain class of affine representations of the Lie algebra of G.

Definition 9.1. A homogeneous space G/K endowed with a G-invariant flat connection D is called a **homogeneous flat manifold** and is denoted by $(G/K, D)$.

Theorem 9.1. *Let G/K be a homogeneous space of a connected Lie group G and let $\mathfrak{g}$ and $\mathfrak{k}$ be the Lie algebras of G and K respectively. Suppose that*

165

*G/K is endowed with a G-invariant flat connection. Then $\mathfrak{g}$ admits an affine representation (f, q) on V (cf. section **1.1**) satisfying the following conditions*

(1) $\dim V = \dim G/K$,
(2) *The mapping $q : \mathfrak{g} \longrightarrow V$ is surjective and the kernel coincides with the Lie algebra $\mathfrak{k}$ of K.*

Conversely, suppose that G is simply connected and that $\mathfrak{g}$ is endowed with an affine representation satisfying the above conditions. Then G/K admits a G-invariant flat connection.

Proof. Suppose that G/K admits a G-invariant flat connection D. For $X \in \mathfrak{g}$ we denote by X^* the vector field on G/K induced by $\exp(-tX)$. Then

$$[X, Y]^* = [X^*, Y^*].$$

Denoting by $\mathcal{L}_{X^*}$ the Lie differentiation with respect to X^*, we set

$$A_{X^*} = \mathcal{L}_{X^*} - D_{X^*}.$$

Then by [Kobayashi and Nomizu (1963, 1969)](I, p235) we have

$$A_{X^*} Y^* = -D_{Y^*} X^*,$$
$$A_{X^*} Y^* - A_{Y^*} X^* = [X^*, Y^*],$$
$$A_{[X^*, Y^*]} = [A_{X^*}, A_{Y^*}].$$

Let V be the tangent space at $o = \{K\}$. We denote by $f(X)$ and $q(X)$ the values of A_{X^*} and $-X^*$ at o respectively, and define mappings f and q by $f : X \longrightarrow f(X)$ and $q : X \longrightarrow q(X)$ respectively. Then by the above equations the pair (f, q) is an affine representation of $\mathfrak{g}$ on V satisfying conditions (1) and (2).

Conversely, suppose that $\mathfrak{g}$ admits an affine representation (f, q) satisfying conditions (1) and (2). Let $\{e_1, \cdots, e_n\}$ be a basis of V and let $\{x^1, \cdots, x^n\}$ be the affine coordinate system on V corresponding to the basis. For $X \in \mathfrak{g}$ we define a vector field X_a on V by

$$X_a = -\sum_i \Big(\sum_j f(X)^i_j x^j + q(X)^i \Big) \frac{\partial}{\partial x^i},$$

where $f(X)^i_j$ and $q(X)^i$ are given by

$$f(X)e_j = \sum_i f(X)^i_j e_i, \quad q(X) = \sum_i q(X)^i e_i.$$

The 1-parameter transformation group generated by X_a is an affine transformation group of V, with linear parts given by

$$\exp(-tf(X)),$$

and translation vector parts given by

$$\sum_{n=1}^{\infty} \frac{(-t)^n}{n!} f(X)^{n-1} q(X).$$

Then

$$[X, Y]_a = [X_a, Y_a],$$

and the set $\mathfrak{g}_a$ given by

$$\mathfrak{g}_a = \{X_a \mid X \in \mathfrak{g}\}$$

is a Lie algebra. Since G acts effectively on G/K, the mapping $X \longrightarrow X_a$ is a Lie algebra isomorphism from $\mathfrak{g}$ to $\mathfrak{g}_a$. Let G_a be the Lie group generated by $\mathfrak{g}_a$. An element s in G_a is an affine transformation of V. We denote by $f(s)$ and $q(s)$ the linear part and the translation vector part of s respectively. Let $\Omega_a = G_a 0 = G_a/K_a$ be the orbit of G_a through the origin 0. Then Ω_a is an open orbit because $q(\mathfrak{g}) = V$. The flat connection D given by the restriction to Ω_a of the standard flat connection of V is G_a-invariant. Since G is simply connected, there exists a covering homomorphism

$$\rho : G \longrightarrow G_a$$

such that $d\rho(X) = X_a$. Since K is the identity component of $\rho^{-1}(K_a)$, ρ induces the universal covering mapping

$$p : G/K \longrightarrow G/\rho^{-1}(K_a) \cong G_a/K_a = \Omega_a.$$

The pull back of the G_a-invariant flat connection D on G_a/K_a by p is a G-invariant flat connection on G/K. $\qquad\square$

Example 9.1. Let $G = GL(n, \mathbf{R})$ and let K be a subgroup of G consisting of all elements such that

$$k = \begin{bmatrix} I_r & k_1 \\ 0 & k_2 \end{bmatrix} \in GL(n, \mathbf{R}), \quad k_1 \in M(r, n-r), \quad k_2 \in M(n-r, n-r),$$

where I_r is the unit matrix of degree r and $M(p, q)$ is the set of all $p \times q$ matrices. Then K is a closed subgroup of G. The homogeneous space G/K is called the **Stiefel manifold**. The Lie algebra of G is given by $\mathfrak{g} = \mathfrak{gl}(n, \mathbf{R}) = M(n, n)\mathrm{D}$ For $X \in \mathfrak{g}$ we set

$$X = \begin{bmatrix} X_{11} & X_{12} \\ X_{21} & X_{22} \end{bmatrix},$$

where $X_{11} \in \mathfrak{gl}(r, \mathbf{R})$, $X_{12} \in M(r, n-r)$, $X_{21} \in M(n-r, r)$, $X_{22} \in \mathfrak{gl}(n-r, \mathbf{R})$. The Lie algebra of K is given by $\mathfrak{k} = \{X \in \mathfrak{g} \mid X_{11} = 0, X_{21} = 0\}$. Put $V = M(n, r)$. For $X \in \mathfrak{g}$ we define an endomorphism $f(X) :$ $V \longrightarrow V$ and an element $q(X) \in V$ by $f(X)v = Xv$ and $q(X) = \begin{bmatrix} X_{11} \\ X_{21} \end{bmatrix}$. Then (f, q) is an affine representation of $\mathfrak{g}$ on V. It is easy to see that $\dim V = \dim G/K = nr$ and the mapping q is surjective and the kernel coincides with $\mathfrak{k}$. Hence by Theorem 9.1 the Stiefel manifold G/K admits a G-invariant flat connection.

Corollary 9.1. *Let G be a connected Lie group with a left-invariant flat connection. Then the Lie algebra $\mathfrak{g}$ of G adimits an operation of multiplication $X \cdot Y$ such that*

(1) $X \cdot Y - Y \cdot X = [X, Y]$,
(2) $[X \cdot Y \cdot Z] = [Y \cdot X \cdot Z]$, *where* $[X \cdot Y \cdot Z] = X \cdot (Y \cdot Z) - (X \cdot Y) \cdot Z$.

Conversely, suppose that the Lie algebra $\mathfrak{g}$ of a simply connected Lie group G is endowed with an operation of multiplication $X \cdot Y$ satisfying the above conditions (1) and (2). Then G admits a left-invariant flat connection.

Proof. Suppose that G is endowed with a left-invariant flat connection. Then there exists an affine representation (f, q) of $\mathfrak{g}$ on V satisfying the conditions of Theorem 9.1. In this case the mapping $q : \mathfrak{g} \longrightarrow V$ is a linear isomorphism. We define an operation of multiplication on $\mathfrak{g}$ by

$$X \cdot Y = q^{-1}(f(X)q(Y)).$$

Then it is easy to see that the multiplication satisfies the above conditions (1) and (2). Conversely, suppose that $\mathfrak{g}$ admits an operation of multiplication satisfying the above conditions (1) and (2). We put

$$f(X)Y = X \cdot Y, \quad q(X) = X.$$

Then (f, q) is an affine representation of $\mathfrak{g}$ on $\mathfrak{g}$ satisfying the conditions of Theorem 9.1. Hence by the theorem the simply connected Lie group G admits a left-invariant flat connection. $\qquad\square$

Definition 9.2. Let V be an algebra over $\mathbf{R}$ with multiplication $x \cdot y$. We put

$$[x \cdot y \cdot z] = x \cdot (y \cdot z) - (x \cdot y) \cdot z.$$

The algebra V is said to be **left symmetric** [Vinberg (1963)] if the following condition is satisfied;

$$[x \cdot y \cdot z] = [y \cdot x \cdot z].$$

Lemma 9.1. *Let V be an algebra over $\mathbf{R}$ with multiplication $x \cdot y$. For $x \in V$ we denote by L_x and R_x the left multiplication and the right multiplication by x respectively. Then the following conditions (1)-(3) are equivalent.*

(1) *V is a left symmetric algebra.*
(2) *$[L_x, L_y] = L_{x \cdot y - y \cdot x} D$*
(3) *$[L_x, R_y] = R_{x \cdot y} - R_y R_x D$*

If V is a left symmetric algebra with multiplication $x \cdot y$, then V is a Lie algebra with respect to the operation of multiplication $[x, y] = x \cdot y - y \cdot x$.

Proof. The proof is straightforward and will be omitted. □

Corollary 9.2. *Let G be a connected Lie group with a left-invariant flat connection D. Using the same notation as in Corollary 9.1, the following conditions (1) and (2) are equivalent.*

(1) *The left-invariant flat connection D is right-invariant.*
(2) *The multiplication $X \cdot Y$ is associative$G$ $[X \cdot Y \cdot Z] = 0 D$*

Proof. Identifying the tangent space of G at the unit element e with the Lie algebra of G we have

$$X \cdot Y = (A_{X*}(-Y^*))_e = (D_{Y*}X^*)_e.$$

For $s \in G$, we denote by l_s and r_s the left translation and the right translation by s respectively. Then $l_{s*}X^* = (Ad(s)X)^*$ and $r_{s*}X^* = X^*$. Since D is l_s-invariant we have

$$l_{s*}(D_{Y*}X^*) = D_{(Ad(s)Y)*}(Ad(s)X)^*,$$

and D is right-invariant if and only if $r_{s*}(D_{Y*}X^*) = D_{r_{s*}Y*}r_{s*}X^* = D_{Y*}X^*$. This condition is equivalent to $l_{s*}r_{s^{-1}*}(D_{Y*}X^*) = l_{s*}(D_{Y*}X^*) = D_{(Ad(s)Y)*}(Ad(s)X)^*$, and hence to

$$Ad(s)(X \cdot Y) = (Ad(s)X) \cdot (Ad(s)Y).$$

The above equation holds if and only if

$$[Z, X \cdot Y] = [Z, X] \cdot Y + X \cdot [Z, Y],$$

which is equivalent to

$$[X \cdot Y \cdot Z] + [Z \cdot X \cdot Y] - [X \cdot Z \cdot Y] = 0.$$

Since the algebra $\mathfrak{g}$ with multiplication $X \cdot Y$ is left symmetric, this condition is equivalent to

$$[X \cdot Y \cdot Z] = 0. \qquad □$$

9.2   Invariant Hessian structures and affine representations

In this section we express Hessian metrics and Koszul forms in terms of the affine representations given in Theorem 9.1. Applying the result, we show that a homogeneous space of a semisimple Lie group does not admit any invariant Hessian structure.

Definition 9.3. A Hessian structure (D, g) on a homogeneous space G/K is said to be an **invariant Hessian structure** if both D and g are G-invariant. A homogeneous space G/K with an invariant Hessian structure (D, g) is called a **homogeneous Hessian manifold** and is denoted by $(G/K, D, g)$.

Let (D, g) be an invariant Hessian structure on G/K. Using the same notation as in the previous section **9.1**, we have

$$(A_{X^*}g)(Y^*, Z^*) = -g(A_{X^*}Y^*, Z^*) - g(Y^*, A_{X^*}Z^*).$$

On the other hand, X^* being a Killing vector field, i.e. $\mathcal{L}_{X^*}g = 0$, we obtain

$$(A_{X^*}g)(Y^*, Z^*) = -(D_{X^*}g)(Y^*, Z^*).$$

These facts together with the Codazzi equation $(D_{X^*}g)(Y^*, Z^*) = (D_{Y^*}g)(X^*, Z^*)$ (cf. Proposition 2.1) yield

$$g(A_{X^*}Y^*, Z^*) + g(Y^*, A_{X^*}Z^*) = g(A_{Y^*}X^*, Z^*) + g(X^*, A_{Y^*}Z^*).$$

This relation implies the following lemma.

Lemma 9.2. *We denote by* $\langle \, , \, \rangle$ *the restriction of the Hessian metric* g *to the origin* o. *We then have*

$$\langle f(X)q(Y), q(Z)\rangle + \langle q(Y), f(X)q(Z)\rangle$$
$$= \langle f(Y)q(X), q(Z)\rangle + \langle q(X), f(Y)q(Z)\rangle. \tag{9.1}$$

Lemma 9.3. *Suppose that the Hessian metric* g *is Koszul type* $g = D\omega$ *where* ω *is* G-*invariant. We then have*

$$\langle q(X), q(Y)\rangle = -\omega_o(f(X)q(Y)),$$

where ω_o *is the restriction of* ω *to* o.

Proof.   Since ω is G-invariant we have

$$0 = (\mathcal{L}_{X^*}\omega)(Y^*) = X^*\omega(Y^*) - \omega([X^*, Y^*]),$$
$$0 = d\omega(X^*, Y^*) = X^*\omega(Y^*) - Y^*\omega(X^*) - \omega([X^*, Y^*]),$$

which imply $Y^*\omega(X^*) = 0$, and so $\omega(X^*)$ is a constant. Hence we obtain

$$g(X^*, Y^*) = (D_{Y^*}\omega)(X^*) = Y^*\omega(X^*) - \omega(D_{Y^*}X^*) = \omega(A_{X^*}Y^*).$$

Taking the value of the above equation at o we obtain the required relation. □

Lemma 9.4. *Let α and β be the first and the second Koszul forms respectively. Then we have*

(1) $\alpha(X^*) = \operatorname{Tr} A_{X^*}$,
(2) $\beta(X^*, Y^*) = \alpha(A_{X^*}Y^*)$.

Proof. Put $X^* = \sum_i \xi^i \dfrac{\partial}{\partial x^i}$ where $\{x^1, \cdots, x^n\}$ is an affine coordinate

system. Since $A_{X^*}\dfrac{\partial}{\partial x^i} = -D_{\partial/\partial x^i}X^* = -\sum_j \dfrac{\partial \xi^j}{\partial x^i}\dfrac{\partial}{\partial x^j}$, we have $A_{X^*}dx^i = $

$\sum_j \dfrac{\partial \xi^i}{\partial x^j}dx^j$. Thus for the volume element $v = F dx^1 \wedge \cdots \wedge dx^n$ of g we obtain

$$A_{X^*}v = F\sum_i dx^1 \wedge \cdots \wedge A_{X^*}dx^i \wedge \cdots \wedge dx^n$$

$$= F\left(\sum_i \dfrac{\partial \xi^i}{\partial x^i}\right) dx^1 \wedge \cdots \wedge dx^n$$

$$= -(\operatorname{Tr} A_{X^*})v.$$

Alternatively, since the volume element v is G-invariant, we have

$$A_{X^*}v = (\mathcal{L}_{X^*} - D_{X^*})v = -D_{X^*}v = -\alpha(X^*)v,$$

and so we obtain

$$\alpha(X^*) = \operatorname{Tr} A_{X^*}.$$

Using this relationship, it follows that $\alpha(X^*) = \sum_i \dfrac{\partial \xi^i}{\partial x^i}$. Since each ξ^i is a polynomial of degree 1 because of the G-invariance of D, $\alpha(X^*)$ is a constant. Therefore

$$(\mathcal{L}_{X^*}\alpha)(Y^*) = X^*(\alpha(Y^*)) - \alpha([X^*, Y^*]) = -\alpha([X, Y]^*)$$
$$= -\operatorname{Tr} A_{[X^*, Y^*]} = -\operatorname{Tr}[A_{X^*}, A_{Y^*}] = 0,$$

that is,

$$\mathcal{L}_{X^*}\alpha = 0.$$

Upon applying this relation we obtain

$$\beta(X^*, Y^*) = (D_{X^*}\alpha)(Y^*) = ((\mathcal{L}_{X^*} - A_{X^*})\alpha)(Y^*)$$
$$= -(A_{X^*}\alpha)(Y^*) = \alpha(A_{X^*}Y^*).$$

$\square$

By Lemma 9.4 we also have the following lemma.

Lemma 9.5. *Let α_o and β_o be the restrictions of the Koszul forms α and β to o respectively. Then*

(1) $\alpha_o(q(X)) = -\operatorname{Tr} f(X)$.
(2) $\beta_o(q(X), q(Y)) = -\alpha_o(f(X)q(Y))$.

Let V^* be the dual space of V and let f^* be the contragredient representation of f;

$$(f^*(X)w^*)(w) = -w^*(f(X)w), \quad \text{for } X \in \mathfrak{g}, \ w \in V \text{ and } w^* \in V^*.$$

We denote by d_{f^*} the coboundary operator for the cohomology of the Lie algebra $\mathfrak{g}$ with coefficients in (V^*, f^*). Define a linear mapping θ from $\mathfrak{g}$ to V^* by

$$(\theta(X))(v) = \langle q(X), v \rangle, \quad \text{for } X \in \mathfrak{g} \text{ and } v \in V.$$

Considering θ as a 1-dimensional (V^*, f^*)-cochain we have

$$((d_{f^*}\theta)(X, Y))(q(Z))$$
$$= \{f^*(X)\theta(Y) - f^*(Y)\theta(X) - \theta([X, Y])\}q(Z)$$
$$= -\langle q(Y), f(X)q(Z)\rangle + \langle q(X), f(Y)q(Z)\rangle - \langle q([X, Y]), q(Z)\rangle$$
$$= -\langle q(Y), f(X)q(Z)\rangle + \langle q(X), f(Y)q(Z)\rangle$$
$$\quad - \langle f(X)q(Y), q(Z)\rangle + \langle f(Y)q(X), q(Z)\rangle.$$

Hence, by Lemma 9.2, condition (9.1) is equivalent to

$$d_{f^*}\theta = 0. \tag{9.2}$$

Theorem 9.2. *Let G/K be a homogeneous space of a semisimple Lie group G. Then G/K does not admit any G-invariant Hessian structure.*

Proof.   Suppose that G/K admits a G-invariant Hessian structure (D, g). We denote by d_f the coboundary operator for the cohomology of the Lie algebra $\mathfrak{g}$ with coefficients in (V, f). Regarding q as a 1-dimensional (V, f)-cochain we have

$$(d_f q)(X, Y) = f(X)q(Y) - f(Y)q(X) - q([X, Y]) = 0,$$

that is, q is a (V, f)-cocycle. Since $H^1(\mathfrak{g}, (V, f)) = \{0\}$ because of the semisimplicity of $\mathfrak{g}$, there exists $e \in V$ such that

$$q = d_f e.$$

The mapping q being surjective we can choose an element $E \in \mathfrak{g}$ such that $q(E) = e$. Then

$$q(X) = f(X)q(E), \quad \text{for } X \in \mathfrak{g}.$$

Since $H^1(\mathfrak{g}, (V^*, f^*)) = \{0\}$ and $d_{f^*}\theta = 0$ by (9.2), there exists $c^* \in V^*$ such that

$$\theta = d_{f^*}c^*.$$

Thus we obtain

$$\langle q(X), q(Y) \rangle = (\theta(X))(q(Y)) = ((d_{f^*}c^*)(X))(q(Y))$$
$$= -c^*(f(X)q(Y)),$$

and in particular

$$\langle q(E), q(X) \rangle = -c^*(f(X)q(E)) = -c^*(q(X)).$$

These relationships imply

$$\langle f(E)q(X), q(Y) \rangle + \langle q(X), f(E)q(Y) \rangle$$
$$= \langle f(X)q(E), q(Y) \rangle + \langle q(E), f(X)q(Y) \rangle$$
$$= \langle q(X), q(Y) \rangle - c^*(f(X)q(Y))$$
$$= 2\langle q(X), q(Y) \rangle.$$

Therefore

$$f(E) + {}^t f(E) = 2I,$$

where I is the identity mapping on V and ${}^t f(E)$ is the adjoint of $f(E)$ with respect to the inner product $\langle \, , \, \rangle$. Considering the trace of the both sides of the above equation we obtain

$$\operatorname{Tr} f(E) = \dim V.$$

On the other hand, since $\mathfrak{g} = [\mathfrak{g}, \mathfrak{g}]$ because of the semisimplicity of $\mathfrak{g}$, we have

$$\operatorname{Tr} f(E) = 0.$$

This is a contradiction. Thus G/K does not admit any G-invariant Hessian structure. $\qquad\square$

9.3　Symmetric spaces with invariant Hessian structures

In this section we prove that a symmetric homogeneous space with an invariant Hessian structure essentially consists of a Euclidean space and a homogeneous self-dual regular convex cone.

Definition 9.4. A homogeneous space G/K of a connected Lie group G is said to be **symmetric** if G admits an involutive automorphism σ satisfying the following conditions.

(1) $\sigma^2(s) = s$, for $s \in G$,
(2) We denote by K_σ the closed subgroup of G consisting of all fixed points of σ, and by K_σ^0 the connected component of K_σ containing the unit element. Then

$$K_\sigma^0 \subset K \subset K_\sigma.$$

Let G/K be a symmetric homogeneous space with an involutive automorphism σ. We denote by $\mathfrak{g}$ the Lie algebra of G and set

$$\mathfrak{k} = \{X \in \mathfrak{g} \mid \sigma_*(X) = X\}, \quad \mathfrak{m} = \{X \in \mathfrak{g} \mid \sigma_*(X) = -X\},$$

where σ_* is the differential of σ. Then we have

$$\mathfrak{g} = \mathfrak{k} + \mathfrak{m}, \tag{9.3}$$

$$[\mathfrak{k}, \mathfrak{k}] \subset \mathfrak{k}, \quad [\mathfrak{k}, \mathfrak{m}] \subset \mathfrak{m}, \quad [\mathfrak{m}, \mathfrak{m}] \subset \mathfrak{k},$$

and $\mathfrak{k}$ is the Lie algebra of K. The above decomposition of $\mathfrak{g}$ is called the **canonical decomposition** for the symmetric homogeneous space G/K.

In the remainder of this section we prove the following theorem and corollaries.

Theorem 9.3. *Let G/K be a symmetric homogeneous space with an invariant Hessian structure. Then we have the following decomposition*

$$G/K = G_0/K_0 \times G_1/K_1 \times \cdots \times G_r/K_r,$$

where the universal covering space of G_0/K_0 is a Euclidean space and the universal covering space of G_i/K_i ($1 \leq i \leq r$) is an irreducible homogeneous self-dual regular convex cone.

Corollary 9.3. *Let G/K be a homogeneous space of a reductive Lie group G. Suppose that G/K admits an invariant Hessian structure. Then the universal covering space of G/K is a direct product of a Euclidean space and a homogeneous self-dual regular convex cone.*

Corollary 9.4. *A compact homogeneous space with an invariant Hessian structure is a Euclidean torus.*

Corollary 9.5. *A homogeneous self-dual regular convex cone is characterized as a simply connected symmetric homogeneous space admitting an invariant Hessian structure with positive definite second Koszul form β.*

Let G/K be a symmetric homogeneous space with an invariant Hessian structure (D, g). Let $\mathfrak{g}$ be the Lie algebra of G and let

$$\mathfrak{g} = \mathfrak{k} + \mathfrak{m}$$

be the canonical decomposition of $\mathfrak{g}$ for the symmetric homogeneous space G/K. By Theorem 9.1, $\mathfrak{g}$ admits an affine representation (f, q) such that the image of the linear mapping $q : \mathfrak{g} \longrightarrow V$ coincides with V and the kernel is $\mathfrak{k}$. Therefore the restriction of q to $\mathfrak{m}$ is a linear isomorphism from $\mathfrak{m}$ onto V. Hence for each $u \in V$ there exists a unique element $X_u \in \mathfrak{m}$ such that

$$q(X_u) = u.$$

We denote by L_u an endomorphism of V given by

$$L_u = f(X_u), \quad \text{for } u \in V,$$

and define a multiplication on V by

$$u * v = L_u v.$$

Then V is a commutative algebra by (9.3).

Lemma 9.6. *Let R be the curvature tensor of g and let R_o be the restriction of R to $o = \{K\}$. Then*

$$R_o(u, v) = -[L_u, L_v].$$

Proof. Identifying $\mathfrak{m}$ with V by the mapping q we have by [KN]

$$R_o(X, Y)Z = -[[X, Y], Z], \quad for\ X,\ Y\ and\ Z \in \mathfrak{m}.$$

With this result we have

$$R_o(u, v)w = q(R_o(X_u, X_v)X_w) = -q([[X_u, X_v], X_w])$$
$$= -f([X_u, X_v])q(X_w) + f(X_w)q([X_u, X_v])$$
$$= -[L_u, L_v]w.$$

$\square$

Lemma 9.7. *Let $W \in \mathfrak{k}$. Then $f(W)$ is a derivation of the algebra V.*

Proof.   Since

$$[W, X_u] \in \mathfrak{m} \quad \text{and} \quad q([W, X_u]) = f(W)q(X_u) - f(X_u)q(W) = f(W)u,$$

we obtain

$$[W, X_u] = X_{f(W)u}.$$

Hence we have

$$
\begin{aligned}
(f(W)u) * v &= f(X_{f(W)u})v = f([W, X_u])v = [f(W), f(X_u)]v \\
&= f(W)f(X_u)v - f(X_u)f(\mathring{W})v \\
&= f(W)(u * v) - u * (f(W)v). \qquad \qquad \square
\end{aligned}
$$

Let α_o and β_o be the restrictions of the Koszul forms α and β to o. By Lemma 9.5 we know

$$\alpha_o(u) = -\operatorname{Tr} L_u, \tag{9.4}$$

$$\beta_o(u, v) = -\alpha_o(u * v) = \operatorname{Tr} L_{u*v}. \tag{9.5}$$

Lemma 9.8.   *We have*

(1) $[[L_u, L_v], L_w] = L_{[u*w*v]}$,
     where $[u * w * v] = u * (w * v) - (u * w) * v$.
(2) $\beta_o(u * v, w) = \beta_o(v, u * w)$.

Proof.   By relations (9.3) we have

$$
\begin{aligned}
[[X_u, X_v], X_w] &\in \mathfrak{m}, \\
q([[X_u, X_v], X_w]) &= f([X_u, X_v])q(X_w) - f(X_w)q([X_u, X_v]) \\
&= [L_u, L_v]w = [u * w * v],
\end{aligned}
$$

which yield

$$[[X_u, X_v], X_w] = X_{[u*w*v]}.$$

Hence

$$[[L_u, L_v], L_w] = f([[X_u, X_v], X_w]) = f(X_{[u*w*v]}) = L_{[u*w*v]}.$$

Using relations (9.5) together with the above result we obtain

$$
\begin{aligned}
\beta_o(u * v, w) - \beta_o(v, u * w) &= \operatorname{Tr} L_{(u*v)*w - v*(u*w)} = -\operatorname{Tr} L_{[v*u*w]} \\
&= -\operatorname{Tr} [[L_v, L_w], L_u] = 0. \qquad \square
\end{aligned}
$$

Lemma 9.9. *We have*

(1) ${}^t f(X) = f(X)$, for $X \in \mathfrak{m}$,
(2) ${}^t f(W) = -f(W)$, for $W \in \mathfrak{k}$,

where ${}^t f(X)$ *is the adjoint operator of* $f(X)$ *with respect to the inner product* $\langle \ , \ \rangle$ *given in Lemma 9.2. In particular the linear Lie algebra* $f(\mathfrak{g})$ *is self-adjoint, that is,* ${}^t f(X) \in f(\mathfrak{g})$ *for all* $f(X) \in f(\mathfrak{g})$.

Proof. By equation (9.1) we have

$$\langle f(Y)q(Z), q(X) \rangle + \langle q(Z), f(Y)q(X) \rangle$$
$$= \langle f(Z)q(Y), q(X) \rangle + \langle q(Y), f(Z)q(X) \rangle.$$

For X, Y and $Z \in \mathfrak{m}$ the above formula is reduced to

$$\langle q(Z), f(X)q(Y) \rangle = \langle f(X)q(Z), q(Y) \rangle,$$

because the algebra V is commutative. This implies assertion (1). For $X, \ Y \in \mathfrak{m}$ and $Z = W \in \mathfrak{k}$ we have

$$\langle f(W)q(Y), q(X) \rangle + \langle q(Y), f(W)q(X) \rangle = 0.$$

Thus assertion (2) is also proved. $\qquad\square$

Lemma 9.10. *The kernel* $\operatorname{Ker} f$ *of* f *is included in* $\mathfrak{m}$.

Proof. Let $Z \in \operatorname{Ker} f \cap \mathfrak{k}$ and $X \in \mathfrak{m}$. Since

$$[Z, X] \in \mathfrak{m} \quad \text{and} \quad q([Z, X]) = f(Z)q(X) - f(X)q(Z) = 0,$$

it follows that $[Z, X] = 0$. We have $[Z, W] \in \operatorname{Ker} f \cap \mathfrak{k}$ for all $W \in \mathfrak{k}$. Thus $\operatorname{Ker} f \cap \mathfrak{k}$ is an ideal of $\mathfrak{g}$ included in $\mathfrak{k}$. Since G acts effectively on G/K, an ideal of $\mathfrak{g}$ included in $\mathfrak{k}$ is reduced to 0. Hence $\operatorname{Ker} f \cap \mathfrak{k} = 0$. By Lemma 9.9 we have $\operatorname{Ker} f = \operatorname{Ker} f \cap \mathfrak{k} + \operatorname{Ker} f \cap \mathfrak{m}$ and so $\operatorname{Ker} f \subset \mathfrak{m}$D $\qquad\square$

Let $\mathfrak{g}(\mathfrak{m})$ be a Lie subalgebra of $\mathfrak{g}$ generated by $\mathfrak{m}$;

$$\mathfrak{g}(\mathfrak{m}) = [\mathfrak{m}, \mathfrak{m}] + \mathfrak{m}.$$

By Lemma 9.9 the Lie algebra $f(\mathfrak{g}(\mathfrak{m}))$ is self-adjoint. Let $G(\mathfrak{m})$ be the connected Lie subgroup of G corresponding to $\mathfrak{g}(\mathfrak{m})$, and put $K(\mathfrak{m}) = G(\mathfrak{m}) \cap K$. Then

$$G/K = G(\mathfrak{m})/K(\mathfrak{m}). \tag{9.6}$$

Let V_0 be the intersection of $\operatorname{Ker} f(X)$ for all $X \in \mathfrak{g}(\mathfrak{m})$. The orthogonal complement V' of V_0 is a $f(\mathfrak{g}(\mathfrak{m}))$-invariant subspace. Hence V' is decomposed into the direct sum of irreducible subspaces

$$V' = V_1 + \cdots + V_r,$$

where V_i are orthogonal to each other. The algebra V is therefore the direct sum of ideals,

$$V = V_0 + V_1 + \cdots + V_r, \tag{9.7}$$
$$V_i * V_i \subset V_i, \qquad V_i * V_j = \{0\}, \quad \text{for } i \neq j.$$

Lemma 9.11. *Set* $\mathfrak{m}_i = \{X \in \mathfrak{m} \mid q(X) \in V_i\}$. *Then*

(1) $\operatorname{Ker} f = \mathfrak{m}_0$,
(2) $[\mathfrak{m}_i, \mathfrak{m}_j] = \{0\}$, *for* $i \neq j$.

Proof. Let $X \in \mathfrak{m}_0$. Then $f(X)q(Y) = f(Y)q(X) = 0$ for all $Y \in \mathfrak{m}$. Hence $X \in \operatorname{Ker} f$. Conversely let $X \in \operatorname{Ker} f$. Then $f(Y)q(X) = f(X)q(Y) = 0$ for all $Y \in \mathfrak{m}$. Since $\mathfrak{g}(\mathfrak{m})$ is generated by $\mathfrak{m}$, we have $f(Y)q(X) = 0$ for all $Y \in \mathfrak{g}(\mathfrak{m})$. Hence $X \in \mathfrak{m}_0$. Thus (1) is proved. Let $X_i \in \mathfrak{m}_i$ and $X_j \in \mathfrak{m}_j$ where $i \neq j$. It follows from relation (9.7) that $f([X_i, X_j])v = f(X_i)f(X_j)v - f(X_j)f(X_i)v = 0$ for all $v \in V$. Thus $[X_i, X_j] \in \operatorname{Ker} f \cap \mathfrak{k}\mathfrak{D}$ Hence $[X_i, X_j] = 0$ by Lemma 9.10. $\square$

Lemma 9.12. *Set* $\mathfrak{g}_i = [\mathfrak{m}_i, \mathfrak{m}_i] + \mathfrak{m}_i$. *Then*

(1) $\mathfrak{g}(\mathfrak{m}) = \mathfrak{g}_0 + \mathfrak{g}_1 + \cdots + \mathfrak{g}_r$,
(2) $\mathfrak{g}_i$ *is an ideal of* $\mathfrak{g}(\mathfrak{m})$.

Proof. Assertion (1) is a consequence of $[\mathfrak{m}_i, \mathfrak{m}_j] = \{0\}$ for $i \neq j$. For the proof of assertion (2) it suffices to show $[[\mathfrak{m}_i, \mathfrak{m}_i], \mathfrak{m}_i] \subset \mathfrak{m}_i$. This follows from $[[\mathfrak{m}_i, \mathfrak{m}_i], \mathfrak{m}_i] \subset \mathfrak{m}$ and $q([\mathfrak{m}_i, \mathfrak{m}_i], \mathfrak{m}_i]) = f([\mathfrak{m}_i, \mathfrak{m}_i])q(\mathfrak{m}_i) \subset f(\mathfrak{g}(\mathfrak{m}))V_i \subset V_i$. $\square$

Lemma 9.13. *For* $X \in \mathfrak{g}_i$ *we denote by* $f_i(X)$ *the restriction of* $f(X)$ *to* V_i. *Then* f_i $(1 \leq i \leq r)$ *is a faithful irreducible representation of* $\mathfrak{g}_i$ *on* V_i.

Proof. Let $X \in \mathfrak{g}_i$ such that $f_i(X) = 0$. Since $f(\mathfrak{g}_i)$ is generated by $f(\mathfrak{m}_i)$ and $f(\mathfrak{m}_i)V_j = \{0\}$ $(j \neq i)$, it follows that $f(X)V_j = 0$. Hence $f(X) = 0$. Given also that $\operatorname{Ker} f \cap \mathfrak{g}_i = \mathfrak{m}_0 \cap \mathfrak{g}_i = \{0\}$, we have $X = 0$. Thus f_i is a faithful representation. Let U_i be a subspace of V_i invariant by $f_i(\mathfrak{g}_i)$. Since $f(\mathfrak{g}_j)$ $(j \neq i)$ is generated by $f(\mathfrak{m}_j)$ and $f(\mathfrak{m}_j)U_i = \{0\}$, we obtain $f(\mathfrak{g}_j)U_i = \{0\}$. Therefore U_i is an $f(\mathfrak{g}(\mathfrak{m}))$-invariant subspace of V_i. Since V_i is an irreducible subspace by $f(\mathfrak{g}(\mathfrak{m}))$ we have $U_i = \{0\}$ or $U_i = V_i$. Hence f_i is an irreducible representation. $\square$

Proposition 9.1. *If the representation* f *of* $\mathfrak{g}$ *on* V *is faithful and irreducible, then the algebra* V *is a compact simple Jordan algebra.*

Proof. We put $U_0 = \{u_0 \in V \mid \beta_o(u_0, v) = 0,$ for all $v \in V\}$. Since $\beta_o(u * u_0, v) = \beta_o(u_0, u * v) = 0$ for $u_0 \in U_0$ and $u, v \in V$, we obtain $f(\mathfrak{m})U_0 \subset U_0$. For $u_0 \in U_0$, $v \in V$ and $W \in \mathfrak{k}$, by Lemma 9.7 and (9.5) we have

$$\beta_o(f(W)u_0, v) = \operatorname{Tr} L_{(f(W)u_0)*v} = \operatorname{Tr} L_{f(W)(u_0*v)} - \operatorname{Tr} L_{u_0*(f(W)v)}$$
$$= \operatorname{Tr} f(X_{f(W)(u_0*v)}) = \operatorname{Tr} f([W, X_{u_0*v}]) = 0.$$

Hence $f(\mathfrak{k})U_0 \subset U_0$, and so $f(\mathfrak{g})U_0 \subset U_0$. Since f is irreducible, it follows that $U_0 = \{0\}$ or $U_0 = VD$

Suppose $U_0 = V$. Let e be the element in V determined by $\langle e, u \rangle = \alpha_o(u)$. Since

$$\langle L_e u, v \rangle = \langle e, u * v \rangle = \alpha_o(u * v) = -\beta_o(u, v) = 0, \quad \text{for all } u \text{ and } v \in V,$$

we have $L_e = 0$ and hence $\langle e, e \rangle = \alpha_o(e) = \operatorname{Tr} L_e = 0$. This implies $e = 0$ and

$$\alpha_o = 0.$$

Since $\mathfrak{g}$ admits a faithful and irreducible representation f, it is known that the Lie algebra $\mathfrak{g}$ is reductive and so it may be decomposed into

$$\mathfrak{g} = \mathfrak{c} + \mathfrak{s},$$

where $\mathfrak{c}$ is the center of $\mathfrak{g}$ and $\mathfrak{s}$ is a semisimple subalgebra of $\mathfrak{g}$ [Bourbaki (1960)]. Let $C \in \mathfrak{c} \cap \mathfrak{k}$. Then $f(C)q(X) = f(C)q(X) - f(X)q(C) = q([C, X]) = 0$ for all $X \in \mathfrak{g}$. Hence $f(C) = 0$ and so $C = 0$. Since $\mathfrak{c} = \mathfrak{c} \cap \mathfrak{k} + \mathfrak{c} \cap \mathfrak{m}$ it follows that

$$\mathfrak{c} \subset \mathfrak{m}.$$

Assume that there exists $C \neq 0 \in \mathfrak{c}$. We shall show that the minimal polynomial $P(x)$ of $f(C)$ is an irreducible polynomial over **R**. Suppose that $P(x) = Q(x)R(x)$, where $Q(x)$ and $R(x)$ are polynomials over **R** whose degrees are less than the degree of $P(x)$. Put $U = Q(f(C))V$. Then U is an $f(\mathfrak{g})$-invariant subspace of V. Hence either $U = \{0\}$ or $U = V$ because f is an irreducible representation. If $U = \{0\}$ then $Q(f(C)) = 0$. However, this is a contradiction because $P(x)$ is a minimal polynomial of $f(C)$. If $U = V$ then $Q(f(C))$ is a linear isomorphism of V. From this result together with $Q(f(C))R(f(C)) = P(f(C)) = 0$ it follows that $R(f(C)) = 0$, which is also a contradiction. Thus $P(x)$ is an irreducible polynomial over **R**. Hence $P(x) = x - \lambda$ where $\lambda \in \mathbf{R}$, or $P(x) = (x - \mu)(x - \bar{\mu})$ where $\mu \in \mathbf{C}$ and $\bar{\mu} \neq \mu$. Since ${}^t f(C) = f(C)$ by Lemma 9.9, the eigenvalues of $f(C)$ are real numbers. Hence $P(x) = x - \lambda$ and $f(C) = \lambda$. From $0 = \alpha_o(q(C)) = \text{Tr} f(C) = \lambda \dim V$ it follows that $\lambda = 0$. Thus, because f is faithful, $C = 0$. This is a contradiction. Therefore $\mathfrak{c} = \{0\}$ and $\mathfrak{g}$ is semisimple. This contradicts Theorem 9.2, and so we have $U_0 = \{0\}$ and so β_o is non-degenerate on V. Therefore, it follows from relations (9.4) and (9.5) and Lemmata 4.2 and 9.8 that V is a semisimple Jordan algebra. Since f is irreducible, the Jordan algebra V is simple. Moreover, by Lemma 9.9 (1) and the following Lemma 9.14, we have that the simple Jordan algebra V is compact.   □

Lemma 9.14. *Let V be a Jordan algebra over* **R**. *Then the following conditions are equivalent* (cf. [Koecher (1962)]).

(1) *The bilinear form given by $(a, b) = \text{Tr} L_{a*b}$ is positive definite.*
(2) *V admits an inner product such that $\langle u * v, w \rangle = \langle v, u * w \rangle$.*

Proof of Theorem 9.3.   Let G_i be a connected Lie subgroup of G corresponding to $\mathfrak{g}_i$ and let $K_i = G_i \cap K(\mathfrak{m})$. It follows from (9.6), (9.7) and Lemma 9.12 that
$$G/K = G(\mathfrak{m})/K(\mathfrak{m}) = G_0/K_0 \times G_1/K_1 \times \cdots \times G_r/K_r.$$
By Lemma 9.6 and 9.11 the universal covering space of G_0/K_0 is a Euclidean space. We shall consider the case G_i/K_i where $i \geq 1$. Let $\tilde{G}_i$ be the universal covering group of G_i and let $\pi : \tilde{G}_i \longrightarrow G_i$ be the covering projection. Denote by $\tilde{K}_i$ the connected component of $\pi^{-1}(K_i)$ containing the unit element. Then $\tilde{G}_i/\tilde{K}_i$ is the universal covering space of G_i/K_i. Since $\tilde{G}_i$ is simply connected, there exists a representation $\boldsymbol{f}_i$ of $\tilde{G}_i$ on V_i such that the differential coincides with f_i. By Proposition 9.1, V_i is a compact Jordan algebra. Therefore, by Lemma 9.14, the bilinear form on V_i defined by $(u, v)_i = \text{Tr} L_{u*v}$ is positive definite. Let e_i be the element in V_i determined by $(e_i, u)_i = \text{Tr} L_u$. Since $(e_i * u, v)_i = (u * e_i, v)_i = (e_i, u * v)_i = \text{Tr} L_{u*v} = (u, v)_i$, for u, $v \in V_i$, it follows that $e_i * u = u * e_i = u$, that is, e_i is the unit element of V_i. Hence $f_i(X)e_i = q(X)$ for all $X \in \mathfrak{m}_i$. Let $W \in \mathfrak{k}_i$. Then $f_i(W)$ is a derivation of V_i, and so $f_i(W)e_i = 0 = q(W)$. These relations imply
$$f_i(X)e_i = q(X), \quad \text{for } X \in \mathfrak{g}_i.$$
An open orbit $\Omega_i = \boldsymbol{f}_i(\tilde{G}_i)e_i$ of $\boldsymbol{f}_i(\tilde{G}_i)$ through e_i is an irreducible homogeneous self-dual regular convex cone [Koecher (1962)] [Vinberg (1960)]. Let $\tilde{H}_i$ be the closed subgroup of $\tilde{G}_i$ defined by $\tilde{H}_i = \{\tilde{h} \in \tilde{G}_i \mid \boldsymbol{f}_i(\tilde{h})e_i = e_i\}$ and let $\tilde{\mathfrak{h}}_i$ be the Lie subalgebra of $\tilde{H}_i$. Since
$$X \in \tilde{\mathfrak{h}}_i \Longleftrightarrow f_i(X)e_i = q(X) = 0 \Longleftrightarrow X \in \text{Lie algebra of } K_i,$$

it follows that $\tilde{K}_i \subset \tilde{H}_i$. We therefore obtain the canonical covering projection $p_i : \tilde{G}_i/\tilde{K}_i \longrightarrow \Omega_i = \tilde{G}_i/\tilde{H}_i$. Since the domain Ω_i is convex, the covering projection p_i is a homeomorphism, and so the irreducible homogeneous self-dual regular convex cone Ω_i is the universal covering space of G_i/K_i. $\square$

Proof of Corollary 9.3.   Let G/K be a homogeneous space of a reductive Lie group G with an invariant Hessian structure. Let $\tilde{G}$ be the universal covering group of G and let $\pi : \tilde{G} \longrightarrow G$ be the covering projection. Denote by $\tilde{K}$ the connected component of $\pi^{-1}(K)$ containing the unit element. By Theorem 8.2 the universal covering space $\tilde{G}/\tilde{K}$ of G/K is a homogeneous convex domain in $\mathbf{R}^n$. Let $\tilde{N}$ be the set of all elements in $\tilde{G}$ which induce the identity transformation on $\tilde{G}/\tilde{K}$. Then $\tilde{N}$ is a normal subgroup of $\tilde{G}$. Put $G^* = \tilde{G}/\tilde{N}$ and $K^* = \tilde{K}/\tilde{N}$. Then $G^*/K^* = \tilde{G}/\tilde{K}$ and G^* acts effectively on G^*/K^*. We may consider G^* as a subgroup of the affine transformation group $A(n)$ of $\mathbf{R}^n$. It is known that the Lie algebra $\mathfrak{g}$ of a reductive Lie group G is decomposed into

$$\mathfrak{g} = \mathfrak{c} + \mathfrak{s},$$

where $\mathfrak{c}$ is the center of $\mathfrak{g}$ and $\mathfrak{s}$ is a semisimple subalgebra of $\mathfrak{g}$. Let C^* and S^* be the connected Lie subgroups of G^* corresponding to $\mathfrak{c}$ and $\mathfrak{s}$ respectively. Then S^* is a closed subgroup of $A(n)$ because S^* is a connected semisimple subgroup of $A(n)$, and we have

$$G^* = C^* S^*.$$

Denoting by $\bar{C}^*$ the closure of C^* in $A(n)$, the group $\bar{C}^* S^*$ coincides with the closure $\bar{G}^*$ of G^* in $A(n)$G

$$\bar{G}^* = \bar{C}^* S^*.$$

The group $\bar{G}^*$ acts on the homogeneous convex domain G^*/K^* and leaves the Hessian structure invariant. Let H_c^* and H_s^* be maximal compact subgroups of $\bar{C}^*$ and S^* respectively. Then the group $H^* = H_c^* H_s^*$ is a maximal compact subgroup of $\bar{G}^* = \bar{C}^* S^*$. Since G^*/K^* is a convex domain, the compact group H^* has a fixed point in G^*/K^*. We assume, without loss of generality, that the fixed point is the origin 0. The isotropy subgroup B^* of $\bar{G}^*$ at the origin 0 is contained in an orthogonal group and is closed in $\bar{G}^*$. Thus B^* is a compact subgroup of $\bar{G}^*$ including H^*. Hence $H^* = B^*$. Since the group $\bar{G}^*$ acts effectively on $\bar{G}^*/H^* = G^*/K^*$, the group H_c^* is reduced to the identity. Hence $B^* = H^* = H_s^*$. Let $\bar{\mathfrak{g}}^*$, $\bar{\mathfrak{c}}^*$, $\mathfrak{s}^*$ and $\mathfrak{h}_s^*$ be the

Lie algebras $\bar{G}^*$, $\bar{C}^*$, S^* and H_s^* respectively. Denote by $\mathfrak{p}_s^*$ the orthogonal complement of $\mathfrak{h}_s^*$ in $\mathfrak{s}^* = \mathfrak{s}$ with respect to the Killing form of $\mathfrak{s}$. Put

$$\mathfrak{h}^* = \mathfrak{h}_s^*, \qquad \mathfrak{p}^* = \bar{\mathfrak{c}}^* + \mathfrak{p}_s^*.$$

Then

$$\bar{\mathfrak{g}}^* = \mathfrak{h}^* + \mathfrak{p}^*,$$
$$[\mathfrak{h}^*, \mathfrak{h}^*] \subset \mathfrak{h}^*, \quad [\mathfrak{h}^*, \mathfrak{p}^*] \subset \mathfrak{p}^*, \quad [\mathfrak{p}^*, \mathfrak{p}^*] \subset \mathfrak{h}^*.$$

Therefore $\bar{G}^*/H^*(= G^*/K^* = \tilde{G}/\tilde{K})$ is a symmetric homogeneous domain with an invariant Hessian structure. Hence, by Theorem 9.3, the universal covering space $\tilde{G}/\tilde{K}$ of G/K is a direct product of a Euclidean space and a homogeneous self-dual regular convex cone.   □

Proof of Corollary 9.4.   The automorphism group G of a compact homogeneous Hessian manifold M is compact. Hence G is a reductive Lie group acting transitively on M. Using the same notation as in Corollary 9.3, by Weyl's theorem, S^* is a compact subgroup. Hence $H^* = S^*$. Since S^* is a normal subgroup of G^* fixing the origin 0, it follows that $S^* \subset K^*$. Since G^* acts effectively, S^* is reduced to the identity, and so $G^* = C^*$. Hence the group $G^* = C^*$ is a compact Abelian group and M is a Euclidean torus.   □

Proof of Corollary 9.5.   We have proved in Theorem 4.6 and Proposition 4.10 that a homogeneous self-dual regular convex cone is a simply connected symmetric homogeneous space admitting an invariant Hessian structure with positive definite second Koszul form. Conversely, suppose that G/K is a simply connected symmetric homogeneous space endowed with an invariant Hessian structure with positive definite second Koszul form. It follows from relations (9.5) and Lemma 9.11 that $V_0 = \{0\}$. Hence G_0/K_0 reduces to a point. Therefore, by Theorem 9.3, G/K is a homogeneous self-dual regular convex cone.   □

Chapter 10

Homogeneous spaces with invariant Hessian structures

In section **4.2** we proved that a homogeneous self-dual regular convex cone is a symmetric space with invariant canonical Hessian structure, and we gave a correspondence between these cones and compact Jordan algebras. In section **9.3** we proved that a simply connected symmetric space with an invariant Hessian structure is the direct product of a Euclidean space and a homogeneous self-dual regular convex cone. In this chapter we generalize these results, and show that a homogeneous space with an invariant Hessian structure essentially consists of a Euclidean space and a homogeneous regular convex cone. In section **10.1** it is shown that a homogeneous Hessian domain admits a simply transitive triangular group. Using an affine representation of the Lie algebra of the triangular group, we reduce our study to that of certain non-associative algebras. In sections **10.2** and **10.3** we state the results of [Vinberg (1963)] which give a correspondence between homogeneous regular convex domains and non-associative algebras, called clans, and a realization of a homogeneous regular convex domain as a real Siegel domain. In section **10.4** we give a correspondence between homogeneous Hessian domains and normal Hessian algebras and apply the structure theorem of normal Hessian algebras for the study of homogeneous Hessian manifolds.

10.1 Simply transitive triangular groups

Let $(G/K, D, g)$ be a homogeneous Hessian manifold. Using the notation of section **8.1**, it follows from Corollary 8.1 (1) that the mapping

$$\exp_o^D : \mathcal{E}_o \longrightarrow G/K$$

is an affine mapping where $o = \{K\}$. Hence $\tilde{g} = (\exp_o^D)^*g$ is a Hessian metric with respect to the standard flat connection $\tilde{D}$ on $\mathcal{E}_o$. By Lemma 8.3, there exists a potential function $\tilde{\varphi}$ such that $\tilde{g} = \tilde{D}d\tilde{\varphi}$. By Lemma 8.1 and 8.4 the domain $\mathcal{E}_o$ is convex, and so $\mathcal{E}_0$ is the universal covering space of G/K with covering projection $\exp_o^D : \mathcal{E}_0 \longrightarrow G/K$ by Corollary 8.1 (2). Thus the universal covering space of a homogeneous Hessian manifold is a homogeneous Hessian domain. Therefore the study of homogeneous Hessian manifolds is essentially reduced to that of homogeneous Hessian domains. For this reason, in the present chapter the Hessian domains considered will be homogeneous unless otherwise stated.

In this section we shall prove that a homogeneous Hessian domain admits a simply transitive triangular group. We first recall Vinberg's theorem on algebraic groups.

Let x_{ij} be a function given by $x_{ij} : [s_{ij}] \in GL(n, \mathbf{R}) \longrightarrow s_{ij} \in \mathbf{R}$. A subgroup G of $GL(n, \mathbf{R})$ is called an **algebraic group** if there exist a family of polynomials $\{p_\lambda(\cdots, x_{ij}, \cdots)\}_{\lambda \in \Lambda}$ of x_{ij} such that

$$G = \{[s_{ij}] \in GL(n, \mathbf{R}) \mid p_\lambda(\cdots, s_{ij}, \cdots) = 0, \text{ for all } \lambda \in \Lambda\}.$$

A subgroup T of $GL(n, \mathbf{R})$ is said to be **triangular** if there exists an element $s \in GL(n, \mathbf{R})$ such that all elements in $s^{-1}Ts$ are upper triangular matrices.

The affine transformation group $A(n)$ of $\mathbf{R}^n$ is naturally identified with a subgroup of $GL(n + 1, \mathbf{R})$. A subgroup G of $A(n)$ is said to be algebraic or triangular if, considering G as a subgroup of $GL(n + 1, \mathbf{R})$ by the above identification, G is algebraic or triangular respectively. The following theorem is due to [Vinberg (1961)].

Theorem 10.1. *Let G be the group of the connected component of an algebraic group. Then there exists a triangular subgroup T and a maximal compact subgroup K of G such that*

$$G = TK, \quad T \cap K = \{e\},$$

where $\{e\}$ is the identity.

Let $(\Omega, D, g = Dd\varphi)$ be a homogeneous Hessian domain in $\mathbf{R}^n$ and let $Aut(\Omega, D, g)$ be the group of all automorphism of $(\Omega, D, g = Dd\varphi)$. In the following we assume that $Aut(\Omega, D, g)$ acts transitively on Ω.

Lemma 10.1. *Let N be the normalizer of $Aut(\Omega, D)$ in $A(n)$. Then N is an algebraic group, and the group of the connected component of N coincides with the group of the connected component of $Aut(\Omega, D)$.*

Proof. For the proof the reader may refer to [Vinberg (1963)]. □

Lemma 10.2. *Each partial derivative* $\dfrac{\partial\varphi}{\partial x^j}$ *is a rational function with respect to the standard affine coordinate system* $\{x^1,\cdots,x^n\}$ *on* $\mathbf{R}^n$.

Proof. Let $\mathfrak{g}$ be the Lie algebra of $Aut(\Omega,D,g)$. For $X\in\mathfrak{g}$ we denote by X^* the vector field on Ω induced by $\exp(-tX)$. Let p be an arbitrary point in Ω. Then we can find $X_1,\cdots,X_n\in\mathfrak{g}$, and a neighbourhood W of p, such that $X_1^*,\cdots,X_n^*$ is a basis of the tangent space at any point of W. We set $\dfrac{\partial}{\partial x^j}=\sum_i \eta_j^i X_i^*$ and $X_i^*=\sum_{ij}\xi_i^j\dfrac{\partial}{\partial x^j}$. Then each ξ_i^j is an affine function of $\{x^1,\cdots,x^n\}$ because $\exp(-tX)\in A(n)$. Since η_j^i is the (i,j)-th component of the inverse matrix of $[\xi_i^j]$, η_j^i is a rational function of $\{x^1,\cdots,x^n\}$. Let $X\in\mathfrak{g}$, then $X^*=\sum_j\xi^j\dfrac{\partial}{\partial x^j}$ is a Killing vector field with respect to $g=Dd\varphi$ and ξ^j is an affine function. Hence, denoting by $\mathcal{L}_{X^*}$ the Lie derivative with respect to X^*, we have

$$
\begin{aligned}
0 &= (\mathcal{L}_{X^*}g)\Big(\frac{\partial}{\partial x^i},\frac{\partial}{\partial x^j}\Big)\\
&= X^*\frac{\partial^2\varphi}{\partial x^i\partial x^j}-g\Big(\Big[X^*,\frac{\partial}{\partial x^i}\Big],\frac{\partial}{\partial x^j}\Big)-g\Big(\frac{\partial}{\partial x^i},\Big[X^*,\frac{\partial}{\partial x^j}\Big]\Big)\\
&= \sum_p\Big\{\xi^p\frac{\partial^3\varphi}{\partial x^i\partial x^j\partial x^p}+\frac{\partial\xi^p}{\partial x^i}\frac{\partial^2\varphi}{\partial x^p\partial x^j}+\frac{\partial\xi^p}{\partial x^j}\frac{\partial^2\varphi}{\partial x^p\partial x^i}\Big\}\\
&= \frac{\partial^2}{\partial x^i\partial x^j}(X^*\varphi),
\end{aligned}
$$

which implies that $X^*\varphi$ is an affine function. Therefore $\dfrac{\partial\varphi}{\partial x^j}=\sum_{ij}\eta_j^i X_i^*\varphi$ is a rational function. □

Proposition 10.1. *The group* $Aut_0(\Omega,D,g)$, *the identity component of* $Aut(\Omega,D,g)$, *coincides with the identity component of an algebraic group.*

Proof. Let $\boldsymbol{f}(s)=[\boldsymbol{f}(s)_j^i]$ and $\boldsymbol{q}(s)=[\boldsymbol{q}(s)^i]$ be the linear part and the translation vector part of $s\in A(n)$ respectively. An element s in $Aut(\Omega,D)$ is contained in $Aut(\Omega,D,g)$ if and only if

$$
\sum_{k,l}\boldsymbol{f}(s)_i^k\boldsymbol{f}(s)_j^l g_{kl}(sp)=g_{ij}(p),\quad\text{for all }p\in\Omega.
$$

By Lemma 10.2 the component $g_{ij} = \dfrac{\partial^2 \varphi}{\partial x^i \partial x^j}$ of g is a rational function. Hence

$$H = \Big\{ s \in A(n) \mid \sum_{k,l} \boldsymbol{f}(s)_i^k \boldsymbol{f}(s)_j^l g_{kl}(sp) = g_{ij}(p), \ \text{ for all } p \in \Omega \Big\}$$

is an algebraic subgroup of $A(n)$ and $Aut(\Omega, D, g) = Aut(\Omega, D) \cap H$. Using Lemma 10.1 and the notation therein, it follows that the group $Aut_0(\Omega, D, g)$ coincides with the group of the connected component of the algebraic group $N \cap H$. □

Proposition 10.2. *The isotropy subgroup K of $Aut_o(\Omega, D, g)$ at $o \in \Omega$ is a maximal compact subgroup of G.*

Proof. The group K is a closed subgroup of $A(n)$ because $Aut_o(\Omega, D, g)$ is a closed subgroup of $A(n)$. We set

$$H_o = \{ h \in A(n) \mid \sum_{p,q} \boldsymbol{f}(h)_i^p \boldsymbol{f}(h)_j^q g_{pq}(o) = g_{ij}(o) \},$$

where $[\boldsymbol{f}(h)_j^i]$ is the linear part of h. Since H_o is a compact group and $K \subset H_o$, K is a compact group. Let K' be a maximal compact subgroup of G including K and let W be a bounded open subset of Ω. Put

$$K'W = \{ k'w \mid k' \in K', \ w \in W \}.$$

Then $K'W$ is a bounded subset invariant by K'. Let o' be the center of gravity of $K'W$. Then o' is, by definition, a fixed point of K' and is contained in Ω by the convexity. Choosing an element $s \in G$ such that $so' = o$, we have $sK's^{-1} \subset K$. Since $sK's^{-1}$ is a maximal compact subgroup of G, we obtain $sK's^{-1} = K$. Therefore K is a maximal compact subgroup of $Aut_o(\Omega, D, g)$. □

It follows from Theorem 10.1 and Propositions 10.1 and 10.2 that there exists a triangular subgroup T and a maximal compact subgroup K of $Aut_o(\Omega, D, g)$ such that

$$Aut_o(\Omega, D, g) = TK, \quad T \cap K = \{e\}.$$

In the proof of Proposition 10.2 we proved that there exists a fixed point $o \in \Omega$ of K such that the isotropy subgroup of $Aut_o(\Omega, D, g)$ at o coincides with K. Therefore the triangular group T acts simply transitively on Ω. Thus we have

Theorem 10.2. *A homogeneous Hessian domain (Ω, D, g) admits a simply transitive triangular subgroup T of $Aut(\Omega, D, g)$.*

10.2 Homogeneous regular convex domains and clans

In this section, according to [Vinberg (1963)] we shall give a bijective correspondence between homogeneous regular convex domains and certain non-
associative algebras, called clans. Let Ω be a homogeneous regular convex
domain in $\mathbf{R}^n$. By the proof of Theorem 8.4, Ω admits a Hessian metric
of Koszul type $(D, g = D\omega)$. Without loss of generality we may assume
that Ω contains the origin 0 of $\mathbf{R}^n$. By Theorem 10.2, there exists a triangular subgroup T of $Aut(\Omega, D, g)$ acting simply transitively on Ω. Let
$f : T \longrightarrow GL(n, \mathbf{R})$ and $q : T \longrightarrow \mathbf{R}^n$ be the mappings which assign each
$s \in T$ to the linear part $f(s)$ and the translation vector part $q(s)$. Let t be
the Lie algebra of T. We denote by $f : \mathfrak{t} \longrightarrow \mathfrak{gl}(n, \mathbf{R})$ and $q : \mathfrak{t} \longrightarrow \mathbf{R}^n$ the
differentials of f and q at the identity respectively.

For $X \in \mathfrak{t}$ we denote by X^* the vector field on Ω induced by $\exp(-tX)$,
then

$$X^* = -\sum_i \left(\sum_j f(X)^i_j x^j + q^i(X)\right) \frac{\partial}{\partial x^i}.$$

Using the notation of section **9.1** we have

$$A_{X^*} Y^* = -\sum_i \left(\sum_{j,k} f(X)^i_j f(Y)^j_k x^k + \sum_j f(X)^i_j q(Y)^j\right) \frac{\partial}{\partial x^i}.$$

Hence

$$X_0^* = -\sum_i q(X)^i \left(\frac{\partial}{\partial x^i}\right)_0,$$

$$(A_{X^*} Y^*)_0 = -\sum_i (f(X)q(Y))^i \left(\frac{\partial}{\partial x^i}\right)_0.$$

Identifying $\sum_i a_i \left(\frac{\partial}{\partial x^i}\right)_0 \in T_0\Omega$ and $[a^i] \in \mathbf{R}^n$ we have

$$X_0^* = -q(X), \qquad (A_{X^*})_0 = f(X). \tag{10.1}$$

The pair (f, q) defined above coincides with the pair given in the section
9.1. Thus (f, q) is an affine representation of t on $V = \mathbf{R}^n$,

(1) f is a linear representation of t on V,
(2) q is a linear isomorphism from t to V,
(3) $q([X, Y]) = f(X)q(Y) - f(Y)q(X)$.

Using the above (1)-(3) we define an operation of multiplication on V by

$$u \cdot v = f(q^{-1}(u))v.$$

Then V is an algebra with this multiplication.

Lemma 10.3. *Let L_u be the operator of the left multiplication by $u \in V$. Then we obtain*

$$[L_u, L_v] = L_{u \cdot v - v \cdot u}.$$

Therefore V is a left symmetric algebra (cf. Definition 9.2 and Lemma 9.1)D

Proof. Since

$$L_u = f(q^{-1}(u)),$$
$$q([q^{-1}(u), q^{-1}(v)]) = f(q^{-1}(u))v - f(q^{-1}(v))u = u \cdot v - v \cdot u,$$

we have

$$\begin{aligned}
L_{u \cdot v - v \cdot u} &= f(q^{-1}(u \cdot v - v \cdot u)) = f([q^{-1}(u), q^{-1}(v)]) \\
&= [f(q^{-1}(u)), f(q^{-1}(v))] = [L_u, L_v].
\end{aligned}$$

$\square$

Proposition 10.3. *If a homogeneous regular convex domain is a cone, then the corresponding left symmetric algebra has the unit element.*

Proof. Let $-e$ be the vertex of the cone. The 1-parameter transformation group of dilations at $-e$ is given by

$$\sigma(t) : x \longrightarrow (\exp t)(x + e) - e,$$

and is included in the center of T. There therefore exists an element E in the center of $\mathfrak{t}$ such that $\sigma(t) = \exp tE$ and

$$f(\exp tE)x = (\exp t)x, \quad q(\exp tE) = (\exp t - 1)e.$$

Hence

$$f(E) = I, \quad q(E) = e,$$

where I is the unit matrix. This implies

$$e \cdot u = f(E)u = u,$$
$$0 = q([E, q^{-1}(u)]) = f(E)u - f(q^{-1}(u))e = u - u \cdot e.$$

Therefore e is the unit element in V.

$\square$

Definition 10.1. A left symmetric algebra V is said to be **normal** if the eigenvalues of L_u are all real numbers for any $u \in V$.

Definition 10.2. A normal left symmetric algebra V is called a **clan** (cf. [Vinberg (1963)]) if V admits a linear function χ on V such that

(1) $\chi(u \cdot v) = \chi(v \cdot u)$, for u and $v \in V$,
(2) $\chi(w \cdot w) > 0$, for $w \neq 0 \in V$.

The clan is denoted by (V, χ).

Theorem 10.3. *Let Ω be a homogeneous regular convex domain. Then we can construct a clan from Ω. If Ω is a cone, then the corresponding clan has the unit element.*

Proof. Since $\mathfrak{t}$ is the Lie algebra of a triangular group T, all the eigenvalues of L_u are real numbers for any $u \in V$. Hence, by Lemma 10.3, V is a normal left symmetric algebra. The homogeneous regular convex domain $\Omega = T0$ admits a Hessian metric of Koszul type $g = D\omega$, where ω is an invariant 1-form (cf. Theorem 8.4). Denoting by χ the value of ω at 0, by Lemma 9.3 the pair (V, χ) is a clan. By Proposition 10.3 the clan has the unit element if Ω is a cone. $\qquad\square$

Conversely, we shall construct a homogeneous regular convex domain from a clan. Let (V, χ) be a clan and let $\{x^1, \cdots, x^n\}$ be the affine coordinate system on V with respect to a basis $\{e_1, \cdots, e_n\}$, i.e. $u = \sum_i x^i(u) e_i$ for $u \in V$. We define a vector field X_u on V by

$$X_u = -\sum_i \Big(\sum_j (L_u)^i_j x^j + x^i(u) \Big) \frac{\partial}{\partial x^i}, \qquad (10.2)$$

where $(L_u)^i_j$ represent the components of L_u with respect to the basis. Then X_u is an infinitesimal affine transformation, and we have

$$[X_u, X_v] = X_{u \cdot v - v \cdot u}.$$

Thus

$$\mathfrak{t}(V) = \{X_u \mid u \in V\}$$

forms a Lie algebra. Let $T(V)$ be the affine transformation group of V generated by $\mathfrak{t}(V)$. For $s \in T(V)$ we denote by $\boldsymbol{f}(s)$ and $\boldsymbol{q}(s)$ the linear part and the translation vector part of s respectively. Then

$$\boldsymbol{f}(\exp X_{-u}) = \exp L_u, \qquad \boldsymbol{q}(\exp X_{-u}) = \sum_{n=1}^{\infty} \frac{1}{n!} (L_u)^{n-1} u. \qquad (10.3)$$

Let f and q be the differentials of $\boldsymbol{f}$ and $\boldsymbol{q}$ at the identity respectively. Then
$$f(X_{-u}) = L_u, \qquad q(X_{-u}) = u. \tag{10.4}$$
Let $\Omega(V)$ be the orbit of $T(V)$ through the origin 0;
$$\Omega(V) = T(V)0 = \{\boldsymbol{q}(s) \mid s \in T(V)\}.$$
Then $\Omega(V)$ is a domain in V because $q : \mathfrak{t}(V) \longrightarrow V$ is a linear isomorphism.

Lemma 10.4. *The group $T(V)$ acts simply transitively on $\Omega(V)$.*

Proof. It is enough to show that the isotropy subgroup B of $T(V)$ at the origin 0 is reduced to the identity element. Let $\mathfrak{b}$ be the Lie algebra of B. It follows from $B = \{s \in T(V) \mid \boldsymbol{q}(s) = 0\}$ that $\mathfrak{b} = \{X_u \in \mathfrak{t}(V) \mid q(X_u) = 0\}$, and so $\mathfrak{b} = \{0\}$. Hence the group B is discrete. Since all the eigenvalues of L_u are real numbers for every $u \in V$, the Lie algebra $\{L_u \mid u \in V\}$ is triangular. Therefore the Lie algebra $\mathfrak{t}(V)$ is also triangular and so the mapping $\exp : \mathfrak{t}(V) \longrightarrow T(V)$ is surjective. Let $s \in B$, then there exists $X_u \in \mathfrak{t}(V)$ such that $s = \exp X_u$. Put $r = \exp \frac{1}{2}X_u$. Since $0 = \boldsymbol{q}(s) = \boldsymbol{q}(r^2) = \boldsymbol{f}(r)\boldsymbol{q}(r) + \boldsymbol{q}(r)$ we obtain $\boldsymbol{f}(r)\boldsymbol{q}(r) = -\boldsymbol{q}(r)$. Suppose $\boldsymbol{q}(r) \neq 0$. Then $\boldsymbol{q}(r)$ is an eigenvector of $\boldsymbol{f}(r)$ corresponding to the eigenvalue -1. However, since $\boldsymbol{f}(r) = \boldsymbol{f}(\exp \frac{1}{2}X_u) = \exp \frac{1}{2}f(X_u) = \exp \frac{1}{2}L_u$ and all the eigenvalues of L_u are real, the eigenvalues of $\boldsymbol{f}(r)$ are all positive real numbers. This is a contradiction. Hence $\boldsymbol{q}(r) = 0$ and $r = \exp \frac{1}{2}X_u \in B$. Repeating the same procedure we obtain $\exp \frac{1}{2^n}X_u \in B$ for all non-negative integers n. Since the group B is a discrete subgroup, we conclude $X_u = 0$ and so s is the identity element. Thus B is reduced to the identity element. $\square$

Lemma 10.5. *Let ω be a 1-form on $\Omega(V)$ defined by*
$$\omega(X_u) = \chi(u) \quad \text{for } u \in V.$$
Then

(1) ω *is a closed 1-form invariant under $T(V)$.*
(2) $D\omega$ *is a positive definite symmetric bilinear form invariant under $T(V)$.*

Proof. The 1-form ω is closed and invariant under $T(V)$ because
$$(d\omega)(X_u, X_v) = X_u(\omega(X_v)) - X_v(\omega(X_u)) - \omega([X_u, X_v])$$
$$= -\chi(u \cdot v - v \cdot u) = 0,$$
$$(\mathcal{L}_{X_u}\omega)(X_v) = X_u(\omega(X_v)) - \omega([X_u, X_v])$$
$$= -\omega(X_{u \cdot v - v \cdot u}) = -\chi(u \cdot v - v \cdot u) = 0,$$

where $\mathcal{L}_{X_u}$ is the Lie derivative by X_u. The invariance of $D\omega$ under $T(V)$ follows from

$$(\mathcal{L}_{X_u} D\omega)(X_v, X_w)$$
$$= X_u(D\omega(X_v, X_w)) - D\omega([X_u, X_v], X_w) - D\omega(X_v, [X_u, X_w])$$
$$= X_u((D_{X_v}\omega)(X_w)) - (D_{X_v}\omega)([X_u, X_w]) - (D_{[X_u, X_v]}\omega)(X_w)$$
$$= (\mathcal{L}_{X_u} D_{X_v}\omega - D_{[X_u, X_v]}\omega)(X_w)$$
$$= \{(\mathcal{L}_{X_u} D_{X_v} - D_{X_v} \mathcal{L}_{X_u} - D_{[X_u, X_v]})\omega\}(X_w)$$
$$= 0,$$

where for the last equation see [Kobayashi and Nomizu (1963, 1969)](I, p.231). To prove that $D\omega$ is positive definite, it is sufficient to show at 0. Since

$$(D\omega)(X_u, X_v) = X_v\omega(X_u) - \omega(D_{X_v} X_u) = -\omega(D_{X_v} X_u),$$

$$D_{X_v} X_u = \sum_j \left\{ \sum_k (L_u L_v)^j_k x^k + x^j (u \cdot v) \right\} \frac{\partial}{\partial x^j},$$

we have

$$(D\omega)_0((X_u)_0, (X_v)_0) = -\omega_0((D_{X_v} X_u)_0) = -\omega_0\left(\sum_j x^j (u \cdot v) \left(\frac{\partial}{\partial x^j} \right)_0 \right)$$

$$= \omega_0((X_{u \cdot v})_0) = \chi(u \cdot v).$$

This implies that $(D\omega)_0$ is positive definite. $\qquad\square$

By Lemma 10.5, ω is a closed 1-form invariant under $T(V)$ and $g = D\omega$ is a Hessian metric of Koszul type on $\Omega(V) = T(V)0$. Therefore, by Theorem 8.5, $\Omega(V) = T(V)0$ is a homogeneous regular convex domain. If the clan (V, χ) has the unit element e, the 1-parameter affine transformation group $\sigma(t)$ generated by $X_e = -\sum_i (x^i + x^i(e)) \frac{\partial}{\partial x^i}$ is given by

$$\sigma(t)x = e^{-t}(x + e) - e,$$

and is a group of dilations at $-e$. Thus $\Omega(V)$ is a cone with vertex $-e$. In summary, we have the following theorem.

Theorem 10.4. *Let (V, χ) be a clan. Then we can construct from the clan (V, χ) a homogeneous regular convex domain $\Omega(V)$. If the clan has the unit element e, then $\Omega(V) = \{(\exp L_u)e \mid u \in V\} - e$ is a cone with vertex $-e$.*

Corollary 10.1. *Let (V, χ) be a clan. The bilinear form on V defined by $\operatorname{Tr} L_{u \cdot v}$ is a positive definite symmetric bilinear form.*

Proof.   Let α and β be the Koszul forms on $\Omega(V)$. Since $\Omega(V)$ is a regular convex domain, the second Koszul form β is positive definite (cf. the proof of Theorem 8.4). Denoting by α_0 and β_0 the restrictions of the Koszul forms to the origin 0, we obtain by Lemma 9.5

$$\alpha_0(u) = \alpha_0(q(X_{-u})) = -\text{Tr} f(X_{-u}) = -\text{Tr} L_u,$$
$$\beta_0(u, v) = -\alpha_0(f(X_{-u})q(X_{-v})) = -\alpha_0(u \cdot v) = \text{Tr} L_{u \cdot v}.$$

Therefore $\text{Tr} L_{u \cdot v} = \beta_0(u, v)$ is positive definite.   □

Example 10.1. Let V be the vector space of all real symmetric matrices of degree n and Ω the set of all positive definite symmetric matrices in V. We have seen in Example 4.1 that Ω is a regular convex cone in V and the group $\{f(s) \mid s \in GL(n, \mathbf{R})\}$ act transitively on Ω, where $f(s)x = sx^ts$ for $x \in V$. Let $T(n, \mathbf{R})$ be the group of all upper triangular matrices with all positive diagonal elements. Then

$$GL^+(n, \mathbf{R}) = T(n, \mathbf{R})SO(n),$$
$$T(n, \mathbf{R}) \cap SO(n) = \{e\}, \text{ where } e \text{ is the unit matrix,}$$

and the group $\{f(s) \mid s \in T(n, \mathbf{R})\}$ acts simply transitively on Ω. Let us consider the cone given by $\Omega - e$. For $s \in GL(n, \mathbf{R})$ we define an affine transformation $a(s)$ of V by

$$a(s) : x \longrightarrow s(x + e)^ts - e.$$

Then $a(s)$ preserves the cone $\Omega - e$ invariant. The linear part $f(s)$ and the translation vector part $q(s)$ of $a(s)$ are given by

$$f(s)x = sx^ts, \qquad q(s) = s^ts - e.$$

The affine transformation group $\{(f(s), q(s)) \mid s \in T(n, \mathbf{R})\}$ acts simply transitively on the regular convex cone $\Omega - e$. The Lie algebra $\mathfrak{t}(n, \mathbf{R})$ of the Lie group $T(n, \mathbf{R})$ is the set of all upper triangular matrices. The differential (f, q) of (f, q) is given by

$$f(X)v = Xv + v\,{}^tX,$$
$$q(X) = X + {}^tX,$$

for $X \in \mathfrak{t}(n, \mathbf{R})$ and $v \in V$. Therefore, putting $\hat{u} = q^{-1}(u)$ for $u = [u_{ij}] \in V$, we obtain

$$\hat{u} = \begin{bmatrix} \frac{1}{2}u_{11} & u_{12} & \cdots & u_{1n} \\ 0 & \frac{1}{2}u_{22} & \cdots & u_{2n} \\ \vdots & \vdots & \ddots & \vdots \\ 0 & 0 & \cdots & \frac{1}{2}u_{nn} \end{bmatrix}.$$

Then the multiplication law on V is given by

$$u \cdot v = f(q^{-1}(u))v = f(\hat{u})v = \hat{u}v + v^t\hat{u},$$

and the unit matrix e is the identity element of the algebra V. Let ψ be the characteristic function on Ω and put $\varphi(x) = \psi(x+e)$. Then $\omega = d\log\varphi$ is a 1-form on $\Omega - e$ invariant under $\{(f(s), q(s)) \mid s \in T(n,\mathbf{R})\}$, and $D\omega$ is positive definite. It follows from (4.2) that

$$\begin{aligned}
\omega_0(q(X)) &= \frac{d}{dt}\Big|_{t=0} \log\varphi(q(\exp(-tX))) \\
&= \frac{d}{dt}\Big|_{t=0} \log\psi(f(\exp(-tX))e) = \frac{d}{dt}\Big|_{t=0} \log\left\{\frac{\psi(e)}{\det f(\exp(-tX))}\right\} \\
&= \frac{d}{dt}\Big|_{t=0} \log\det\exp tf(X) = \operatorname{Tr} f(X).
\end{aligned}$$

Using these expressions we have

$$\omega_0(u) = \operatorname{Tr} f(q^{-1}(u)) = n\operatorname{Tr} q^{-1}(u) = \frac{n}{2}\operatorname{Tr} u,$$

$$\langle u, v \rangle = \omega_0(u \cdot v) = \frac{n}{2}\operatorname{Tr} u \cdot v = \frac{n}{2}\operatorname{Tr} uv.$$

Therefore (V, ω_0) is a clan with the unit element e.

10.3 Principal decompositions of clans and real Siegel domains

For a clan V we constructed in Theorem 10.4 the homogeneous regular convex domain $\Omega(V)$. In this section we shall prove that a clan V admits the principal decomposition $V = V_0 + N$, and that the domain $\Omega(V)$ is realized as a real Siegel domain by using this decomposition [Vinberg (1963)].

Let (V, χ) be a clan and let $\langle u, v \rangle = \chi(u \cdot v)$ the inner product on V. We define an element $e_0 \in V$ by

$$\langle e_0, u \rangle = \chi(u) \quad \text{for all } u \in V.$$

Then we have the following lemma.

Lemma 10.6. *The element e_0 is an idempotent, i.e. $e_0 \cdot e_0 = e_0$, and the operator R_{e_0} of the right multiplication by e_0 is symmetric with respect to the inner product $\langle u, v \rangle$ and satisfies*

$$R_{e_0}^2 = R_{e_0}.$$

Proof.   We have

$$\langle u, R_{e_0} v \rangle - \langle v, R_{e_0} u \rangle = \chi(u \cdot (v \cdot e_0) - v \cdot (u \cdot e_0))$$
$$= \chi([L_u, L_v]e_0) = \chi(L_{u \cdot v - v \cdot u} e_0)$$
$$= \langle u \cdot v - v \cdot u, e_0 \rangle = \chi(u \cdot v - v \cdot u) = 0,$$

for all u and $v \in V$. This means that R_{e_0} is symmetric with respect to the inner product $\langle u, v \rangle$. Hence

$$\langle e_0 \cdot e_0, u \rangle = \langle R_{e_0} e_0, u \rangle = \langle e_0, R_{e_0} u \rangle = \chi(u \cdot e_0) = \langle e_0, u \rangle,$$

for all $u \in V$, and so

$$e_0 \cdot e_0 = e_0.$$

By Lemma 9.1 (3) we obtain

$$[L_{e_0}, R_{e_0}] = R_{e_0} - R_{e_0}^2.$$

Put $P = R_{e_0} - R_{e_0}^2$. Since $[R_{e_0}, P] = 0$ we have $\operatorname{Tr} P^2 = \operatorname{Tr}[L_{e_0}, R_{e_0}]P = \operatorname{Tr} L_{e_0}[R_{e_0}, P] = 0$. Since P is symmetric and $\operatorname{Tr} P^2 = 0$, we have $P = 0$ and so $R_{e_0}^2 = R_{e_0}$. $\square$

Since R_{e_0} is an orthogonal projection by Lemma 10.6, we obtain an orthogonal decomposition

$$V = V_0 + N,$$

where $V_0 = \{x \in V \mid R_{e_0} x = x\}$ and $N = \{y \in V \mid R_{e_0} y = 0\}$.

The decomposition $V = V_0 + N$ is said to be the **principal decomposition** of the clan, and the element e_0 is called the **principal idempotent**. If V has the unit element e, then e is the principal idempotent because $\langle e, u \rangle = \chi(e \cdot u) = \chi(u)$.

Lemma 10.7. *Let e_0 be the principal idempotent of a clan V. Then*

$$L_{e_0} = \frac{1}{2}(R_{e_0} + I),$$

where I is the identity mapping on V.

Proof.   By Lemma 9.1 (2) we have

$$\langle L_{e_0} u, v \rangle + \langle u, L_{e_0} v \rangle = \chi(L_{e_0 \cdot u} v + L_u L_{e_0} v) = \chi(L_{u \cdot e_0} v + L_{e_0} L_u v)$$
$$= \langle R_{e_0} u, v \rangle + \langle u, v \rangle,$$

which implies

$$L_{e_0} + {}^t L_{e_0} = R_{e_0} + I,$$

where ${}^t L_{e_0}$ is the adjoint operator of L_{e_0}. Put

$$K = \frac{1}{2}(R_{e_0} + I) - L_{e_0} = \frac{1}{2}({}^t L_{e_0} - L_{e_0}).$$

Since

$$\left[L_{e_0}, \frac{1}{2}(R_{e_0} + I)\right] = \frac{1}{2}[L_{e_0}, R_{e_0}] = \frac{1}{2}(R_{e_0} - R_{e_0}^2) = 0,$$

and since all the eigenvalues of L_{e_0} and $\frac{1}{2}(R_{e_0} + I)$ are real numbers, it follows that all the eigenvalues of K are real numbers. Since ${}^t K = -K$, the eigenvalues of K are 0 or purely imaginary. Therefore all the eigenvalues of K are 0 and so $K = 0$. □

Lemma 10.8. *Let $L_{e_0} u = \lambda_u u$ and $L_{e_0} v = \lambda_v v$ where λ_u and $\lambda_v \in \mathbf{R}$. Then*

$$L_{e_0}(u \cdot v) = (-\lambda_u + \lambda_v + 1)u \cdot v.$$

Proof. By Lemma 10.7 we have

$$
\begin{aligned}
L_{e_0}(u \cdot v) &= e_0 \cdot (u \cdot v) = (e_0 \cdot u) \cdot v + [e_0 \cdot u \cdot v] \\
&= \lambda_u(u \cdot v) + [u \cdot e_0 \cdot v] = \lambda_u(u \cdot v) + u \cdot (e_0 \cdot v) - (u \cdot e_0) \cdot v \\
&= \lambda_u(u \cdot v) + \lambda_v(u \cdot v) - (2\lambda_u - 1)u \cdot v \\
&= (-\lambda_u + \lambda_v + 1)u \cdot v
\end{aligned}
$$

□

By Lemma 10.8 we obtain the following proposition.

Proposition 10.4. *Let $V = V_0 + N$ be the principal decomposition of a clan (V, χ) and let e_0 be the principal idempotent of V. Then*

$$V_0 \cdot V_0 \subset V_0, \qquad V_0 \cdot N \subset N,$$
$$N \cdot V_0 = \{0\}, \qquad N \cdot N \subset V_0.$$

In particular V_0 is a subalgebra with the unit element e_0.

Definition 10.3. A clan (V, χ) is said to be **elementary** if $\dim V_0 = 1$.

Lemma 10.9. *Let $V = V_0 + N$ be the principal decomposition of a clan (V, χ). Then*

(1) $y \cdot z = z \cdot y$ *for y and $z \in N$.*
(2) $\exp L_a(y \cdot z) = (\exp L_a y) \cdot (\exp L_a z)$ *for $a \in V_0$ and $y, z \in N$.*

Proof.   Let y and $z \in N$. Since $R_{e_0}(y \cdot z - z \cdot y) = L_{y \cdot z - z \cdot y} e_0 = [L_y, L_z] e_0$
$= L_y R_{e_0} z - L_z R_{e_0} y = 0$, we obtain $y \cdot z - z \cdot y \in N$. On the other hand,
by Proposition 10.4, we have $y \cdot z - z \cdot y \in V_0$. Hence $y \cdot z - z \cdot y = 0$.
Again by Proposition 10.4, it follows that $L_a(y \cdot z) - y \cdot (L_a z) = [L_a, L_y] z =$
$L_{a \cdot y - y \cdot a} z = L_{a \cdot y} z = (L_a y) \cdot z$. Re-arranging, $L_a(y \cdot z) = (L_a y) \cdot z + y \cdot (L_a z)$.
This implies (2).   □

Theorem 10.5. *Let $V = V_0 + N$ be the principal decomposition of a clan*
(V, χ) with the principal idempotent e_0. Then the homogeneous regular
convex domain $\Omega(V)$ constructed from the clan (V, χ) is expressed by

$$\Omega(V) = \left\{ x + y \mid x \in V_0, \ y \in N, \ x - \frac{1}{2} y \cdot y \in \Omega(V_0) \right\},$$

where $\Omega(V_0)$ is the homogeneous regular convex cone with vertex $-e_0$ cor-
responding to the clan (V_0, χ).

Proof.   Since the clan (V_0, χ) has the unit element e_0, the corresponding
homogeneous regular convex domain $\Omega(V_0)$ is a cone with vertex $-e_0$. We
set

$$\tilde{\Omega}(V) = \left\{ x + y \mid x \in V_0, \ y \in N, \ x - \frac{1}{2} y \cdot y \in \Omega(V_0) \right\}.$$

Let $a \in V_0$ and $x + y \in \tilde{\Omega}(V)$. It follows from (10.2) that

$$\exp X_{-a}(x + y) = \exp L_a(x + y) + \sum_{n=0}^{\infty} \frac{1}{(n+1)!} (L_a)^n a$$

$$= \exp L_a x + \sum_{n=0}^{\infty} \frac{1}{(n+1)!} (L_a)^n a + \exp L_a y$$

$$= \exp X_{-a} x + \exp L_a y.$$

By Lemma 10.9 we obtain

$$\exp X_{-a} x - \frac{1}{2} (\exp L_a y) \cdot (\exp L_a y)$$

$$= \exp X_{-a} x - \frac{1}{2} \exp L_a (y \cdot y)$$

$$= \exp L_a x + \sum_{n=0}^{\infty} \frac{1}{(n+1)!} (L_a)^n a - \frac{1}{2} \exp L_a (y \cdot y)$$

$$= \exp L_a \left(x - \frac{1}{2} y \cdot y \right) + \sum_{n=0}^{\infty} \frac{1}{(n+1)!} (L_a)^n a$$

$$= \exp X_{-a} \left(x - \frac{1}{2} y \cdot y \right) \in \Omega(V_0).$$

Since $\exp L_a y \in N$ and $\exp X_{-a} x \in V_0$ we have

$$\exp X_{-a}(x + y) = \exp L_a y + \exp X_{-a} x \in \tilde{\Omega}(V).$$

Hence the affine transformation $\exp X_{-a}$ leaves $\tilde{\Omega}(V)$ invariant. Let $b \in N$ and $x + y \in \tilde{\Omega}(V)$. It follows from Proposition 10.4 that

$$\exp X_{-b}(x + y) = \exp L_b(x + y) + \sum_{n=0}^{\infty} \frac{1}{(n+1)!}(L_b)^n b$$

$$= (x + y + b \cdot y) + \left(b + \frac{1}{2}b \cdot b\right) = \left(x + b \cdot y + \frac{1}{2}b \cdot b\right) + (y + b).$$

Since $x + b \cdot y + \dfrac{1}{2}b \cdot b \in V_0$ and $y + b \in N$ by Proposition 10.4, and since

$$x + b \cdot y + \frac{1}{2}b \cdot b - \frac{1}{2}(y + b) \cdot (y + b) = x - \frac{1}{2}y \cdot y \in \Omega(V_0)$$

by Lemma 10.9 (1), we obtain

$$\exp X_{-b}(x + y) \in \tilde{\Omega}(V).$$

Hence the affine transformation $\exp X_{-b}$ leaves $\tilde{\Omega}(V)$ invariant. Therefore $\tilde{\Omega}(V)$ is invariant under the group $T(V) = \{\exp X_u \mid u \in V\}$. Let $x + y \in \tilde{\Omega}(V)$ be an arbitrary point. Then

$$\exp X_{-y}(x + y) = x - \frac{1}{2}y \cdot y \in \Omega(V_0).$$

Since $T(V_0) = \{\exp X_{u_0} \mid u_0 \in V_0\}$ acts transitively on $\Omega(V_0)$ there exists $u_0 \in V_0$ such that

$$\exp X_{u_0}\left(x - \frac{1}{2}y \cdot y\right) = 0.$$

Hence $\exp X_{u_0} \exp X_{-y}(x + y) = 0$. Therefore $\tilde{\Omega}(V) = T(V)0 = \Omega(V)$. $\square$

Definition 10.4. Let V be a finite-dimensional real vector space. Suppose that V is decomposed into the direct sum $V = V_0 + W$ of two subspaces V_0 and W. Suppose further that there exists an open regular convex cone Ω_0 with vertex $-v_0$ in V_0 and a symmetric bilinear function $F : W \times W \longrightarrow V_0$ satisfying the following conditions,

(1) $F(w, w) \in \overline{\Omega}_0 + v_0$ for all $w \in W$,
(2) $F(w, w) = 0$ if and only if $w = 0$.

Then a domain in V defined by

$$\{u + w \mid u \in V_0, \ w \in W, \ u - F(w, w) \in \Omega_0\}$$

is called a **real Siegel domain** and is denoted by $\mathcal{S}(\Omega_0, F)$.

Lemma 10.10. *A real Siegel domain* $\mathcal{S}(\Omega_0, F)$ *is a regular convex domain.*

Proof. Let $\lambda(u + w) + (1 - \lambda)(u' + w')$ be a point in the segment of a line joining any two points $u + v$ and $u' + v' \in \mathcal{S}(\Omega_0, F)$ where $0 \le \lambda \le 1$. It follows from the convexity of $\Omega_0 + v_0$ and condition (1) for F that

$$\lambda u + (1 - \lambda)u' - F(\lambda w + (1 - \lambda)w', \lambda w + (1 - \lambda)w')$$
$$= \lambda(u - F(w, w)) + (1 - \lambda)(u' - F(w', w'))$$
$$+ \lambda(1 - \lambda)F(w - w', w - w') \in \Omega_0 + v_0.$$

Thus $\mathcal{S}(\Omega_0, F)$ is a convex set. Suppose that a full line $u + w + t(a + b)$ through a point $u + w \in \mathcal{S}(\Omega_0, F)$ in a direction $a + b \in V_0 + W$ is contained in $\mathcal{S}(\Omega_0, F)$. Then

$$u + ta - F(w + tb, w + tb)$$
$$= u - F(w, w) + ta - 2tF(w, b) - t^2 F(b, b) \in \Omega_0 + v_0.$$

Since $\Omega_0 + v_0$ is a cone with vertex 0, it follows

$$\frac{1}{t^2}\{u - F(w, w) + ta - 2tF(w, b) - t^2 F(b, b)\} \in \Omega_0 + v_0,$$

for all $t \in \mathbf{R}$. Letting $t \longrightarrow \infty$, we see $-F(b, b) \in \overline{\Omega}_0 + v_0$, that is $F(b, b) \in -(\overline{\Omega}_0 + v_0)$. On the other hand by condition (1) for F we know $F(b, b) \in \overline{\Omega}_0 + v_0$. Hence $F(b, b) = 0$. By condition (2) we have $b = 0$. This means

$$u - F(w, w) + ta \in \Omega_0,$$

for all $t \in \mathbf{R}$. Since Ω_0 contains no straight line we obtain $a = 0$. Therefore $\mathcal{S}(\Omega_0, F)$ does not contain any full straight line.   □

Lemma 10.11. *With the same notation as in* Theorem 10.5 *we have*

(1) $y \cdot y = \overline{\Omega(V_0)} + e_0$ *for* $y \in N$.
(2) $y \cdot y = 0$ *if and only if* $y = 0$.

Proof. Since $\left(\frac{1}{2}y \cdot y - e_0\right) - \frac{1}{2}(\pm y) \cdot (\pm y) = -e_0 \in \overline{\Omega(V_0)}$, it follows from Theorem 10.5 that

$$\left(\frac{1}{2}y \cdot y - e_0\right) \pm y \in \overline{\Omega(V)}.$$

Since $\overline{\Omega(V)}$ is a convex domain we obtain

$$\frac{1}{2}y \cdot y - e_0 = \frac{1}{2}\left\{\left(\frac{1}{2}y \cdot y - e_0 + y\right) + \left(\frac{1}{2}y \cdot y - e_0 - y\right)\right\} \in \overline{\Omega(V)}.$$

Therefore $\frac{1}{2}y \cdot y - e_0 \in \overline{\Omega(V)} \cap V_0 = \overline{\Omega(V_0)}$ and $\frac{1}{2}y \cdot y \in \overline{\Omega(V_0)} + e_0 D$ Hence we obtain (1). Suppose $y \cdot y = 0$. Then $\langle y, y \rangle = \chi(y \cdot y) = 0$, and so $y = 0D$ Thus (2) is also proved.   □

By Theorem 10.5 and Lemma 10.11 we obtain the following corollary.

Corollary 10.2. *Let V be a clan and let $V = V_0 + N$ be the principal decomposition. Then the homogeneous regular convex domain $\Omega(V)$ corresponding to V is a real Siegel domain $S(\Omega(V_0), F)$ where $\Omega(V_0)$ is the cone corresponding to V_0 and F is the symmetric bilinear function given by*

$$F : N \times N \ni (y, z) \longrightarrow \frac{1}{2} y \cdot z \in V_0.$$

Proposition 10.5. *An n-dimensional elementary clan V corresponds to a real Siegel domain given by*

$$x^n > \frac{1}{2} \sum_{i=1}^{n-1} (x^i)^2 - 1,$$

(cf. Proposition 3.8).

Proof. Let (V, χ) be an n-dimensional elementary clan with principal decomposition $V = V_0 + N$ and principal idempotent e_0. Choose an orthogonal basis $\{e_1, \cdots, e_{n-1}\}$ of N such that $\langle e_i, e_j \rangle = \chi(e_i \cdot e_j) = \delta_{ij} \chi(e_0)$. Then

$$e_i \cdot e_j = \delta_{ij} e_0.$$

Let $\{x^1, \cdots, x^{n-1}, x^n\}$ be an affine coordinate system on V with respect to the orthogonal basis $\{e_1, \cdots, e_{n-1}, e_n\}$ of V where $e_n = e_0$. Then

$$\Omega_0 = \{x^n e_n \mid x^n > -1\},$$

$$F(y, z) = \frac{1}{2} y \cdot z = \frac{1}{2} \left(\sum_{i=1}^{n-1} y^i z^i \right) e_0, \quad \text{for } y = \sum_{i=1}^{n-1} y^i e_i, \ z = \sum_{i=1}^{n-1} z^i e_i \in N.$$

Therefore

$$\Omega(V) = \left\{ x^n e_n + \sum_{i=1}^{n-1} x^i e_i \ \Big| \ x^n - \frac{1}{2} \sum_{i=1}^{n-1} (x^i)^2 > -1 \right\}.$$

Conversely, let Ω be a domain in $\mathbf{R}^n$ defined by

$$x^n > \frac{1}{2} \sum_{i=1}^{n-1} (x^i)^2 - 1.$$

Let us consider a direct sum decomposition of $\mathbf{R}^n$ such that

$$\mathbf{R}^n = V_0 + N,$$
$$V_0 = \{x \in \mathbf{R}^n \mid x^i = 0, \ 1 \le i \le n - 1\},$$
$$N = \{x \in \mathbf{R}^n \mid x^n = 0\}.$$

We denote an element in $\mathbf{R}^n$ by $\begin{bmatrix} a \\ b \end{bmatrix}$ where $a \in \mathbf{R}^{n-1}$ and $b \in \mathbf{R}$. We define an open regular convex cone Ω_0 in V_0 with vertex $-e_0$ and a bilinear mapping $F : N \times N \longrightarrow V_0$ by

$$\Omega_0 = \{x \in V_0 \mid x^n > -1\},$$

$$F(y, z) = \begin{bmatrix} 0 \\ {}^t y z \end{bmatrix}.$$

Then F and Ω_0 satisfy the conditions of Definition 10.4 and Ω is a real Siegel domain $\mathcal{S}(\Omega_0, F)$. For $\lambda \in \mathbf{R}^+$ we define an affine transformation of $\mathbf{R}^n$ by

$$x^i \longrightarrow \lambda x^i, \quad \text{for } 1 \leq i \leq n - 1,$$

$$x^n \longrightarrow \lambda^2 x^n + \lambda^2 - 1.$$

Then the affine transformation leaves Ω invariant. The linear part $\boldsymbol{f}(\lambda)$ and the translation vector part $\boldsymbol{q}(\lambda)$ of the affine transformation are expressed by

$$\boldsymbol{f}(\lambda) = \begin{bmatrix} \lambda I_{n-1} & 0 \\ 0 & \lambda^2 \end{bmatrix}, \quad \boldsymbol{q}(\lambda) = \begin{bmatrix} 0 \\ \lambda^2 - 1 \end{bmatrix},$$

where I_{n-1} is the unit matrix of degree $n - 1$. For $a = [a^i] \in N$ we define an affine transformation of $\mathbf{R}^n$ by

$$x^i \longrightarrow x^i + a^i, \quad \text{for } 1 \leq i \leq n - 1,$$

$$x^n \longrightarrow x^n + \sum_{i=1}^{n-1} a^i x^i + \frac{1}{2} \sum_{i=1}^{n-1} (a^i)^2.$$

Then the affine transformation leaves Ω invariant. Identifying N with $\mathbf{R}^{n-1}$, the linear part $\boldsymbol{f}(a)$ and the translation vector part $\boldsymbol{q}(a)$ of the affine transformation are expressed by

$$\boldsymbol{f}(a) = \begin{bmatrix} I_{n-1} & 0 \\ {}^t a & 1 \end{bmatrix}, \quad \boldsymbol{q}(a) = \begin{bmatrix} a \\ \frac{1}{2} {}^t aa \end{bmatrix}.$$

By $(s, x) \in G \times \Omega \longrightarrow \boldsymbol{f}(s)x + \boldsymbol{q}(s) \in \Omega$ the Lie group $G = \mathbf{R}^+ \times N$ acts simply transitively on Ω. The Lie algebra $\mathfrak{g}$ of the Lie group $G = \mathbf{R}^+ \times N$ is identified with $\mathbf{R} + N$. Since

$$\boldsymbol{f}(\exp tr) = \boldsymbol{f}(e^{tr}) = \begin{bmatrix} e^{tr} I_{n-1} & 0 \\ 0 & e^{2tr} \end{bmatrix}, \quad \boldsymbol{q}(\exp tr) = \begin{bmatrix} 0 \\ e^{2tr} - 1 \end{bmatrix},$$

for $r \in \mathbf{R} \subset \mathfrak{g}$, the differentials f of $\boldsymbol{f}$ and q of $\boldsymbol{q}$ are given by

$$f(r) = \begin{bmatrix} r I_{n-1} & 0 \\ 0 & 2r \end{bmatrix}, \quad q(r) = \begin{bmatrix} 0 \\ 2r \end{bmatrix}.$$

Since

$$f(\exp ta) = f(ta) = \begin{bmatrix} I_{n-1} & 0 \\ t\,{}^t a & 1 \end{bmatrix}, \qquad q(\exp ta) = q(ta) = \begin{bmatrix} ta \\ \frac{t^2}{2}\,{}^t aa \end{bmatrix},$$

for $a \in N = \mathbf{R}^{n-1} \subset \mathfrak{g}$, the differentials f of f and q of q are expressed by

$$f(a) = \begin{bmatrix} 0 & 0 \\ {}^t a & 0 \end{bmatrix}, \qquad q(a) = \begin{bmatrix} a \\ 0 \end{bmatrix}.$$

These imply

$$q : r + a \in \mathfrak{g} = \mathbf{R} + N \longrightarrow \begin{bmatrix} a \\ 2r \end{bmatrix} \in \mathbf{R}^n.$$

Therefore

$$q^{-1}(u) = \frac{1}{2} u^n + \begin{bmatrix} u^1 \\ \vdots \\ u^{n-1} \\ 0 \end{bmatrix} \in \mathfrak{g} = \mathbf{R} + N,$$

$$L_u = f(q^{-1}(u)) = \begin{bmatrix} \frac{1}{2} u^n I_{n-1} & 0 \\ u^1 \cdots u^{n-1} & u^n \end{bmatrix}, \quad \text{for } u = [u^i] \in \mathbf{R}^n.$$

Thus

$$u \cdot v = L_u v = \begin{bmatrix} \frac{1}{2} u^n v^1 \\ \vdots \\ \frac{1}{2} u^n v^{n-1} \\ {}^t uv \end{bmatrix}, \quad \text{for } u = [u^i] \text{ and } v = [v^i] \in \mathbf{R}^n.$$

Putting $\varphi(x) = -\log\left\{x^n + 1 - \frac{1}{2}\sum_{i=1}^{n-1}(x^i)^2\right\}$ the 1-form $\omega = d\varphi$ is invariant under the group $G = \mathbf{R}^+ \times N$, and the symmetric bilinear form $D\omega$ is positive-definite (cf. Proposition 3.8). For $X = r + a \in \mathfrak{g} = \mathbf{R} + N$ we have

$$q(\exp tX) = q((e^{tr})(ta)) = f(e^{tr})\, q(ta) + q(e^{tr})$$

$$= \begin{bmatrix} 0 \\ e^{2tr}\left(\frac{t^2}{2}\,{}^t aa + 1\right) - 1 \end{bmatrix}.$$

Hence

$$\omega_0(q(X)) = \frac{d}{dt}\Big|_{t=0} \varphi(\, q(\exp(-tX)))$$

$$= \frac{d}{dt}\Big|_{t=0} \log\left\{e^{2tr}\left(\frac{t^2}{2}\,{}^t aa + 1\right)\right\} = 2r.$$

Thus $\omega_0(u) = u^n$ for $u = [u^i] \in \mathbf{R}^n$ and

$$\langle u, v \rangle = \omega_0(u \cdot v) = {}^t uv, \text{ for } u \text{ and } v \in \mathbf{R}^n.$$

Therefore $(\mathbf{R}^n, \omega_0)$ is a clan. Let $e_0 = \begin{bmatrix} 0 \\ 1 \end{bmatrix} \in V_0$. Then

$$\langle u, e_0 \rangle = \omega_0(u \cdot e_0) = \omega_0(u^n e_0) = u^n = \omega_0(u),$$
$$L_{e_0} v_0 = v_0, \quad \text{for } v_0 \in V_0,$$
$$L_{e_0} u = \frac{1}{2} u, \quad \text{for } u \in N.$$

Hence e_0 is the principal idempotent of the clan $(\mathbf{R}^n, \omega_0)$, and $\mathbf{R}^n = V_0 + N$ is the principal decomposition. Thus the clan is elementary. $\qquad \square$

Let (V, χ) be a clan and let $\Omega(V)$ be the corresponding homogeneous regular convex domain. We identify V with the tangent space $T_0 \Omega(V)$ of $\Omega(V)$ at 0 by the correspondence $u \longleftrightarrow -(X_u)_0$. Denoting by γ_u and A_u the values of $-\gamma_{X_u}$ and $-A_{X_u}$ at 0 we have

$$\gamma_u v = (\gamma_{X_u} X_v)_0 = \sum \gamma^k_{ij}(0) u^i v^j \left(\frac{\partial}{\partial x^k} \right)_0,$$
$$A_u v = (A_{X_u} X_v)_0 = -(D_{X_v} X_u)_0 = -\sum (u \cdot v)^i \left(\frac{\partial}{\partial x^i} \right)_0$$
$$= -u \cdot v.$$

We put

$$A^\nabla_{X_u} = \mathcal{L}_{X_u} - \nabla_{X_u}.$$

Then by [Kobayashi and Nomizu (1963, 1969)](I, p.235) we know

$$A^\nabla_{X_u} X_v = -\nabla_{X_v} X_u.$$

Denoting by A^∇_u the value of $-A^\nabla_{X_u}$ at 0 we obtain

$$A^\nabla_u v = (A^\nabla_{X_u} X_v)_0 = -(\nabla_{X_v} X_u)_0$$
$$= -\sum_{ij} v^i \left\{ \nabla_{\partial/\partial x^i} \left(\sum (L_u)^j_q x^q + u^j \right) \frac{\partial}{\partial x^j} \right\}_0$$
$$= -\sum (L_u)^j_i v^i \left(\frac{\partial}{\partial x^j} \right)_0 - \sum v^j \gamma^k_{ij}(0) \left(\frac{\partial}{\partial x^k} \right)_0$$
$$= -u \cdot v - \gamma_u v = -(L_u + \gamma_u)v.$$

Lemma 10.12. *For $u \in V$ we have*

(1) $A_u = -L_u,$

(2) $\gamma_u = -\dfrac{1}{2}(L_u + {}^t L_u)$,

(3) $A_u^\nabla = -\dfrac{1}{2}(L_u - {}^t L_u)$.

(4) *The sectional curvature K for a plane spanned by u and $v \in V$ is given by*

$$K = \frac{\langle \gamma_u v, \gamma_u v \rangle - \langle \gamma_u u, \gamma_v v \rangle}{\langle u, u \rangle \langle v, v \rangle - \langle u, v \rangle^2}.$$

Proof. By Lemma 2.1 we have

$$2g(\gamma_{X_u} X_v, X_w) = g(A_{X_u} X_v, X_w) + g(X_v, A_{X_u} X_w),$$

which implies assertion (2). Assertion (3) follows from $A_u^\nabla = -(L_u + \gamma_u)$ and (2). By Proposition 2.3, we obtain (4). □

Proposition 10.6. *Let (V, χ) be a clan and let $\Omega(V)$ be the homogeneous regular convex domain corresponding to the clan. For the canonical Hessian structure (D, g) on $\Omega(V)$ we have*

(1) *If the difference tensor $\gamma = \nabla - D$ is ∇-parallel, then*

$$[A_u^\nabla, \gamma_v] = \gamma_{A_u^\nabla v}.$$

(2) *If the curvature tensor R of ∇ is ∇-parallel, then*

$$[A_u^\nabla, [\gamma_v, \gamma_w]] = [\gamma_{A_u^\nabla v}, \gamma_w] + [\gamma_v, \gamma_{A_u^\nabla w}].$$

Proof. Since X_u is an infinitesimal affine transformation with respect to ∇ and D, we have (cf. [Kobayashi and Nomizu (1963, 1969)])

$$[\mathcal{L}_{X_u}, \nabla_{X_v}] = \nabla_{\mathcal{L}_{X_u} X_v}, \quad [\mathcal{L}_{X_u}, D_{X_v}] = D_{\mathcal{L}_{X_u} X_v}.$$

Hence

$$[\mathcal{L}_{X_u}, \gamma_{X_v}] = \gamma_{\mathcal{L}_{X_u} X_v}.$$

If γ is ∇-parallel, then, by Lemma 2.2, we have

$$[\nabla_{X_u}, \gamma_{X_v}] = \gamma_{\nabla_{X_u} X_v}.$$

Therefore

$$[A_{X_u}^\nabla, \gamma_{X_v}] = [\mathcal{L}_{X_u} - \nabla_{X_u}, \gamma_{X_v}] = \gamma_{\mathcal{L}_{X_u} X_v} - \gamma_{\nabla_{X_u} X_v}$$

$$= \gamma_{A_{X_u}^\nabla X_v},$$

which implies assertion (1).

Suppose that R is ∇-parallel. Since X_u is an infinitesimal affine transformation with respect to ∇ we have

$$\mathcal{L}_{X_u} R = 0.$$

Hence

$$
\begin{aligned}
0 &= (A^{\nabla}_{X_u} R)(X_v, X_w) X_z \\
&= A^{\nabla}_{X_u}(R(X_v, X_w) X_z) - R(A^{\nabla}_{X_u} X_v, X_w) X_z \\
&\quad - R(X_v, A^{\nabla}_{X_u} X_w) X_z - R(X_v, X_w) A^{\nabla}_{X_u} X_z \\
&= (A^{\nabla}_{X_u}[\gamma_{X_v}, \gamma_{X_w}] - [\gamma_{A^{\nabla}_{X_u} X_v}, \gamma_{X_w}] - [\gamma_{X_v}, \gamma_{A^{\nabla}_{X_u} X_w}] - [\gamma_{X_v}, \gamma_{X_w}] A^{\nabla}_{X_u}) X_z \\
&= ([A^{\nabla}_{X_u}, [\gamma_{X_v}, \gamma_{X_w}]] - [\gamma_{A^{\nabla}_{X_u} X_v}, \gamma_{X_w}] - [\gamma_{X_v}, \gamma_{A^{\nabla}_{X_u} X_w}]) X_z,
\end{aligned}
$$

and so assertion (2) has also been proved.   □

Proposition 10.7. *Let Ω be a homogeneous regular convex domain and let V be the corresponding clan. Then the following statements (1)-(3) are equivalent.*

(1) *Ω is affine isomorphic to the domain given by*

$$
\left\{ (x^1, \cdots, x^n) \in \mathbf{R}^n \ \middle| \ x^n > \frac{1}{2} \sum_{i=1}^{n-1} (x^i)^2 \right\}.
$$

(2) *The sectional curvature with respect to the second Koszul form β on Ω is a negative constant.*

(3) *V is an elementary clan.*

Proof. By Proposition 10.5 the statements (1) and (3) are equivalent. We shall here show that statement (2) implies (3). Let $V = V_0 + N$ be the principal decomposition of the clan V with the principal idempotent e_0. By Lemmata 10.6 and 10.7 we know ${}^t L_{e_0} = L_{e_0} $D Hence $\gamma_{e_0} = -L_{e_0}$ by Lemma 10.12. Suppose that there exists $v_0 \in V_0$ independent to e_0. Since $L_{e_0} v_0 = v_0$ we obtain

$$
\begin{aligned}
\langle \gamma_{e_0} v_0, \gamma_{e_0} v_0 \rangle - \langle \gamma_{e_0} e_0, \gamma_{v_0} v_0 \rangle &= \langle v_0, v_0 \rangle + \langle e_0, \gamma_{v_0} v_0 \rangle \\
&= \langle v_0, v_0 \rangle + \langle \gamma_{e_0} v_0, v_0 \rangle = 0.
\end{aligned}
$$

By Lemma 10.12, this implies that the sectional curvature K for the space spanned by e_0 and v_0 vanishes, which contradicts (2). Therefore $V_0 = \mathbf{R} e_0$ and V is an elementary clan. We shall prove that statement (2) follows from (3). Let $V = \mathbf{R} e_0 + N$ be the principal decomposition of the elementary clan V. Let $\{e_1, \cdots, e_{n-1}\}$ be an orthonormal basis of N with respect to the inner product $\langle u, v \rangle = \operatorname{Tr} L_{u \cdot v}$. Putting $e_n = c e_0$ where $c = \langle e_0, e_0 \rangle^{-\frac{1}{2}}$ we obtain

$$
e_i \cdot e_j = c \delta_{ij} e_n, \quad e_n \cdot e_i = \frac{c}{2} e_i, \quad e_i \cdot e_n = 0, \quad \text{for } 1 \le i, j \le n-1,
$$

$$
e_n \cdot e_n = c e_n.
$$

Hence

$$u \cdot v = \frac{c}{2} \sum_{i=1}^{n-1} (u_n v_i) e_i + c \Big(\sum_{i=1}^{n-1} u_i v_i + u_n v_n \Big) e_n,$$

for $u = \sum_{i=1}^{n} u_i e_i$ and $v = \sum_{i=1}^{n} v_i e_i$. Since

$$\langle \gamma_u v, e_i \rangle = \frac{1}{2} \{ \langle u \cdot v, e_i \rangle + \langle v, u \cdot e_i \rangle \} = \frac{c}{2} (u_n v_i + v_n u_i), \quad \text{for } 1 \le i \le n-1,$$

$$\langle \gamma_u v, e_n \rangle = \frac{1}{2} \{ \langle u \cdot v, e_n \rangle + \langle v, u \cdot e_n \rangle \} = \frac{c}{2} \Big(\sum_{i=1}^{n-1} u_i v_i + 2 u_n v_n \Big),$$

we have

$$\gamma_u v = \frac{c}{2} \Big\{ \sum_{i=1}^{n-1} (u_n v_i + v_n u_i) e_i + \Big(\sum_{i=1}^{n-1} u_i v_i + 2 u_n v_n \Big) e_n \Big\}.$$

Using this expression we obtain

$$\langle \gamma_u v, \gamma_u v \rangle - \langle \gamma_u u, \gamma_v v \rangle$$

$$= \frac{c^2}{4} \Big\{ \sum_{i=1}^{n-1} (u_n v_i + v_n u_i)^2 + \Big(\sum_{i=1}^{n-1} u_i v_i + 2 u_n v_n \Big)^2$$

$$- \Big(\sum_{i=1}^{n-1} u_i^2 + 2 u_n^2 \Big) \Big(\sum_{i=1}^{n-1} v_i^2 + 2 v_n^2 \Big) - \sum_{i=1}^{n-1} u_n v_n u_i v_i \Big\}$$

$$= -\frac{c^2}{4} \Big\{ \Big(\sum_{i=1}^{n-1} u_i^2 \Big) \Big(\sum_{i=1}^{n-1} v_i^2 \Big) - \Big(\sum_{i=1}^{n-1} u_i v_i \Big)^2 + \sum_{i=1}^{n-1} (u_n v_i - v_n u_i)^2 \Big\}$$

$$= -\frac{c^2}{4} \{ \langle u, u \rangle \langle v, v \rangle - \langle u, v \rangle^2 \}.$$

Hence, by Lemma 10.12 (4) the sectional curvature K is a constant of value $-\frac{1}{4} \langle e_0, e_0 \rangle^{-1} = -(2(\dim V + 1))^{-1} D$ □

We shall give a differential geometric characterization of homogeneous self-dual regular convex cones among homogeneous regular convex domains. We proved in Theorem 4.11 that the difference tensor $\gamma = \nabla - D$ on a homogeneous self-dual regular convex cone is ∇-parallel. Here, we shall now show the converse.

Theorem 10.6. *Let (V, χ) be a clan satisfying the condition*

(C) $$[A_u^\nabla, \gamma_v] = \gamma_{A_u^\nabla v}.$$

We define a new operation of multiplication on V by

$$u * v = \gamma_u v.$$

Then V is a compact Jordan algebra with this multiplication.

Lemma 10.13. *Under the same condition of the above theorem we have*

(1) *The clan has the unit element e.*
(2) $\operatorname{Tr} \gamma_{\gamma_u v} = \operatorname{Tr} L_{L_u v}$.
(3) $[[\gamma_u, \gamma_v], \gamma_w] = \gamma_{[\gamma_u, \gamma_v]w}$.

Proof. By the condition (C) and Lemma 10.12 we have

$$0 = \big\langle ([A_u^\nabla, \gamma_u] - \gamma_{A_u^\nabla u})u, u \big\rangle = -3\langle \gamma_u u, A_u^\nabla u \rangle$$
$$= \frac{3}{4}(\langle L_u u, L_u u \rangle - \langle {}^t L_u u, {}^t L_u u \rangle),$$

that is,

$$\langle L_u u, L_u u \rangle = \langle {}^t L_u u, {}^t L_u u \rangle.$$

Let $V = V_0 + N$ be the principal decomposition of V. For $u \in N$ we have $\langle L_u u, L_u u \rangle = \langle {}^t L_u u, {}^t L_u u \rangle = \langle u, L_u {}^t L_u u \rangle = 0$ because $L_u V \subset V_0$ and $V = V_0 + N$ is an orthogonal decomposition. Hence $u \cdot u = 0$. Since $\langle u, u \rangle = \chi(u \cdot u) = 0$ we have $u = 0$. Therefore $V = V_0$ and V has the unit element e. Again by condition (C) and Lemma 10.12, we have $0 = \operatorname{Tr}[A_u^\nabla, \gamma_v] = \operatorname{Tr} \gamma_{A_u^\nabla v} = -\operatorname{Tr} \gamma_{(L_u + \gamma_u)v}$. Hence

$$\operatorname{Tr} \gamma_{\gamma_u v} = -\operatorname{Tr} \gamma_{L_u v} = \operatorname{Tr} L_{L_u v}.$$

From Proposition 2.3 (1) we have

$$[\gamma_{X_u}, \gamma_{X_v}] = -R^\nabla(X_u, X_v) = -[A_{X_u}^\nabla, A_{X_v}^\nabla] + A_{[X_u, X_v]}^\nabla.$$

Applying this equation and condition (C) we obtain

$$[[\gamma_u, \gamma_v], \gamma_w] = \gamma_{\{-[A_u^\nabla, A_v^\nabla] - A_{u \cdot v - v \cdot u}^\nabla\}w} = \gamma_{[\gamma_u, \gamma_v]w}. \qquad \square$$

Proof of Theorem 10.6. By Lemma 10.13 (3) we have

$$[[\gamma_u, \gamma_v], \gamma_w] = \gamma_{[u * w * v]},$$

where $[u * w * v] = u * (v * w) - v * (u * w)$. Again by Lemma 10.13 (2) a symmetric bilinear form defined by

$$(u, v) = \operatorname{Tr} \gamma_{u * v} = \operatorname{Tr} L_{u \cdot v} = \langle u, v \rangle$$

is positive-definite. By Lemma 4.2 these facts imply that V is a compact Jordan algebra with respect to the multiplication $u * v = \gamma_u v$. □

Proposition 10.8. *Let V be a clan satisfying the condition* (C) *in Theorem 10.6, and let $\Omega(V)$ be the homogeneous regular convex cone corresponding to the clan V. We denote by $\Omega(V, *)$ the homogeneous self-dual regular convex cone corresponding to the compact Jordan algebra $(V, *)$ with multiplication $u * v = \gamma_u v$ (cf. Theorem 4.9 and 10.6). Then the homogeneous regular convex cone $\Omega(V) + e$ coincides with $\Omega(V, *)$.*

Proof. It is known that $\exp \gamma_u$ is contained in the linear automorphism group $Aut(\Omega(V, *))$ of $\Omega(V, *)$, and that

$$\Omega(V, *) = \{\exp \gamma_v \mid v \in V\}e$$

$$= \left\{ \exp v = \sum_{n=0}^{\infty} \frac{1}{n!} \overbrace{v * \cdots * v}^{n \text{ terms}} \mid v \in V \right\},$$

(cf. Theorem 4.9). Since condition (C) implies $A_u^\nabla (v * w) = (A_u^\nabla v) * w + v * A_u^\nabla w$, we obtain

$$\exp A_u^\nabla (v * w) = (\exp A_u^\nabla v) * (\exp A_u^\nabla w).$$

Hence, for any $\exp v \in \Omega(V, *)$, we have

$$\exp A_u^\nabla \exp v = \exp A_u^\nabla \left(\sum_n \frac{1}{n!} v * \cdots * v \right)$$

$$= \sum_n \frac{1}{n!} \{(\exp A_u^\nabla v) * \cdots * (\exp A_u^\nabla v)\}$$

$$= \exp(\exp A_u^\nabla v) \in \Omega(V, *),$$

which implies

$$\exp A_u^\nabla \in Aut(\Omega(V, *)).$$

Therefore A_u^∇ and γ_u are contained in the Lie algebra $\mathfrak{g}$ of $Aut(\Omega(V, *))$, and so $L_u = \gamma_u + A_u^\nabla \in \mathfrak{g}$. Since $\Omega(V) + e = \{\exp L_u \mid u \in V\}e$ by Theorem 10.4, we have

$$\Omega(V) + e \subset \Omega(V, *).$$

By Lemma 10.13 (2), both characteristic functions of $\Omega(V) + e$ and $\Omega(V, *)$ coincide. It follows from the property of the characteristic function stated in Proposition 4.3 that

$$\Omega(V) + e = \Omega(V, *).$$

□

Theorem 10.7. *Let Ω be a homogeneous regular convex domain. Then the following conditions are equivalent.*

(1) Ω *is a self-dual cone.*
(2) *The difference tensor $\gamma = \nabla - D$ is ∇-parallel.*

Proof. In Theorem 4.12 we proved that statement (1) implies (2). Suppose that statement (2) holds. Then, by Proposition 10.6 (1), the corresponding clan satisfies condition (C) in Theorem 10.6. Therefore, by Proposition 10.8, Ω is a self-dual cone. □

10.4 Homogeneous Hessian domains and normal Hessian algebras

By using the same approach as in section **10.2**, we shall give a correspondence between homogeneous Hessian domains and normal Hessian algebras, and we state without proof a structure theorem of normal Hessian algebras. From the theorem we will see that the second Koszul forms plays an important role in the structure theory of homogeneous Hessian manifolds.

Let (Ω, D, g) be a homogeneous Hessian domain, then there exists a triangular group T acting simply transitively on Ω. In section **10.2** we showed that the flat connection D invariant by T induces an affine representation (f, q) on V of the Lie algebra $\mathfrak{t}$ of T, and defined an operation of multiplication on V by

$$u \cdot v = f(q^{-1}(u))v.$$

Then the algebra V with this multiplication is a normal left symmetric algebra. By Lemma 9.2 we obtain the lemma.

Lemma 10.14. *Let $\langle \, , \, \rangle$ be the restriction of the Hessian metric g to the origin 0. Then we have*

$$\langle u \cdot v, w \rangle + \langle v, u \cdot w \rangle = \langle v \cdot u, w \rangle + \langle u, v \cdot w \rangle. \tag{10.5}$$

Definition 10.5. A normal left symmetric algebra V with an inner product $\langle \, , \, \rangle$ satisfying the condition (10.5) is called a **normal Hessian algebra** and is denoted by $(V, \langle \, , \, \rangle)$.

It should be remarked that a clan (V, χ) is a normal Hessian algebra $(V, \langle \, , \, \rangle)$ where $\langle u, v \rangle = \chi(u \cdot v)$.

By Lemma 10.14 we have the following theorem.

Theorem 10.8. *For a homogeneous Hessian domain* (Ω, D, g) *there corresponds a normal Hessian algebra.*

Conversely, we shall show that a normal Hessian algebra $(V, \langle \ , \ \rangle)$ induces a homogeneous Hessian domain. We use the same method and notation as in section **10.2**. Let $\{e_1, \cdots, e_n\}$ be an orthonormal basis of V. Since $T(V)$ acts simply transitively on $\Omega(V) = T(V)0$, we can define a Riemannian metric g on $\Omega(V)$ by

$$g_{ij}(s0) = \sum_{p=1}^{n} \boldsymbol{f}(s^{-1})_i^p \, \boldsymbol{f}(s^{-1})_j^p, \quad \text{for } s \in T(V), \tag{10.6}$$

where $[\, \boldsymbol{f}(s^{-1})_j^i\,]$ is the matrix representation of $\boldsymbol{f}(s^{-1})$ with respect to $\{e_1, \cdots, e_n\}$.

Lemma 10.15. *The pair* (D, g) *is a Hessian structure on* $\Omega(V)$ *invariant by* $T(V)$.

Proof. By definition, g is invariant under $T(V)$. For $u \in V$ we denote by $\exp t X_u$ the 1-parameter affine transformation group generated by

$$X_u = -\sum_i (\sum_j (L_u)^i_j x^j + x^i(u)) \frac{\partial}{\partial x^i}.$$

Then we have

$$\frac{d}{dt}\bigg|_{t=0} \boldsymbol{f}(\exp t X_u) = -L_u, \qquad \frac{d}{dt}\bigg|_{t=0} \boldsymbol{q}(\exp t X_u) = -u.$$

For a fixed $s \in T(V)$, we define a linear isomorphism $u \in V \longrightarrow u' \in V$ by $s^{-1} \exp t X_u s = \exp t X_{u'}$. Then

$$L_{u'} = \boldsymbol{f}(s)^{-1} L_u \boldsymbol{f}(s),$$
$$u' = \boldsymbol{f}(s)^{-1} L_u \boldsymbol{q}(s) + \boldsymbol{f}(s)^{-1} u = L_{u'} \boldsymbol{f}(s)^{-1} \boldsymbol{q}(s) + \boldsymbol{f}(s)^{-1} u.$$

Denoting by $\mathcal{L}_X$ the Lie differentiation with respect to a vector field X on $\Omega(V)$, we set

$$A_X = \mathcal{L}_X - D_X,$$

and then

$$A_X Y = -D_Y X.$$

Hence

$$(A_{X_u} X_v)_x = -\sum_i (L_u L_v x + L_u v)^i \left(\frac{\partial}{\partial x^i}\right)_x.$$

Since the derivation A_{X_u} maps a function to 0, and given also that $\mathcal{L}_{X_u}g = 0$, we have

$$(D_{X_u}g)(X_v, X_w) = -(A_{X_u}g)(X_v, X_w)$$
$$= g(A_{X_u}X_v, X_w) + g(X_v, A_{X_u}X_w).$$

Therefore

$$g(s0)((A_{X_u}X_v)_{s0}, (X_w)_{s0})$$
$$= \sum_{i,j,p} \mathbf{f}(s^{-1})_i^p \mathbf{f}(s^{-1})_j^q (L_u L_v s0 + L_u v)^i (L_w s0 + w)^j$$
$$= \sum_p \{\mathbf{f}(s^{-1})(L_u L_v \mathbf{q}(s) + L_u v)\}^p w'^p$$
$$= \sum_p \{L_{u'} L_{v'} \mathbf{f}(s^{-1}) \mathbf{q}(s) + L_{u'} \mathbf{f}(s)^{-1}v\}^p w'^p$$
$$= \sum_p (u' \cdot v')^p w'^p$$
$$= \langle u' \cdot v', w' \rangle.$$

These results and (10.5) imply that (D, g) satisfies the Codazzi equation

$$(D_{X_u}g)(X_v, X_w) = (D_{X_v}g)(X_u, X_w),$$

so by Proposition 2.1 (2) the pair (D, g) is therefore a Hessian structure.

$\square$

The following theorem summarizes the above results.

Theorem 10.9. *A normal Hessian algebra $(V, \langle \ , \ \rangle)$ induces a homogeneous Hessian domain $\Omega(V)$ on which a triangular group $T(V)$ acts simply transitively.*

We state without proof a structure theorem of normal Hessian algebras. For the detailed proof the reader may refer to [Shima (1980)].

Theorem 10.10 (Structure theorem of normal Hessian algebras).
Let $(V, \langle \ , \ \rangle)$ be a normal Hessian algebra. Then we have

(1) *Let $C = \{c \in V \mid c \cdot c = 0\}$. Then C is a vector subspace of V.*
(2) *The orthogonal complement U of C is a clan.*
(3) *The decomposition $V = C + U$ satisfies*

$$C \cdot C = \{0\}, \qquad C \cdot U \subset C,$$
$$U \cdot C \subset C, \qquad U \cdot U \subset U.$$

Corollary 10.3. *With same notation as used in* Theorem 10.10, *we denote by* $\Omega(V)$ *the homogeneous Hessian domain corresponding to* V, *and by* $\Omega(U)$ *the homogeneous regular convex domain corresponding to the clan* U. *Then we have*

(1) $\Omega(V) = C + \Omega(U)$.
(2) *The restriction of the Hessian metric* g *on* C *is the Euclidean metric.*

Proof. We set $\tilde{\Omega}(V) = C + \Omega(U)$. For $a \in C$ and $x \in \Omega(U)$ we have

$$\exp X_{-u}(a + x) = \exp L_u a + \exp X_{-u}x \in C + \Omega(U), \quad u \in U,$$
$$\exp X_{-c}(a + x) = (a + c + c \cdot x) + x \in C + \Omega(U), \quad c \in C,$$

and so $T(V)\tilde{\Omega}(V) = \tilde{\Omega}(V)$. Since $\Omega(U) = T(U)0$, for $x \in \Omega(U)$ there exists $u_0 \in U$ such that

$$\exp X_{-u_0}x = 0.$$

For $a \in C$ we put $c_0 = -\exp L_{u_0}a \in C$. Then

$$\exp X_{-c_0} \exp X_{-u_0}(a + x) = 0.$$

This implies $\tilde{\Omega}(V) = T(V)0 = \Omega(V)$, and (1) is provedD For $c \in C$ we put $s = \exp X_{-c}$. Then $s0 = c$, and $f(s^{-1})$ is the identity mapping on C. Hence by (10.6)

$$g(c)_{ij} = \delta_{ij}.$$

Therefore the restriction of g on C is a Euclidean metric. □

Corollary 10.4. *With the same notation as in* Theorems 10.10, *the ideal* C *is a maximal vector subspace of* V *contained in* $\Omega(V)$.

Corollary 10.5. *Let* V *be a normal Hessian algebra and let* $\Omega(V)$ *be the homogeneous Hessian domain corresponding to* V. *Denoting by* β_0 *the restriction of the second Koszul form* β *on* $\Omega(V)$ *at the origin* 0 *we have*

(1) β_0 *is positive semi-definite.*
(2) *The null space* $\{w \in V \mid \beta_0(w, v) = 0, \text{ for all } v \in V\}$ *of* β_0 *is a maximal vector subspace of* V *contained in* $\Omega(V)$.

Proof. By Lemma 9.5 we know

$$\beta_0(u, v) = \operatorname{Tr} L_{u \cdot v}, \quad \text{for } u, v \in V.$$

Hence it follows from Theorem 10.10 (2) that

$$\beta_0(a, v) = 0, \quad \text{for } a \in C, v \in V.$$

For $a, b \in C$ and $u \in V$ we have

$$\langle u \cdot a, b \rangle + \langle a, u \cdot b \rangle = \langle a \cdot u, b \rangle + \langle u, a \cdot b \rangle = \langle a \cdot u, b \rangle.$$

Therefore

$$L_u|_C + {}^t L_u|_C = R_u|_C,$$

where the symbol $|_C$ means the restriction of the corresponding operator on C. In particular we obtain

$$ {}^t R_u|_C = R_u|_C,$$

$$\operatorname{Tr} L_u|_C = \frac{1}{2} \operatorname{Tr} R_u|_C.$$

Since U is a clan, we know by Corollary 10.1

$$\operatorname{Tr} L_{u \cdot u}|_U > 0, \quad \text{for } u \neq 0 \in U.$$

It follows from Lemma 9.1 (3) that

$$\begin{aligned}
\operatorname{Tr} L_{u \cdot u}|_C &= \frac{1}{2} \operatorname{Tr} R_{u \cdot u}|_C \\
&= \frac{1}{2} \operatorname{Tr} (R_u R_u + [L_u, R_u])|_C \\
&= \frac{1}{2} \operatorname{Tr} (R_u R_u)|_C = \frac{1}{2} \operatorname{Tr} (R_u \, {}^t R_u)|_C \geq 0.
\end{aligned}$$

Hence

$$\beta_0(u, u) = \operatorname{Tr} L_{u \cdot u} = \operatorname{Tr} L_{u \cdot u}|_C + \operatorname{Tr} L_{u \cdot u}|_U > 0, \quad \text{for } u \neq 0 \in U.$$

Therefore β_0 is positive semi-definite, and the null space of β_0 coincides with C. On the other hand, by Corollary 10.4, C is a maximal vector subspace of V contained in $\Omega(V)$. □

By Corollaries 8.1, 8.2, 10.3, 10.4 and 10.5 we obtain

Corollary 10.6. *Let $(G/K, D, g)$ be a homogeneous Hessian manifold and let β be the second Koszul form on G/K. Then we have*

(1) *β is positive semi-definite.*
(2) *Let $u \neq 0$ be a tangent vector at $x \in G/K$. Then the geodesic $\exp_x^D t u$ with respect to D is complete if and only if $\beta_x(u, v) = 0$ for all tangent vectors v at x.*
(3) *The following conditions are equivalent.*

    (a) *$\beta = 0$.*
    (b) *The Levi-Civita connection of g coincides with D.*

(c) *D is complete.*

(4) *The following conditions are equivalent.*

(a) *There is no complete geodesic with respect to D.*

(b) *β is positive definite.*

(c) *The universal covering space of $(G/K, D, g)$ is a homogeneous regular convex domain.*

Chapter 11

Homogeneous spaces with invariant projectively flat connections

In section **9.1** we derived the correspondence between an invariant flat connection on a homogeneous space G/K, and a certain affine representation of the Lie algebra of G. Generalizing this result, in section **11.1** we characterize an invariant projectively flat connection on G/K by an affine representation of the central extention of the Lie algebra $\mathfrak{g}$ of G. In course of the proof we establish that G/K is endowed with an invariant projectively flat connection if and only if G/K admits a G-equivariant central affine hypersurface immersion. In section **11.2** we show that symmetric spaces of semisimple Lie groups with invariant projectively flat connections correspond bijectively to central simple Jordan algebras. We prove in section **11.3** that a homogeneous space G/K carries an invariant Codazzi structure of constant curvature if and only if G/K admits a G-equivariant immersion of codimension 1 into a certain homogeneous Hessian domain.

11.1 Invariant projectively flat connections

We first recall the following fundamental facts on projectively flat connections (Definition 1.8).

Theorem 11.1. *Let D be a torsion-free connection with symmetric Ricci tensor Ric_D. The connection D is projectively flat if and only if the following conditions hold.*

(1) *The curvature tensor R_D is given by*
$$R_D(X,Y)Z = \frac{1}{n-1}\{\mathrm{Ric}_D(Y,Z)X - \mathrm{Ric}_D(X,Z)Y\}, \quad n = \dim\ M.$$

(2) *The Ricci tensor satisfies the Codazzi equation,*
$$(D_X\mathrm{Ric}_D)(Y,Z) = (D_Y\mathrm{Ric}_D)(X,Z).$$

Theorem 11.2. *A connection induced by a central affine hypersurface immersion is projectively flat.*

For details the reader may refer to [Nomizu and Sasaki (1994)].

Let G/K be a homogeneous space of a connected Lie group G, and let $\mathfrak{g}$ and $\mathfrak{k}$ be the Lie algebras of G and K respectively. We define the **central extension $\tilde{\mathfrak{g}}$ of the Lie algebra $\mathfrak{g}$** by

$$\tilde{\mathfrak{g}} = \mathfrak{g} \oplus \mathbf{R}E, \qquad [\tilde{\mathfrak{g}}, E] = \{0\}.$$

Theorem 11.3. *Suppose that a homogeneous space G/K admits an invariant projectively flat connection, then there exists an affine representation $(\tilde{f}, \tilde{q})$ of the central extension $\tilde{\mathfrak{g}} = \mathfrak{g} \oplus \mathbf{R}E$ of $\mathfrak{g}$ on $\tilde{V}$ such that*

(1) $\dim \tilde{V} = \dim G/K + 1 C$
(2) $\tilde{q} : \tilde{\mathfrak{g}} \longrightarrow \tilde{V}$ *is surjective and the kernel coincides with $\mathfrak{k}$,*
(3) $\tilde{f}(E)$ *is the identity mapping on $\tilde{V}$ and $\tilde{q}(E) \neq 0 D$*

Conversely, suppose that the Lie group G is simply connected, and that the central extension $\tilde{\mathfrak{g}}$ of the Lie algebra $\mathfrak{g}$ of G admits an affine representation $(\tilde{f}, \tilde{q})$ satisfying the above conditions. Then there exists an invariant projectively flat connection on G/K.

Corollary 11.1. *Let G/K be a homogeneous space of a simply connected Lie group G. Then the following conditions are equivalent.*

(1) *There exists an invariant projectively flat connection on G/K.*
(2) *G/K admits an equivariant central affine hypersurface immersion.*

Proof of Theorem 11.3.   Suppose that G/K admits an invariant projectively flat connection D. Using the same notations as in the section **9.1** we have by [Kobayashi and Nomizu (1963, 1969)](I, p.235).

$$[X, Y]^* = [X^*, Y^*],$$
$$A_{X^*}Y^* = -D_{Y^*}X^*,$$
$$A_{X^*}Y^* - A_{Y^*}X^* = [X^*, Y^*],$$
$$A_{[X^*, Y^*]} = [A_{X^*}, A_{Y^*}] - R_D(X^*, Y^*).$$

Let V be the tangent space of G/K at $o = \{K\}$. We denote by $f(X)$ and $q(X)$ the restriction of A_{X^*} and $-X^*$ at o respectively. Then

$$q([X, Y]) = f(X)q(Y) - f(Y)q(X), \qquad (11.1)$$
$$f([X, Y]) = [f(X), f(Y)] - R_D(X^*, Y^*)_o,$$
$$\mathrm{Ker}\, q = \mathfrak{k}.$$

Since D is projectively flat, by Theorem 11.1 we have

$$R_D(X^*, Y^*)Z^* = Q(Y^*, Z^*)X^* - Q(X^*, Z^*)Y^*,$$

where $Q = \dfrac{1}{n-1} \operatorname{Ric}_D$. Therefore

$$f([X,Y])q(Z) \tag{11.2}$$
$$= [f(X), f(Y)]q(Z) - Q_o(q(Y), q(Z))q(X) + Q_o(q(X), q(Z))q(Y).$$

Since Q is invariant by G, we obtain

$$(D_{X^*}Q)(Y^*, Z^*)$$
$$= X^*(Q(Y^*, Z^*)) - Q(D_{X^*}Y^*, Z^*) - Q(Y^*, D_{X^*}Z^*)$$
$$= Q(\mathcal{L}_{X^*}Y^*, Z^*) + Q(Y^*, \mathcal{L}_{X^*}Z^*) - Q(D_{X^*}Y^*, Z^*) - Q(Y^*, D_{X^*}Z^*)$$
$$= Q(A_{X^*}Y^*, Z^*) + Q(Y^*, A_{X^*}Z^*),$$

and, by Theorem 11.1 Q satisfies the Codazzi equation

$$(D_{X^*}Q)(Y^*, Z^*) = (D_{Y^*}Q)(X^*, Z^*).$$

Thus we have

$$Q_o(f(X)q(Y), q(Z)) + Q_o(q(Y), f(X)q(Z)) \tag{11.3}$$
$$= Q_o(f(Y)q(X), q(Z)) + Q_o(q(X), f(Y)q(Z)).$$

Let us consider an extended vector space $\tilde{V}$ of V

$$\tilde{V} = V \oplus \mathbf{R}e.$$

For $X \in \mathfrak{g}$ we define a linear mapping $\tilde{f}(X) : \tilde{V} \longrightarrow \tilde{V}$ by

$$\tilde{f}(X)q(Z) = f(X)q(Z) - Q_o(q(X), q(Z))e,$$
$$\tilde{f}(X)e = q(X).$$

Then

$$[\tilde{f}(X), \tilde{f}(Y)]q(Z)$$
$$= [f(X), f(Y)]q(Z) - Q_o(q(Y), q(Z))q(X) + Q_o(q(X), q(Z))q(Y)$$
$$\quad -\{Q_o(q(X), f(Y)q(Z)) - Q_o(q(Y), f(X)q(Z))\}e$$
$$= f([X,Y])q(Z) - \{Q_o(f(X)q(Y), q(Z)) - Q_o(f(Y)q(X), q(Z))\}e$$
$$= f([X,Y])q(Z) - Q_o(q([X,Y]), q(Z))e$$
$$= \tilde{f}([X,Y])q(Z),$$

and

$$[\tilde{f}(X), \tilde{f}(Y)]e = \tilde{f}(X)q(Y) - \tilde{f}(Y)q(X)$$
$$= f(X)q(Y) - f(Y)q(X) = q([X,Y])$$
$$= \tilde{f}([X,Y])e,$$

by equations (11.1), (11.2) and (11.3). These imply that $(\tilde{f}, q)$ is an affine representation of $\mathfrak{g}$ on $\tilde{V}$;

$$\tilde{f}([X,Y]) = [\tilde{f}(X), \tilde{f}(Y)],$$
$$q([X,Y]) = \tilde{f}(X)q(Y) - \tilde{f}(Y)q(X), \quad \text{for } X, Y \in \mathfrak{g}.$$

We define an extended affine representation $(\tilde{f}, \tilde{q})$ of $\tilde{\mathfrak{g}}$ on $\tilde{V}$ by

$$\tilde{f}(\tilde{X}) = \begin{cases} \tilde{f}(X), & \tilde{X} = X \in \mathfrak{g} \\ I_{\tilde{V}}, & \tilde{X} = E, \end{cases}$$

$$\tilde{q}(\tilde{X}) = \begin{cases} q(X), & \tilde{X} = X \in \mathfrak{g} \\ e, & \tilde{X} = E, \end{cases}$$

where $I_{\tilde{V}}$ is the identity mapping on $\tilde{V}$. Then the affine representation $(\tilde{f}, \tilde{q})$ satisfies conditions (1)-(3).

Conversely, suppose that there exists an affine representation $(\tilde{f}, \tilde{q})$ of $\tilde{\mathfrak{g}}$ satisfying conditions (1)-(3). Let $\{x^1, \cdots, x^{n+1}\}$ be the affine coordinate system on $\tilde{V}$ with respect to a basis $\{e_1, \cdots, e_{n+1}\}$, i.e. $u = \sum_i x^i(u)e_i$ for $u \in \tilde{V}$. For $\tilde{X} \in \tilde{\mathfrak{g}}$ we define an infinitesimal affine transformation $\tilde{X}_a$ on $\tilde{V}$ by

$$\tilde{X}_a = -\sum_i (\sum_j \tilde{f}(\tilde{X})^i_j x^j + \tilde{q}(\tilde{X})^i) \frac{\partial}{\partial x^i},$$

where $\tilde{f}(\tilde{X})^i_j$ and $\tilde{q}(\tilde{X})^i$ are the components of $\tilde{f}(\tilde{X})$ and $\tilde{q}(\tilde{X})$ with respect to the basis respectively. Since $[\tilde{X}, \tilde{Y}]_a = [\tilde{X}_a, \tilde{Y}_a]$, the space $\tilde{\mathfrak{g}}_a$ defined by

$$\tilde{\mathfrak{g}}_a = \{\tilde{X}_a \mid \tilde{X} \in \tilde{\mathfrak{g}}\}$$

forms a Lie algebra. Since G acts effectively on G/K, the mapping $\tilde{X} \longrightarrow \tilde{X}_a$ is a Lie algebra isomorphism from $\tilde{\mathfrak{g}}$ to $\tilde{\mathfrak{g}}_a$. Let $\tilde{G}_a$ be the Lie group generated by $\tilde{\mathfrak{g}}_a$. For an affine transformation $s \in \tilde{G}_a$ of $\tilde{V}$, we denote by $\tilde{f}(s)$ and $\tilde{q}(s)$ the linear transformation part and the parallel translation vector part of s respectively. The orbit $\tilde{\Omega}_a$ of $\tilde{G}_a$ through the origin 0,

$$\tilde{\Omega}_a = \tilde{G}_a 0 = \tilde{G}_a/\tilde{K}_a,$$

is an open orbit because $\tilde{q}(\tilde{\mathfrak{g}}) = \tilde{V}$. Let G_a be the connected Lie subgroup of $\tilde{G}_a$ generated by $\mathfrak{g}_a = \{X_a \mid X \in \mathfrak{g}\}$, and let M_a the orbit of G_a through the origin

$$M_a = G_a 0 = G_a/K_a.$$

Using a vector field $E_a = -\sum_i (x^i + \tilde{q}(E)^i) \frac{\partial}{\partial x^i}$ transversal to M_a, we define an induced connection D_a and an affine fundamental form h_a on M_a by

$$\tilde{D}_{X_a} Y_a = D_{a X_a} Y_a + h_a(X_a, Y_a) E_a, \quad \text{for } X, Y \in \mathfrak{g},$$

where $\tilde{D}$ is the standard flat connection on $\tilde{V}$. Then the immersion of M_a into $\tilde{V}$ is a centro-affine immersion centred at $-\tilde{q}(E)$. Therefore the induced connection D_a is projectively flat by Theorem 11.2, and is invariant by G_a because $\tilde{D}$ and E_a are invariant under $\tilde{G}_a$. Since G is simply connected, there exists a covering homomorphism

$$\rho : G \longrightarrow G_a$$

such that $\rho_*(X) = X_a$ for $X \in \mathfrak{g}$. Because K is the identity component of $\rho^{-1}(K_a)$, ρ induces the universal covering mapping

$$p : G/K \longrightarrow G/\rho^{-1}(K_a) \cong G_a/K_a = M_a.$$

Hence we can define an invariant projectively flat connection on G/K using the above covering mapping p and the invariant projectively flat connection D_a on M_a. $\qquad\square$

Proof of Corollary 11.1. In course of the proof of Theorem 11.3, we proved that assertion (1) implies (2). By Theorem 11.2, assertion (1) follows from (2). $\qquad\square$

In Corollary 9.1 we gave the correspondence between Lie groups with left-invariant flat connections and left symmetric algebras. As an application of Theorem 11.3, we now extend this result.

Corollary 11.2. *Suppose that a Lie group G is endowed with a left-invariant projectively flat connection. Then we can define a left symmetric multiplication $\tilde{X} \cdot \tilde{Y}$ on the central extention $\tilde{\mathfrak{g}} = \mathfrak{g} \oplus \mathbf{R}E$ of $\mathfrak{g}$ satisfying the following conditions*

(1) $\tilde{X} \cdot \tilde{Y} - \tilde{Y} \cdot \tilde{X} = [\tilde{X}, \tilde{Y}]$,
(2) $\tilde{X} \cdot E = E \cdot \tilde{X} = \tilde{X}$.

Conversely, if a Lie group G is simply connected and the central extention $\tilde{\mathfrak{g}} = \mathfrak{g} \oplus \mathbf{R}E$ of the Lie algebra $\mathfrak{g}$ of G admits a left symmetric multiplication satisfying the above conditions, then G admits a left-invariant projectively flat connection.

Proof. Suppose that G is endowed with a left-invariant projectively flat connection. By Theorem 11.3 there exists an affine representation $(\tilde{f}, \tilde{q})$ of $\tilde{\mathfrak{g}}$ on $\tilde{V}$. Then $\tilde{q} : \tilde{\mathfrak{g}} \longrightarrow \tilde{V}$ is a linear isomorphism. We define an operation of multiplication on $\tilde{\mathfrak{g}}$ by

$$\tilde{X} \cdot \tilde{Y} = \tilde{q}^{-1}(\tilde{f}(\tilde{X})\tilde{q}(\tilde{Y})).$$

It is easy to see that $\tilde{\mathfrak{g}}$ is a left symmetric algebra satisfying conditions (1) and (2). Conversely, assume that there exists a left symmetric multiplication on $\tilde{\mathfrak{g}}$ satisfying (1) and (2). Then, defining $\tilde{f}$ and $\tilde{q}$ by

$$\tilde{f}(\tilde{X})\tilde{Y} = \tilde{X} \cdot \tilde{Y}, \quad \tilde{q}(\tilde{X}) = \tilde{X},$$

the pair $(\tilde{f}, \tilde{q})$ is an affine representation of $\tilde{\mathfrak{g}}$ on $\tilde{\mathfrak{g}}$ satisfying conditions (1), (2) and (3) of Theorem 11.3. Therefore G admits a left-invariant projectively flat connection. $\qquad\square$

Example 11.1. The Lie group $SL(n, \mathbf{R})$ is considered as an equivariant central affine hypersurface in $\mathfrak{gl}(n, \mathbf{R})$ centered at 0. Hence the induced connection is invariant and projectively flat. Let $\mathfrak{sl}(n, \mathbf{R})$ be the Lie algebra of $SL(n, \mathbf{R})$ and let E be the unit matrix of degree n. Then the central extension of $\mathfrak{sl}(n, \mathbf{R})$ is given by $\mathfrak{gl}(n, \mathbf{R}) = \mathfrak{sl}(n, \mathbf{R}) \oplus \mathbf{R}E$. The operation of the multiplication $X \cdot Y$ on the left symmetric algebra $\mathfrak{gl}(n, \mathbf{R})$ coincides with ordinary matrix multiplication.

Example 11.2. Let (V, χ) be a clan with unit element e. We set $\tilde{\mathfrak{g}} = \{L_u \mid u \in V\}$ and $\mathfrak{g} = \{L_u \mid u \in V, \ \chi(u) = 0\}$. Then $\tilde{\mathfrak{g}}$ and $\mathfrak{g}$ are linear Lie algebras because $\chi(u \cdot v - v \cdot u) = 0$ and $\tilde{\mathfrak{g}} = \mathfrak{g} \oplus \mathbf{R}L_e$. For L_u and $L_v \in \tilde{\mathfrak{g}}$ we define

$$L_u \cdot L_v = L_{u \cdot v}.$$

Then $L_u \cdot L_v - L_v \cdot L_u = L_{u \cdot v - v \cdot u} = [L_u, L_v]$ and $L_u \cdot L_e = L_e \cdot L_u = L_u$. Hence, by Corollary 11.2, the simply connected Lie group corresponding to $\mathfrak{g}$ admits a left-invariant projectively flat connection.

11.2 Symmetric spaces with invariant projectively flat connections

In this section we prove that symmetric homogeneous spaces with invariant projectively flat connections correspond bijectively to central simple Jordan algebras.

Let G/K be a symmetric homogeneous space of a semisimple Lie group G, and let

$$\mathfrak{g} = \mathfrak{k} + \mathfrak{m},$$
$$[\mathfrak{k}, \mathfrak{k}] \subset \mathfrak{k}, \quad [\mathfrak{k}, \mathfrak{m}] \subset \mathfrak{m}, \quad [\mathfrak{m}, \mathfrak{m}] \subset \mathfrak{k},$$

be the canonical decomposition for the symmetric homogeneous space G/K where $\mathfrak{g}$ and $\mathfrak{k}$ are the Lie algebras of G and K respectively.

Suppose that G/K admits an invariant projectively flat connection. Let $\tilde{\mathfrak{g}} = \mathfrak{g} \oplus \mathbf{R}E$ be the central extension of $\mathfrak{g}$ defined in section **11.1**. Putting $\tilde{\mathfrak{k}} = \mathfrak{k}$ and $\tilde{\mathfrak{m}} = \mathfrak{m} \oplus \mathbf{R}E$ we have

$$\tilde{\mathfrak{g}} = \tilde{\mathfrak{k}} + \tilde{\mathfrak{m}},$$
$$[\tilde{\mathfrak{k}}, \tilde{\mathfrak{k}}] \subset \tilde{\mathfrak{k}}, \quad [\tilde{\mathfrak{k}}, \tilde{\mathfrak{m}}] \subset \tilde{\mathfrak{m}}, \quad [\tilde{\mathfrak{m}}, \tilde{\mathfrak{m}}] \subset \tilde{\mathfrak{k}}.$$

It follows from Theorem 11.3 that $\tilde{\mathfrak{g}}$ admits an affine representation $(\tilde{f}, \tilde{q})$ satisfying the conditions in the theorem. Since $\tilde{q} : \tilde{\mathfrak{m}} \longrightarrow \tilde{V}$ is a linear isomorphism, for each $u \in \tilde{V}$ there exists a unique element $X_u \in \tilde{\mathfrak{m}}$ such that $\tilde{q}(X_u) = u$. We define an operation of multiplication on $\tilde{V}$ by

$$u * v = \tilde{f}(X_u)v.$$

Then the algebra $\tilde{V}$ with multiplication $u * v$ is commutative and has a unit element $e = \tilde{q}(E)$. In fact

$$u * v - v * u = \tilde{f}(X_u)\tilde{q}(X_v) - \tilde{f}(X_v)\tilde{q}(X_u) = \tilde{q}([X_u, X_v]) = 0,$$
$$e * u = \tilde{f}(E)u = u.$$

In the same way as in the proof of Lemma 9.7, the following lemma may be proved.

Lemma 11.1. *Let $W \in \tilde{\mathfrak{k}}$. Then $\tilde{f}(W)$ is a derivation of the algebra $\tilde{V}$.*

Denoting by L_u the operator of multiplication by $u \in \tilde{V}$ we define a symmetric bilinear form τ on $\tilde{V}$ by

$$\tau(u, v) = \operatorname{Tr} L_{u*v}.$$

In the same way that we proved Lemma 9.8, the following lemma may be proved.

Lemma 11.2.

(1) $[[L_u, L_v], L_w] = L_{[u*w*v]}$, *where* $[u * w * v] = u * (w * v) - (u * w) * v$.
(2) $\tau(u * v, w) = \tau(v, u * w)$.

Lemma 11.3. *τ is non-degenerate.*

Proof. Let $\tilde{V}_0$ be the null space of τ,

$$\tilde{V}_0 = \{v_0 \in \tilde{V} \mid \tau(v_0, v) = 0, \quad v \in \tilde{V}\},$$

and let $v_0 \in \tilde{V}_0$, $v \in \tilde{V}$ and $W \in \tilde{\mathfrak{k}}$. Since

$$\tilde{q}([W, X_{v_o*v}]) = \tilde{f}(W)\tilde{q}(X_{v_o*v}) - \tilde{f}(X_{v_o*v})\tilde{q}(W) = \tilde{f}(W)(v_0 * v),$$

by Lemma 11.1 we obtain

$$[\tilde{f}(W), \tilde{f}(X_{v_0*v})] = \tilde{f}(X_{\tilde{f}(W)(v_0*v)}) = L_{\tilde{f}(W)(v_0*v)}$$
$$= L_{(\tilde{f}(W)v_0)*v} + L_{v_0*(\tilde{f}(W)v)}.$$

Hence

$$0 = \mathrm{Tr}\,[\tilde{f}(W), \tilde{f}(X_{v_0*v})] = \mathrm{Tr}\,L_{(\tilde{f}(W)v_0)*v} + \mathrm{Tr}\,L_{v_0*(\tilde{f}(W)v)}$$
$$= \tau(\tilde{f}(W)v_0, v),$$

and so

$$\tilde{f}(\tilde{\mathfrak{k}})\tilde{V}_0 \subset \tilde{V}_0.$$

Let $v_0 \in \tilde{V}_0$, $v \in \tilde{V}$ and $X \in \tilde{\mathfrak{m}}$. Then

$$\tau(\tilde{f}(X)v_0, v) = \tau(\tilde{q}(X) * v_0, v) = \tau(v_0, \tilde{q}(X) * v) = 0,$$

and so

$$\tilde{f}(\tilde{\mathfrak{m}})\tilde{V}_0 \subset \tilde{V}_0.$$

These imply $\tilde{f}(\tilde{\mathfrak{g}})\tilde{V}_0 \subset \tilde{V}_0$, in particular

$$\tilde{f}(\mathfrak{g})\tilde{V}_0 \subset \tilde{V}_0.$$

Since the representation $\tilde{f}$ of the semisimple Lie algebra $\mathfrak{g}$ is completely reducible, there exists a complementary subspace $\tilde{V}_1$ invariant under $\tilde{f}(\mathfrak{g})$;

$$\tilde{V} = \tilde{V}_0 \oplus \tilde{V}_1, \qquad \tilde{f}(\mathfrak{g})\tilde{V}_1 \subset \tilde{V}_1.$$

Since $\tilde{f}(E) = I_{\tilde{V}}$ we have

$$\tilde{f}(\tilde{\mathfrak{g}})\tilde{V}_i \subset \tilde{V}_i.$$

Hence

$$\tilde{V} * \tilde{V}_i \subset \tilde{V}_i.$$

We set $e = e_0 + e_1$, where $e_i \in \tilde{V}_i$. Then

$$L_{e_i} v_j = \delta_{ij} v_j, \quad \text{for } v_j \in \tilde{V}_j,$$

where δ_{ij} is Kronecker's delta. Therefore

$$\dim \tilde{V}_0 = \mathrm{Tr}\,L_{e_0} \mid \tilde{V}_0 = \mathrm{Tr}\,L_{e_0} = \mathrm{Tr}\,L_{e_0*e_0} = \tau(e_0, e_0) = 0.$$

This implies that τ is non-degenerate. $\qquad \square$

The following lemma follows from Lemmata 4.2, 11.2 and 11.3.

Lemma 11.4. $\tilde{V}$ *is a semisimple Jordan algebra.*

Lemma 11.5. *The representation $\tilde{f}$ of $\tilde{\mathfrak{g}}$ on $\tilde{V}$ is faithful.*

Proof. Let $\mathrm{Ker}_{\mathfrak{g}}\tilde{f}$ be the kernel of $\tilde{f}$ on $\mathfrak{g}$;

$$\mathrm{Ker}_{\mathfrak{g}}\tilde{f} = \{X \in \mathfrak{g} \mid \tilde{f}(X) = 0\}.$$

We denote by $d_{\tilde{f}}$ the coboundary operator for the cohomology of coefficients on the representation $(\tilde{V}, \tilde{f})$ of $\mathfrak{g}$. Regarding $\tilde{q}$ as 1-dimensional $(\tilde{V}, \tilde{f})$-cochain we obtain

$$(d_{\tilde{f}}\tilde{q})(X, Y) = \tilde{f}(X)\tilde{q}(Y) - \tilde{f}(Y)\tilde{q}(X) - \tilde{q}([X, Y]) = 0.$$

Since $\mathfrak{g}$ is a semisimple Lie algebra, the cohomology $H^1(\mathfrak{g}, (\tilde{V}, \tilde{f}))$ vanishes, and there exists $\tilde{e} \in \tilde{V}$ such that $\tilde{q} = d_{\tilde{f}}\tilde{e}$, that is

$$\tilde{q}(X) = \tilde{f}(X)\tilde{e}, \quad \text{for } X \in \mathfrak{g}.$$

This implies $\mathrm{Ker}_{\mathfrak{g}}\tilde{f} \subset \tilde{\mathfrak{k}} = \mathfrak{k}$, and so $\mathrm{Ker}_{\mathfrak{g}}\tilde{f}$ is an ideal of $\mathfrak{g}$ contained in $\mathfrak{k}$. Since G acts effectively on G/K we have

$$\mathrm{Ker}_{\mathfrak{g}}\tilde{f} = \{0\}.$$

Suppose $\tilde{f}(\tilde{X}) = 0$ where $\tilde{X} = X + cE$ and $X \in \mathfrak{g}$. Then $\mathrm{Tr}\,\tilde{f}(X) = \mathrm{Tr}\,(-c\tilde{f}(E)) = -c\dim\tilde{V}$. It follows from the semisimplicity of $\mathfrak{g}$ that $\mathfrak{g} = [\mathfrak{g}, \mathfrak{g}]$ and so $\mathrm{Tr}\,\tilde{f}(X) = 0$. These results yield $c = 0$ and $X \in \mathrm{Ker}_{\mathfrak{g}}\tilde{f} = \{0\}$. Hence $\tilde{X} = 0$. Thus the representation $\tilde{f}$ is faithful. $\qquad\square$

Let $\mathfrak{m}(\tilde{V})$ be a vector subspace spanned by $\{L_v \mid v \in \tilde{V}\}$ and let $\mathfrak{k}(\tilde{V})$ be a vector subspace spanned by $\{[L_u, L_v] \mid u, v \in \tilde{V}\}$. Then

$$\tilde{f}(\tilde{\mathfrak{m}}) = \mathfrak{m}(\tilde{V}).$$

We set

$$\mathfrak{g}(\tilde{V}) = \mathfrak{k}(\tilde{V}) + \mathfrak{m}(\tilde{V}). \tag{11.4}$$

Then $\mathfrak{g}(\tilde{V})$ is a Lie algebra. Since any element in $\tilde{f}(\tilde{\mathfrak{k}})$ is, by Lemma 11.1, a derivation of the Jordan algebra $\tilde{V}$, and any derivation of a semisimple Jordan algebra is an inner derivation [Braun and Koecher (1966)], we obtain

$$\tilde{f}(\tilde{\mathfrak{k}}) = \mathfrak{k}(\tilde{V}).$$

Therefore

$$\tilde{f} : \tilde{\mathfrak{g}} = \tilde{\mathfrak{k}} + \tilde{\mathfrak{m}} \longrightarrow \mathfrak{g}(\tilde{V}) = \mathfrak{k}(\tilde{V}) + \mathfrak{m}(\tilde{V}) \tag{11.5}$$

is an isomorphism as Lie algebra including decomposition.

Definition 11.1. Let $\mathbf{A}$ be a Jordan algebra with multiplication $u * v$.

(1) The **center** $Z(\mathbf{A})$ of the Jordan algebra $\mathbf{A}$ is by definition

$$Z(\mathbf{A}) = \{u \in \mathbf{A} \mid [u * v * w] = [v * u * w] = [v * w * u] = 0, \text{ for all } v, w \in \mathbf{A}\},$$

where $[u * v * w] = u * (v * w) - (u * v) * w$.

(2) A Jordan algebra $\mathbf{A}$ with unit element e is said to be **central simple** if $\mathbf{A}$ is simple and $Z(\mathbf{A}) = \mathbf{R}e$.

Lemma 11.6. *The Jordan algebra $\tilde{V}$ is central simple.*

Proof. Let $c \in Z(\tilde{V})$ and $u \in \tilde{V}$. Then $[L_c, L_u] = 0$ because $0 = [c * v * u] = [L_c, L_u]v$ for all $v \in \tilde{V}$. This result together with equation (11.4) implies that L_c is an element of the center of $\mathfrak{g}(\tilde{V})$. It follows from (11.5) that the center of $\mathfrak{g}(\tilde{V})$ is the image of the center $\mathbf{R}E$ of $\tilde{\mathfrak{g}}$ under $\tilde{f}$. Hence $L_c \in \tilde{f}(\mathbf{R}E) = \mathbf{R}L_e\mathbf{D}$ Thus we have $Z(\tilde{V}) = \mathbf{R}e$. Since $\tilde{V}$ is a semisimple Jordan algebra, $\tilde{V}$ is decomposed into the direct sum of simple ideals $\tilde{V}_i\mathbf{G}$

$$\tilde{V} = \tilde{V}_1 \oplus \cdots \oplus \tilde{V}_k.$$

Let $e = e_1 + \cdots + e_k$ be the decomposition of the unit element e of $\tilde{V}$ where $e_i \in \tilde{V}_i$. Suppose $\tilde{V}_1 \neq \{0\}$. Then e_1 is the unit element of $\tilde{V}_1$. For any $c_1 \neq 0 \in Z(\tilde{V}_1)$ we obtain

$$[c_1 * \tilde{V}_1 * \tilde{V}_1] = \{0\},$$
$$[c_1 * \tilde{V}_i * \tilde{V}_j] = \{0\}, \quad \text{for } i \neq 1 \text{ and } j \neq 1.$$

Hence

$$[c_1 * \tilde{V} * \tilde{V}] = \{0\}.$$

In the same way we have

$$[\tilde{V} * c_1 * \tilde{V}] = \{0\}, \quad [\tilde{V} * \tilde{V} * c_1] = \{0\}.$$

Therefore $c_1 \in Z(\tilde{V}) = \mathbf{R}e$ and $c_1 = ae$, $a \neq 0$. This means $\tilde{V}_i = \{0\}$ for all $i \neq 1$. Thus $\tilde{V}$ is simple. $\square$

The above results are summarized in the following theorem.

Theorem 11.4. *Let G/K be a symmetric homogeneous space of a semisimple Lie group G on which G acts effectively. Suppose that G/K admits an invariant projectively flat connection. Then there exists a central simple Jordan algebra $\tilde{V}$ satisfying the following properties.*

(1) $\tilde{V}$ is decomposed into a direct sum of vector spaces,

$$\tilde{V} = V \oplus \mathbf{R}e,$$

where e is the unit element of $\tilde{V}$.

(2) Let $\mathfrak{m}(V)$ be a vector subspace spanned by $\{L_v \mid v \in V\}$ and let $\mathfrak{k}(V)$ be a vector subspace spanned by $\{[L_u, L_v] \mid u, v \in V\}$. Then $\mathfrak{g}(V) = \mathfrak{k}(V) + \mathfrak{m}(V)$ is a Lie algebra, and is isomorphic to the Lie algebra $\mathfrak{g}$ of G including the canonical decomposition $\mathfrak{g} = \mathfrak{k} + \mathfrak{m}$ for the symmetric homogeneous space G/K.

Conversely, we shall construct from a central simple Jordan algebra a semisimple symmetric homogeneous space with an invariant projectively flat connection.

Theorem 11.5. *Let $\tilde{V}$ be a central simple Jordan algebra. We set $V = \{v \in \tilde{V} \mid \mathrm{Tr}\, L_v = 0\}$ and denote by $\mathfrak{m}(V)$, $\mathfrak{k}(V)$ vector subspaces spanned by $\{L_v \mid v \in V\}$, $\{[L_u, L_v] \mid u, v \in V\}$ respectively. We put $\mathfrak{g}(V) = \mathfrak{k}(V) + \mathfrak{m}(V)$. Then $\mathfrak{g}(V)$ and $\mathfrak{k}(V)$ are linear Lie algebras. Let $G(V)$ be the simply connected Lie group with Lie algebra $\mathfrak{g}(V)$ and let $K(V)$ be the connected Lie subgroup of $G(V)$ corresponding to $\mathfrak{k}(V)$. Then $G(V)$ is a semisimple Lie group and $G(V)/K(V)$ is a symmetric homogeneous space with an invariant projectively flat connection.*

Proof. By [Braun and Koecher (1966)] we know that $\mathfrak{g}(V)$ is a semisimple Lie algebra and

$$[\mathfrak{k}(V), \mathfrak{m}(V)] \subset \mathfrak{m}(V), \quad [\mathfrak{m}(V), \mathfrak{m}(V)] \subset \mathfrak{k}(V).$$

Denoting by $\mathfrak{m}(\tilde{V})$ a vector subspace spanned by $\{L_v \mid v \in \tilde{V}\}$, and by $\mathfrak{k}(\tilde{V})$ a vector subspace spanned by $\{[L_u, L_v] \mid u, v \in \tilde{V}\}$, we set

$$\mathfrak{g}(\tilde{V}) = \mathfrak{k}(\tilde{V}) + \mathfrak{m}(\tilde{V}). \tag{11.6}$$

Then $\mathfrak{g}(\tilde{V})$ is a Lie algebra and

$$\mathfrak{g}(\tilde{V}) = \mathfrak{g}(V) + \mathbf{R}I_{\tilde{V}},$$

where $I_{\tilde{V}}$ is the identity mapping on $\tilde{V}$ [Braun and Koecher (1966)]. We define a linear representation $\tilde{f}$ of $\mathfrak{g}(\tilde{V})$ on $\tilde{V}$ by $\tilde{f}(\tilde{X}) = \tilde{X}$ and a linear mapping $\tilde{q}$ from $\mathfrak{g}(\tilde{V})$ to $\tilde{V}$ by $\tilde{q}(W + L_{\tilde{v}}) = \tilde{v}$ where $W \in \mathfrak{k}(\tilde{V})$ and $L_{\tilde{v}} \in \mathfrak{m}(\tilde{V})$. Then $(\tilde{f}, \tilde{q})$ is an affine representation of $\mathfrak{g}(\tilde{V})$ on $\tilde{V}$ and satisfies the conditions (1)-(3) of Theorem 11.3. Hence $G(V)/K(V)$ is a symmetric homogeneous space with an invariant projectively flat connection. $\square$

Example 11.3. (Quadratic hypersurfaces) We set

$$J = \begin{bmatrix} -I_p & 0 \\ 0 & I_{n-p} \end{bmatrix},$$

$$\tilde{J} = \begin{bmatrix} -I_p & 0 \\ 0 & I_{n+1-p} \end{bmatrix} = \begin{bmatrix} J & 0 \\ 0 & 1 \end{bmatrix},$$

where I_p is the unit matrix of degree p. A quadratic hypersurface M_p^n is by definition the connected component of $\{x \in \mathbf{R}^{n+1} \mid {}^t x \tilde{J} x = 1\}$ containing ${}^t e = [0, \cdots, 0, 1]$. Then M_p^n is a central affine hypersurface centered at the origin 0 and the induced connection on M_p^n is projectively flat. M_0^n is a sphere and M_n^n is a component of a two-sheeted hyperboloid. Let $SO(p, n-p)$ be a linear Lie group preserving J,

$$SO(p, n-p) = \{s \in SL(n, \mathbf{R}) \mid {}^t s J s = J\}.$$

The group $SO(p, n+1-p)$ acts transitively on M_p^n. The isotropy subgroup at e is $\left\{ \begin{bmatrix} s & 0 \\ 0 & 1 \end{bmatrix} \mid s \in SO(p, n-p) \right\}$ and is identified with $SO(p, n-p)$. Therefore we have $M_p^n = SO(p, n+1-p)/SO(p, n-p)$. The central affine hypersurface immersion M_p^n is $SO(p, n+1-p)$-equivariant. Hence the induced projectively flat connection on $SO(p, n+1-p)/SO(p, n-p)$ is invariant under $SO(p, n+1-p)$. We define an involutive automorphism σ of $SO(p, n+1-p)$ by $\sigma(s) = HsH^{-1}$ where $H = \begin{bmatrix} I_n & 0 \\ 0 & -1 \end{bmatrix}$. Then $M_p^n = SO(p, n+1-p)/SO(p, n-p)$ is a symmetric homogeneous space with involution σ. The Lie algebra $\mathfrak{o}(p, n-p)$ of $SO(p, n-p)$ is given by

$$\mathfrak{o}(p, n-p) = \{A \in \mathfrak{gl}(n, \mathbf{R}) \mid {}^t A J + J A = 0\},$$

and the Lie algebra $\mathfrak{o}(p, n+1-p)$ of $SO(p, n+1-p)$ is given by

$$\mathfrak{o}(p, n+1-p) = \left\{ \begin{bmatrix} A & a \\ -{}^t(Ja) & 0 \end{bmatrix} \mid A \in \mathfrak{o}(p, n-p),\ a \in \mathbf{R}^n \right\}.$$

Since the differential σ_* of σ is expressed by

$$\sigma_* \begin{bmatrix} A & a \\ -{}^t(Ja) & 0 \end{bmatrix} = \begin{bmatrix} A & -a \\ {}^t(Ja) & 0 \end{bmatrix},$$

the canonical decomposition of $\mathfrak{g} = \mathfrak{o}(p, n+1-p)$ for the symmetric space $SO(p, n+1-p)/SO(p, n-p)$ is given by

$$\mathfrak{g} = \mathfrak{k} + \mathfrak{m},$$

$$\mathfrak{k} = \left\{ \begin{bmatrix} A & 0 \\ 0 & 0 \end{bmatrix} \mid A \in \mathfrak{o}(p, n-p) \right\},$$

$$\mathfrak{m} = \left\{ \begin{bmatrix} 0 & a \\ -{}^t(Ja) & 0 \end{bmatrix} \mid a \in \mathbf{R}^n \right\}.$$

We set

$$\tilde{\mathfrak{g}} = \mathfrak{g} \oplus \mathbf{R}E, \qquad \tilde{\mathfrak{m}} = \mathfrak{m} \oplus \mathbf{R}E,$$

where $E = I_{n+1}$, and define an affine representation $(\tilde{f}, \tilde{q})$ of $\tilde{\mathfrak{g}}$ on $\mathbf{R}^{n+1}$ by $\tilde{f}(\tilde{X}) = \tilde{X}$ and $\tilde{q}(\tilde{X}) = \tilde{X}e$. Then the affine representation $(\tilde{f}, \tilde{q})$ satisfies the conditions of Theorem 11.3. Then for $\tilde{u} = {}^t[u_1, \cdots, u_n, u_{n+1}] \in \mathbf{R}^{n+1}$ we obtain

$$X_{\tilde{u}} = \begin{bmatrix} u_{n+1}I_n & u \\ -{}^t(Ju) & u_{n+1} \end{bmatrix}, \qquad u = {}^t[u_1, \cdots, u_n].$$

Hence

$$\begin{bmatrix} u \\ u_{n+1} \end{bmatrix} * \begin{bmatrix} v \\ v_{n+1} \end{bmatrix} = X_{\tilde{u}}\tilde{v} = \begin{bmatrix} u_{n+1}v + v_{n+1}u \\ -{}^tuJv + u_{n+1}v_{n+1} \end{bmatrix}.$$

The central simple Jordan algebra corresponding to $SO(p, n + 1 - p)/SO(p, n - p)$ coincides with the Jordan algebra defined in Example 4.4. [Braun and Koecher (1966)](p.193)

Example 11.4. Let $\tilde{V}$ be the vector space of all symmetric matrices of degree n and let Ω_p^n be the set of all element in $\tilde{V}$ with signature $(p, n - p)$. Then $GL^+(n, \mathbf{R})$ acts on Ω_p^n by $GL^+(n, \mathbf{R}) \times \Omega_p^n \ni (s, x) \longrightarrow sx{}^ts \in \Omega_p^n$. A matrix J defined in Example 11.3 is contained in Ω_p^n. For any $x \in \Omega_p^n$, by Sylvester's law of inertia there exists $s \in GL^+(n.\mathbf{R})$ such that $sx{}^ts = J$. Hence $GL^+(n, \mathbf{R})$ acts transitively on Ω_p^n. Since the isotropy subgroup of $GL^+(n, \mathbf{R})$ at J is $SO(p, n - p)$, we have

$$\Omega_p^n = GL^+(n, \mathbf{R})/SO(p, n - p).$$

Let $M_p^n = \{x \in \Omega_p^n \mid |\det x| = 1\}$. Then M_p^n is a central affine hypersurface in $\tilde{V}$ centered at 0. Hence the induced connection on M_p^n is projectively flat. The group $SL(n, \mathbf{R})$ acts transitively on M_p^n by

$$(s, x) \in SL(n, \mathbf{R}) \times M_p^n \longrightarrow sx{}^ts \in M_p^n.$$

The central affine hypersurface immersion is $SL(n, \mathbf{R})$-equivariant. Hence the induced projectively flat connection on M_p^n is invariant under $SL(n, \mathbf{R})$. The homogeneous space $M_p^n = SL(n, \mathbf{R})/SO(p, n - p)$ is symmetric with respect to the involutive automorphism $\sigma : s \longrightarrow J{}^ts^{-1}J$. Since the differential σ_* of σ is given by $\sigma_*(X) = -J{}^tXJ$, the canonical decomposition of the Lie algebra $\mathfrak{g} = \mathfrak{sl}(n, \mathbf{R})$ for the symmetric space is given by

$$\mathfrak{g} = \mathfrak{k} + \mathfrak{m},$$

$$\mathfrak{k} = \{A \in \mathfrak{g} \mid J{}^tAJ = -A\} = \mathfrak{o}(p, n - p),$$

$$\mathfrak{m} = \{A \in \mathfrak{g} \mid J{}^tAJ = A\}.$$

Let $\tilde{\mathfrak{g}} = \mathfrak{g} \oplus \mathbf{R}E = \mathfrak{gl}(n, \mathbf{R})$ where $E = I_n$. We define an affine representation $(\tilde{f}, \tilde{q})$ of $\tilde{\mathfrak{g}}$ on $\tilde{V}$ by $\tilde{f}(X)v = -{}^tXv - vX$ and $\tilde{q}(X) = -{}^tXJ - JX$ where $X \in \tilde{\mathfrak{g}}$ and $v \in \tilde{V}$. Then $(\tilde{f}, \tilde{q})$ satisfies the conditions of Theorem 11.3. For $u \in \tilde{V}$ there exists a unique $X_u \in \mathfrak{m}$ such that $\tilde{q}(X_u) = u$. Then $X_u = -\dfrac{1}{2}Ju$. Thus the operation of multiplication of the central simple Jordan algebra $\tilde{V}$ corresponding to $SL(n, \mathbf{R})/SO(p, n-p)$ is given by

$$u * v = \tilde{f}(X_u)v = \frac{1}{2}(uJv + vJu).$$

In the case of $p = 0$, the multiplication is reduced to $u * v = \dfrac{1}{2}(uv + vu)$ and the Jordan algebra $\tilde{V}$ coincides with the space of all real symmetric matrices of degree n (cf. Theorem 4.12).

11.3   Invariant Codazzi structures of constant curvature

A pair (D, g) of a torsion-free connection D and a non-degenerate metric g on a manifold M is called a **Codazzi structure** if it satisfies the Codazzi equation,

$$(D_X g)(Y, Z) = (D_Y g)(X, Z).$$

A manifold M provided with a Codazzi structure (D, g) is said to be a **Codazzi manifold** and is denoted by (M, D, g). A Codazzi structure (D, g) is said to be of a **constant curvature** c if the curvature tensor R_D of D is given by

$$R_D(X, Y)Z = c\{g(Y, Z)X - g(X, Z)Y\},$$

(cf. Definition 2.8 and 2.9). A homogeneous space G/K endowed with an invariant Codazzi structure (D, g) is called a homogeneous Codazzi manifold, and is denoted by $(G/K, D, g)$.

In this section we study an invariant Codazzi structure of a constant curvature c. By Proposition 2.9, an invariant Codazzi structure of constant curvature $c = 0$ is an invariant Hessian structure, which has been extensively studied in chapters **9** and **10**. We shall prove that a homogeneous space with an invariant Codazzi structure of a non-zero constant curvature $c \neq 0$ is obtained by an equivariant immersion of codimension 1 into a certain homogeneous space with an invariant Hessian structure.

Theorem 11.6. *Let G/K be a simply connected homogeneous space of a simply connected Lie group G. Suppose that G/K is endowed with an invariant Codazzi structure (D, g) of constant curvature $c \neq 0$. Then there* *exists a G-equivariant immersion ρ of codimension 1 from G/K into a homogeneous space $\hat{G}/\hat{K}$ with an invariant Hessian structure $(\hat{D}, \hat{g})$ satisfying the following conditions:*

Let $\hat{\mathfrak{g}}$ be the Lie algebra of $\hat{G}$. For $\hat{X} \in \hat{\mathfrak{g}}$ we denote by $\hat{X}^*$ the vector field on $\hat{G}/\hat{K}$ induced by $\exp(-t\hat{X})$. Then there exists an element $\hat{E}$ in the center of $\hat{\mathfrak{g}}$ such that

(1) $\hat{E}^*$ is transversal to G/K, and the pair (D, g) coincides with the induced connection and the affine fundamental form of an affine immersion $(\rho, -c\hat{E}^*)$,

(2) $\hat{D}_{\hat{X}^*} \hat{E}^* = -\hat{X}^*, \quad$ for $\hat{X} \in \hat{\mathfrak{g}}$,

(3) $\hat{g}(\hat{E}^*, \hat{E}^*) = -\dfrac{1}{c}.$

Proof. Since the Codazzi structure (D, g) is of a constant curvature $c \neq 0$, we have

$$R_D(X, Y)Z = c\{g(Y, Z)X - g(X, Z)Y\},$$

and so the Ricci tensor Ric_D of D is given by

$$\mathrm{Ric}_D = c(n - 1)g,$$

where n is the dimension of G/K. Hence

$$R_D(X, Y)Z = \frac{1}{n-1}\{\mathrm{Ric}_D(Y, Z)X - \mathrm{Ric}_D(X, Z)Y\},$$
$$(D_X \mathrm{Ric}_D)(Y, Z) = (D_Y \mathrm{Ric}_D)(X, Z).$$

Therefore, by Theorem 11.1, D is an invariant projectively flat connection on G/K. Also by Theorem 11.1, the central extension $\tilde{\mathfrak{g}}$ of the Lie algebra $\mathfrak{g}$ of G admits an affine representation. Here we recall the proof of Theorem 11.3 and its accompanying notation. Let $(\tilde{f}, \tilde{q})$ be the affine representation of the central extension $\tilde{\mathfrak{g}} = \mathfrak{g} \oplus \mathbf{R}E$ on $\tilde{V} = V \oplus \mathbf{R}e$ where V is the tangent space of G/K at $o = \{K\}$. We denote by $\{e_1, \cdots, e_n\}$ a basis of V and by $\{x^1, \cdots, x^{n+1}\}$ the affine coordinate system on $\tilde{V}$ defined by $x^{n+1}(e_i) = 0$, $x^{n+1}(e) = 1$ and $x^j(e_i) = \delta_i^j$, $x^j(e) = 0$ for $1 \leq i, j \leq n$. For $\tilde{X} \in \tilde{\mathfrak{g}}$ we define an infinitesimal affine transformation $\tilde{X}_a$ on $\tilde{V}$ by

$$\tilde{X}_a = -\sum_i \left(\sum_j \tilde{f}(\tilde{X})_j^i x^j + \tilde{q}(\tilde{X})^i\right) \frac{\partial}{\partial x^i},$$

where $\tilde{f}(\tilde{X})_j^i$ and $\tilde{q}(\tilde{X})^i$ are the components of $\tilde{f}(\tilde{X})$ and $\tilde{q}(\tilde{X})$ with respect to the basis. Then the vector space $\tilde{\mathfrak{g}}_a$ defined by

$$\tilde{\mathfrak{g}}_a = \{\tilde{X}_a \mid \tilde{X} \in \tilde{\mathfrak{g}}\}$$

is a Lie algebra because of $[\tilde{X}, \tilde{Y}]_a = [\tilde{X}_a, \tilde{Y}_a]$. Since G acts effectively on G/K, the mapping $\tilde{\mathfrak{g}} \ni \tilde{X} \longrightarrow \tilde{X}_a \in \tilde{\mathfrak{g}}_a$ is a Lie algebra isomorphism. Let $\tilde{G}_a$ be the Lie group generated by $\tilde{\mathfrak{g}}_a$. The orbit $\tilde{\Omega}_a$ of $\tilde{G}_a$ through the origin 0

$$\tilde{\Omega}_a = \tilde{G}_a 0 = \tilde{G}_a / \tilde{K}_a$$

is an open orbit because $\tilde{q}(\tilde{\mathfrak{g}}) = \tilde{V}$. Let G_a be the connected Lie subgroup of $\tilde{G}_a$ generated by $\mathfrak{g}_a = \{X_a \mid X \in \mathfrak{g}\}$, and let M_a the orbit of G_a through the origin

$$M_a = G_a 0 = G_a / K_a.$$

Since $E_a = -\sum_i (x^i + \tilde{q}(E)^i) \dfrac{\partial}{\partial x^i}$ is transversal to M_a, we define an induced connection D_a and an affine fundamental form h_a on M_a by

$$\tilde{D}_{X_a} Y_a = D_{a X_a} Y_a + h_a(X_a, Y_a) E_a, \quad \text{for } X, Y \in \mathfrak{g},$$

where $\tilde{D}$ is the standard flat connection on $\tilde{V}$. Then the immersion of M_a into $\tilde{V}$ is a central affine immersion centered at $-\tilde{q}(E)$. Therefore the induced connection D_a is projectively flat by Theorem 11.2, and is invariant by G_a because both $\tilde{D}$ and E_a are invariant under $\tilde{G}_a$. Since G is simply connected, there exists a covering homomorphism

$$\rho : G \longrightarrow G_a.$$

Let p be a G-equivariant immersion of G/K into $\tilde{V}$ defined by

$$p : G/K \ni sK \longrightarrow \rho(s)0 \in M_a = G_a 0 \subset \tilde{V}.$$

For $X \in \mathfrak{g}$ we denote by X^* the vector field on G/K induced by $\exp(-tX)$. Then

$$p_*(X^*) = X_a.$$

Let D' and h' be the induced connection and the affine fundamental form on G/K with respect to the affine immersion $(p, -cE_a)$,

$$\tilde{D}_{X^*} p_*(Y^*) = p_*(D'_{X^*} Y^*) + h'(X^*, Y^*)(-cE_a), \quad \text{for } X, Y \in \mathfrak{g}.$$

Then

$$p_*(D'_{X^*} Y^*) = D_{a X_a} Y_a,$$

$$h'(X^*, Y^*) = -\frac{1}{c} h_a(X_a, Y_a).$$

Since

$$p(sx) = \rho(s)p(x), \quad \rho(s)_* E_a = E_a, \quad \text{for } s \in G,$$

and since D_a and h_a are invariant by G_a, both D' and h' are invariant under G. It is straightforward to see that

$$(\tilde{D}_{X_a} Y_a)_0 = p_*((D_{X^*} Y^*)_o) + g(X^*, Y^*)_o(-ce),$$

hence we have $D' = D$ and $h' = g$, and so

$$\tilde{D}_{X_a} Y_a = p_*((D_{X^*} Y^*)) + g(X^*, Y^*)(-cE_a).$$

Put $\hat{G} = \tilde{G}_a$, $\hat{K} = \rho(K)$ and $\hat{\mathfrak{g}} = \tilde{\mathfrak{g}}_a$. Then we have a covering map from $\hat{G}/\hat{K}$ to the open orbit $\tilde{\Omega}_a = \tilde{G}_a 0$ of $\tilde{G}_a$,

$$\pi : \hat{G}/\hat{K} \longrightarrow \tilde{\Omega}_a = \tilde{G}_a 0.$$

Since the open orbit $\tilde{\Omega}_a = \tilde{G}_a 0$ admits a $\tilde{G}_a$-invariant flat affine connection $\tilde{D}$, there exists a $\hat{G}$-invariant flat affine connection $\hat{D}$ on $\hat{G}/\hat{K}$ induced by $\tilde{D}$ and π. For $X \in \mathfrak{g}$ we denote by $\hat{X}^*$ the vector field on $\hat{G}/\hat{K}$ induced by $\exp(-tX)$. Let $\hat{p}$ be a G-equivariant immersion from G/K to $\hat{G}/\hat{K}$ given by

$$\hat{p} : G/K \ni sK \longrightarrow \rho(s)\hat{K} \in \hat{G}/\hat{K}.$$

Since $p = \pi \circ \hat{p}$, $\hat{p}_*(X^*) = \hat{X}^*$ and $\pi_*(\hat{X}^*) = X_a$, we have

$$\begin{aligned}
\pi_*(\hat{D}_{\hat{X}^*} \hat{Y}^*) &= \tilde{D}_{\pi_*(\hat{X}^*)} \pi_*(\hat{Y}^*) = \tilde{D}_{X_a} Y_a \\
&= p_*((D_{X^*} Y^*)) + g(X^*, Y^*)(-cE_a) \\
&= \pi_* \left\{ \hat{p}_*(D_{X^*} Y^*) + g(X^*, Y^*)(-c\hat{E}^*) \right\}.
\end{aligned}$$

Thus the Gauss formula for the immersion $(\hat{p}, -c\hat{E})$ is given by

$$\begin{aligned}
\hat{D}_{X^*} \hat{p}(Y^*) &= \hat{D}_{\hat{p}_*(X^*)} \hat{p}_*(Y^*) = \hat{D}_{\hat{X}^*} \hat{Y}^* \\
&= \hat{p}_*(D_{X^*} Y^*) + g(X^*, Y^*)(-c\hat{E}^*),
\end{aligned}$$

which proves assertion (1).

We shall show that $\hat{G}/\hat{K}$ admits an invariant Hessian structure $(\hat{D}, \hat{g})$ satisfying conditions (2) and (3). From the Codazzi equation $(D_X \text{Ric}_D)(Y, Z) = (D_Y \text{Ric}_D)(X, Z)$ we obtain

$$\begin{aligned}
Q_o(f(X)q(Y), q(Z)) &+ Q_o(q(Y), f(X)q(Z)) \qquad (11.7) \\
&= Q_o(f(Y)q(X), q(Z)) + Q_o(q(X), f(Y)q(Z)).
\end{aligned}$$

Let us define a non-degenerate symmetric bilinear form $\tilde{Q}_o$ on $\tilde{V}$ by

$$\tilde{Q}_o(u, v) = \begin{cases} Q_o(u, v) & u, v \in V \\ 0 & u \in V, v = e \\ -1 & u = v = e. \end{cases}$$

Let $\tilde{X} = X + xE, \tilde{Y} = Y + yE, \tilde{Z} = Z + zE$, where $X, Y, Z \in \mathfrak{g}$ and $x, y, z \in \mathbf{R}$. Then we have

$$\tilde{Q}_o(\tilde{f}(\tilde{X})\tilde{q}(\tilde{Y}), \tilde{q}(\tilde{Z}))$$

$$= \tilde{Q}_o(f(X)q(Y) + xq(Y) + yq(X) + \{xy - Q_o(q(X), q(Y))\}e, q(Z) + ze)$$

$$= Q_o(f(X)q(Y), q(Z))$$

$$+ xQ_o(q(Y), q(Z)) + yQ_o(q(X), q(Z)) + zQ_o(q(X), q(Y)) - xyz.$$

The above expression together with equation (11.7) imply

$$\tilde{Q}_o(\tilde{f}(\tilde{X})\tilde{q}(\tilde{Y}), \tilde{q}(\tilde{Z})) + \tilde{Q}_o(\tilde{q}(\tilde{Y}), \tilde{f}(X)\tilde{q}(\tilde{Z})) \qquad (11.8)$$

$$= \tilde{Q}_o(\tilde{f}(\tilde{Y})\tilde{q}(\tilde{X}), \tilde{q}(\tilde{Z})) + \tilde{Q}_o(\tilde{q}(\tilde{X}), \tilde{f}(\tilde{Y})\tilde{q}(\tilde{Z})),$$

for $\tilde{X}, \tilde{Y}$ and $\tilde{Z} \in \tilde{\mathfrak{g}}$. We now define a non-degenerate bilinear form $\hat{g}_{\hat{o}}$ on the tangent space $\hat{G}/\hat{K}$ at $\hat{o} = \hat{K}$ by

$$\hat{g}_{\hat{o}}(\hat{X}_{\hat{o}}^*, \hat{Y}_{\hat{o}}^*) = \frac{1}{c}\tilde{Q}_o(\pi_*(\hat{X}_{\hat{o}}^*), \pi_*(\hat{Y}_{\hat{o}}^*)).$$

Then $\hat{g}_{\hat{o}}$ defines a $\hat{G}$-invariant non-degenerate metric $\hat{g}$ on $\hat{G}/\hat{K}$. It follows from equation (11.8) that $\hat{g}$ satisfies the Codazzi equation. Hence $(\hat{D}, \hat{g})$ is a Hessian structure on $\hat{G}/\hat{K}$. Since

$$\pi_*(\hat{D}_{\hat{X}*}\hat{E}^*) = \tilde{D}_{\pi_*(\hat{X}*)}\pi_*(\hat{E}^*) = \tilde{D}_{X_a}E_a = -X_a = -\pi_*(\hat{X}^*),$$

we have

$$\hat{D}_{\hat{X}*}\hat{E}^* = -\hat{X}^*,$$

which proves assertion (2). Assertion (3) follows from

$$\hat{g}_{\hat{o}}(\hat{E}^*, \hat{E}^*) = \frac{1}{c}Q_o(\pi_*(\hat{E}^*), \pi_*(\hat{E}^*)) = \frac{1}{c}Q_o(e, e) = -\frac{1}{c}. \qquad \square$$

Conversely, we have the following theorem.

Theorem 11.7. *Let* $(\hat{G}/\hat{K}, \hat{D}, \hat{g})$ *be a homogeneous Hessian manifold. Suppose that there exist an element* $\hat{E}$ *in the center of the Lie algebra* $\hat{\mathfrak{g}}$ *of* $\hat{G}$, *and a non-zero constant* c *such that*

(1) $\hat{D}_{\hat{X}*}\hat{E}^* = -\hat{X}^*$, *where* $\hat{X}^*$ *is a vector field on* $\hat{G}/\hat{K}$ *induced by* $\exp(-t\hat{X})$ *for* $\hat{X} \in \hat{\mathfrak{g}}$.

(2) $\hat{g}(\hat{E}^*, \hat{E}^*) = -\dfrac{1}{c}$.

Then there exists a homogeneous submanifold G/K *of* $\hat{G}/\hat{K}$ *of codimension 1 such that* $\hat{E}^*$ *is transversal to* G/K, *and that the pair* (D, g) *of the induced connection* D *and the affine fundamental form* g *with respect to* $\hat{D}$ *and* $\hat{E}^*$ *is a G-invariant Codazzi structure of constant curvature* c.

Proof. Let $\hat{V}$ be the tangent space of $\hat{G}/\hat{K}$ at $\hat{o} = \hat{K}$. We put

$$\hat{A}_{\hat{X}*}\hat{Y}^* = -\hat{D}_{\hat{Y}*}\hat{X}^*,$$

and denote by $\hat{f}(\hat{X})$ and $\hat{q}(\hat{X})$ the values of $\hat{A}_{\hat{X}*}$ and $\hat{X}^*$ at $\hat{o}$ respectively. Then by Theorem 9.1 the pair $(\hat{f}, \hat{q})$ is an affine representation of $\hat{\mathfrak{g}}$ on $\hat{V}$. Let $\langle \, , \, \rangle$ be the restriction of $\hat{g}$ at $\hat{o}$. By Lemma 9.2 we have

$$\langle \hat{f}(\hat{X})\hat{q}(\hat{Y}), \hat{q}(\hat{Z}) \rangle + \langle \hat{q}(\hat{Y}), \hat{f}(\hat{X})\hat{q}(\hat{Z}) \rangle$$
$$= \langle \hat{f}(\hat{Y})\hat{q}(X), \hat{q}(\hat{Z}) \rangle + \langle \hat{q}(\hat{X}), \hat{f}(\hat{Y})\hat{q}(\hat{Z}) \rangle.$$

Since $\hat{D}_{\hat{X}*}\hat{E}^* = -\hat{X}^*$, we have

$$\hat{f}(\hat{E}) = I_{\hat{V}},$$

where $I_{\hat{V}}$ is the identity mapping on $\hat{V}$. We define a subspace $\mathfrak{g}$ of $\hat{\mathfrak{g}}$ by

$$\mathfrak{g} = \{X \in \hat{\mathfrak{g}} \mid \langle \hat{q}(X), \hat{q}(\hat{E}) \rangle = 0\}.$$

Then $\mathfrak{g}$ is a subalgebra of $\hat{\mathfrak{g}}$. In fact, for $X, Y \in \mathfrak{g}$ we have

$$\langle \hat{q}([X, Y]), \hat{q}(\hat{E}) \rangle = \langle \hat{f}(X)\hat{q}(Y) - \hat{f}(Y)\hat{q}(X), \hat{q}(\hat{E}) \rangle$$
$$= \langle \hat{q}(X), \hat{f}(Y)\hat{q}(\hat{E}) \rangle - \langle \hat{q}(Y), \hat{f}(X)\hat{q}(\hat{E}) \rangle$$
$$= 0.$$

Since $\langle \hat{q}(\hat{E}), \hat{q}(\hat{E}) \rangle = -\dfrac{1}{c}$ we have

$$\hat{\mathfrak{g}} = \mathfrak{g} \oplus \mathbf{R}\hat{E}.$$

Let G be the connected Lie subgroup of $\hat{G}$ corresponding to $\mathfrak{g}$. Let G/K be the orbit of G through $\hat{o}$. Then $K = \hat{K}$. For $X \in \mathfrak{g}$ and $s \in G$ we have

$$\hat{g}_{s\hat{o}}(X^*_{s\hat{o}}, \hat{E}^*_{s\hat{o}}) = \hat{g}_{s\hat{o}}((s_*)_{\hat{o}}(Ad(s^{-1})X)^*_{\hat{o}}, (s_*)_{\hat{o}}(Ad(s^{-1})\hat{E})^*_{\hat{o}})$$
$$= \hat{g}_{\hat{o}}(Ad(s^{-1}X)^*_{\hat{o}}, \hat{E}^*_{\hat{o}})$$
$$= \langle \hat{q}(Ad(s^{-1})X), \hat{q}(\hat{E}) \rangle$$
$$= 0.$$

This implies that $\hat{E}^*_{s\hat{o}}$ is orthogonal to the tangent space of G/K at $s\hat{o}$. Using the flat affine connection $\hat{D}$ on $\hat{G}/\hat{K}$ and the transversal vector field $\hat{E}^*$, we define the induced connection D and the affine fundamental form h on G/K by

$$\hat{D}_{X*}Y^* = D_{X*}Y^* + h(X^*, Y^*)\hat{E}^*,$$

where $X, Y \in \mathfrak{g}$. Then

$$\hat{g}(\hat{D}_{X*}Y^*, \hat{E}^*) = -\frac{1}{c}h(X^*, Y^*).$$

Alternatively, we have

$$\hat{g}(\hat{D}_{X^*}Y^*, \hat{E}^*)$$
$$= X^*(\hat{g}(Y^*, \hat{E}^*)) - (\hat{D}_{X^*}\hat{g})(Y^*, \hat{E}^*) - \hat{g}(Y^*, \hat{D}_{X^*}\hat{E}^*)$$
$$= -(\hat{D}_{X^*}\hat{g})(Y^*, \hat{E}^*) + \hat{g}(X^*, Y^*)$$
$$= -(\hat{D}_{\hat{E}^*}\hat{g})(X^*, Y^*) + \hat{g}(X^*, Y^*)$$
$$= -\hat{E}^*(\hat{g}(X^*, Y^*)) + \hat{g}(\hat{D}_{\hat{E}^*}X^*, Y^*) + \hat{g}(X^*, \hat{D}_{\hat{E}^*}Y^*) + \hat{g}(X^*, Y^*)$$
$$= -\hat{g}(X^*, Y^*).$$

Thus we obtain

$$h(X^*, Y^*) = c\hat{g}(X^*, Y^*).$$

Since

$$\hat{D}_{X^*}\hat{D}_{Y^*}Z^* = D_{X^*}D_{Y^*}Z^* - h(Y^*, Z^*)X^*$$
$$+\{h(X^*, D_{Y^*}Z^*) + X^*h(Y^*, Z^*)\}\hat{E}^*,$$

we have

$$R_{\hat{D}}(X^*, Y^*)Z^* = \hat{D}_{X^*}\hat{D}_{Y^*}Z^* - \hat{D}_{Y^*}\hat{D}_{X^*}Z^* - \hat{D}_{[X^*, Y^*]}Z^*$$
$$= R_D(X^*, Y^*)Z^* - h(Y^*, Z^*)X^* + h(X^*, Z^*)Y^*$$
$$+\{(D_{X^*}h)(Y^*, Z^*) - (D_{Y^*}h)(X^*, Z^*)\}\hat{E}^*,$$

which implies

$$(D_{X^*}h)(Y^*, Z^*) = (D_{Y^*}h)(X^*, Z^*),$$
$$R_D(X^*, Y^*)Z^* = h(Y^*, Z^*)X^* - h(X^*, Z^*)Y^*.$$

Therefore, denoting by g the restriction of $\hat{g}$ to G/K, the pair (D, g) is a G-invariant Codazzi structure of constant curvature c. □

Corollary 11.3. *Let Ω be a homogeneous regular convex cone and let ψ be the characteristic function. Then each level surface $\psi^{-1}(c)$ of ψ admits an invariant Codazzi structure of constant curvature with value $-(\dim \Omega)^{-1}$.*

Proof.   We may assume that the vertex of Ω is the origin 0. Let $\hat{G}$ be the linear automorphism group of Ω and let G be a closed subgroup defined by $G = \{s \in \hat{G} \mid \det s = 1\}$. For any two points p and q of $\psi^{-1}(c)$ there exists $\hat{s} \in \hat{G}$ such that $q = \hat{s}p$. Since

$$c = \psi(q) = \psi(\hat{s}p) = \frac{\psi(p)}{\det \hat{s}} = \frac{c}{\det \hat{s}}$$

by (4.2), it follows that $\hat{s} \in G$. Hence G acts transitively on $\psi^{-1}(c)$. Since the one parameter group e^t is contained in $\hat{G}$ the linear Lie algebra $\hat{\mathfrak{g}}$ of $\hat{G}$

contains the identity transformation $\hat{E}$ of Ω. Let $(\hat{D}, \hat{g} = \hat{D}d\log\psi)$ be the canonical Hessian structure on Ω. Using the same notation as in Theorem 11.7, we have

$$\hat{D}_{\hat{X}^*}\hat{E}^* = -\hat{X}^*, \quad \text{for } \hat{X} \in \hat{\mathfrak{g}},$$

because $\hat{E}^* = -\sum x^i \dfrac{\partial}{\partial x^i}$. Since $d\log\psi$ is invariant under $\hat{G}$, denoting by $\mathcal{L}_{\hat{E}^*}$ the Lie derivative with respect to $\hat{E}^*$, we obtain

$$0 = \mathcal{L}_{\hat{E}^*}d\log\psi = (d\iota_{\hat{E}^*} + \iota_{\hat{E}^*}d)d\log\psi$$
$$= d((d\log\psi)(\hat{E}^*)).$$

Therefore $(d\log\psi)(\hat{E}^*)$ is a constant. Hence

$$\hat{g}(\hat{X}^*, \hat{E}^*) = (\hat{D}_{\hat{X}^*}d\log\psi)(\hat{E}^*)$$
$$= \hat{X}^*((d\log\psi)(\hat{E}^*)) - (d\log\psi)(\hat{D}_{\hat{X}^*}\hat{E}^*)$$
$$= (d\log\psi)(\hat{X}^*),$$

and so

$$\hat{g}(\hat{E}^*, \hat{E}^*) = (d\log\psi)(\hat{E}^*).$$

Since

$$(\hat{E}^*\psi)(x) = \frac{d}{dt}\bigg|_{t=0} \psi((\exp(-t\hat{E}))x)$$
$$= \frac{d}{dt}\bigg|_{t=0} \psi(x)\big(\det(\exp(-t\hat{E}))\big)^{-1}$$
$$= \frac{d}{dt}\bigg|_{t=0} e^{nt}\psi(x) = n\psi(x), \quad n = \dim\Omega,$$

we have

$$\hat{g}(\hat{E}^*, \hat{E}^*) = \dim\Omega.$$

Therefore our assertion follows from Theorem 11.7. $\qquad\qquad\square$

Bibliography

Amari, S. (1985). Differential-geometrical methods in statistics, Springer Lecture Notes in Statistics **28**.

Amari, S. and Nagaoka, H (2000). Methods of information geometry, Translation of Mathematical Monographs **191**, AMS, Oxford, Univ. Press.

Agaoka, Y. (1982). Invariant flat projective structures on homogeneous space, Hokkaido Math. J. **11**, pp. 125–172.

Borel, A. (1954). Kählerian coset spaces of semi-simple Lie groups, Proc. Nat. Acad. Sci. USA **40**, pp. 1147–1151.

Bott, R. and Chern, S. S. (1965). Hermitian vector bundles and the equidistribution of the zeroes of their holomorphic sections, Acta Math. **114**, pp. 71-112.

Bourbaki, N. (1960). Éléments de Mathématique, Groupes et Algébres de Lie, Chapitre I, Hermann, Paris.

Braun, H. and Koecher, M. (1966). Jordan Algebren, Springer.

Calabi, E. (1954). The space of Kähler metrics, Proc. Int. Congr. Math. Amsterdam **2**, pp. 206-207.

Calabi, E. (1955). On Kähler manifolds with vanishing canonical class, Algebraic Geometry and Topology, A Symposium in Honor of S. Lefschetz, Princeton Univ. Press, pp. 78-89.

Calabi, E. (1958). Improper affine hypersurfaces of convex type and a generalization of a theorem by K. Järgens, Michigan Math. J. **5**, pp. 105-126.

Cartan, E. (1935). Sur les domaines bornés de l'espace de n variable complexes, Abh. Math. Sem. Hamburg **11**, pp. 116-162.

Chen, B. Y. and Ogiue, K. (1975). Some characterizations of complex space forms in terms of Chern classes, Quart. J. Math. Oxford **26**, pp. 459-464.

Cheng, S. Y and Yau, S. T. (1982). The real Monge-Ampére equation and affine flat structures, Proc. the 1980 Beijing symposium of differential geometry and differential equations, Science Press, Beijing, China, Gordon and Breach, Science Publishers, Inc., New York, pp. 339-370.

Cheng, S. Y and Yau, S. T. (1986). Complete affine hypersurfaces, Part **I**. The completeness of affine metrics, Comm. Pure Appl. Math. **39**(6), pp. 839-866.

Chern, S. S. (1978). Affine minimal hypersurfaces, Proc. Japan-U.S. Seminar, Tokyo, pp. 17-30.

Delanoë, P. (1989). Remarques sur les variëtës localement hessiennes, Osaka J. Math. **26**, pp. 65-69.

Dombrowski, P. (1962). On the geometry of the tangent bundles, J. Reine Angew. Math. **210**, pp. 73-88.

Eguchi, S.(1992). Geometry of minimum contrast, Hiroshima Math. J. **22**, pp. 631-647.

Faraut, J. and Korányi, A. (1994). Analysis on symmetric cones, Oxford Science Publications.

Fried, D., Goldman, W. and Hirsch, M. W. (1981). Affine manifolds with nilpotent holonomy, Comm. Math. Hel. **56**, pp. 487-523.

Griffiths, P. and Harris, J. (1978). Principles of Algebraic Geometry, John Wiley and Sons.

Gindikin, S. G., Vinberg, E. B. and Pyateskii-Shapiro, I. I. (1967). Homogeneous Kähler manifolds, Collect. I. M. E..

Hao, J. H. and Shima, H. (1994). Level surfaces of non-degenerate functions in $\mathbf{R}^{n+1}$, Geometriae Dedicata **50**, pp. 193-204.

Helmstetter, J. (1979). Radical d'une algèbre symétrique a gauche, Ann. Inst. Fourier Grenoble **29**, pp. 17-35.

Jörgens (1954). Über die Lösungen der Differentialgleichung $rt - s^2 = 1$, Math. Ann. **127**, pp. 130-134.

Kaup, W. (1968). Hyperbolische Raüme, Ann. Inst. Fourier **18**, pp. 303-330.

Kim, H. (1986). Complete left-invariant affine structures on nilpotent Lie groups, J. Differential Geom. **24**, pp. 373-394.

Kobayashi, K. and Nomizu, K. (1963, 1969). Foundations of Differential Geometry, vol. **I** and **II**, John Wiley and Sons, New York.

Kobayashi, K. (1972). Transformation groups in differential geometry, Springer-Verlag.

Kobayashi, K. (1987). Differential Geometry of Complex Vector Bundles, Publ. Math. Soc. Japan No.**15**, Iwanami-Princeton Univ. Press.

KobayashiCK. (1997, 1998). Complex Geometry vol. **1** and **2** (in Japanese), Iwanami Shoten Publishers.

KodairaCK. (1986). Complex Manifolds and Deformation of Complex Structures, Springer-Verlag.

Koecher, M. (1957). Positivitätsbereiche im $\mathbf{R}^m$, Amer. J. Math. **79**, pp. 575-596.

Koecher, M. (1962). Jordan algebras and their applications, Lectures notes, Univ. of Minnesota, Minneapolis.

Koszul, J. L. (1955). Sur la forme hermitienne canonique des espaces homogènes complexes, Canad. J. Math. **7**, pp. 562-576.

Koszul, J. L. (1961). Domaines bornés homogènes et orbites de groupes de transformations affines, Bull. Soc. Math. France 89, pp. 515-533.

Koszul, J. L. (1962). Ouverts convexes homogènes des espaces affines, Math. Z. **79**, pp. 254-259.

Koszul, J. L. (1965). Variétés localement plates et convexité, Osaka J. Maht. **2**, pp. 285-290.

Koszul, J. L. (1968a). Déformations des connexions localement plates, Ann. Inst. Fourier **18**, pp. 103-114.

Koszul, J. L. (1968b). Connexions hyperboliques et déformations, Sympos. Math. **11**, pp. 357-361.

Kurose, T. (1990). Dual connections and affine geometry, Math. Z. **203**, pp. 115-121.

Miyaoka, Y. (1977). On the Chern numbers of surfaces of general type, Inventiones math. **42**, pp. 225-237.

Murakami, S. (1952). On the automorphisms of a real semi-simple Lie algebra, J. Math. Soc. Japan, **4**, pp. 103-133.

Morrow, J. and Kodaira, K. (1971). Complex Manifolds, Holt, Rinehart and Winston, Inc. New York.

Mizuhara, A. and Shima, H. (1999). Invariant projectively flat connections and its applications, Lobachevskii J. Math., **4**, pp. 99-107.

Noguchi, M. (1992). Geometry of statistical manifolds, Differential Geometry and its Applications **2**, pp. 197-222.

Nomizu, N. (1981). Introduction to Modern Differential Geometry (in Japanese), Shokabo Publishers.

Nomizu, K. and Sasaki, T. (1994). Affine Differential Geometry, Cambridge Univ. Press.

Nomizu, K. and Simon, U. (1992). Notes on conjugate connections, Geometry and Topology of Submanifolds, IV ed. by F. Dillen and L. Verstraelen, World Scientific, Singapore, pp. 152-172.

Ochiai, T. (1966). A lemma on open convex cones, J. Fac. Sci. Univ. Tokyo, **12**, pp. 231-234.

Ohara, A. and Amari, S. (1994). Differential geometric structures of stable feedback systems with dual connections, Kybernetika, **30**, pp. 369-386.

Pogorelov, A.V. (1978). The Minkowski Multidimensional Problem, Winston and Sons, Washington.

Pyatetskii-Shapiro, I.I. (1959). On a problem of E. Cartan, Dokl. Akad. Nauk SSSR, **124**, pp. 272-273.

Pyatetskii-Shapiro, I.I. (1969). Automorphic functions and the geometry of classical domains, Gordon and Breach.

Rothaus, O. S. (1960). Domains of positivity, Abh. Math. Sem. Univ. Hamburg, **24**, pp. 189-235.

Rothaus, O. S. (1966). The construction of homogeneous convex cones, Ann. of Math., **83**, pp. 358-376.

Sasaki, T. (1985). A note on characteristic functions and projectively invariant metrics on a bounded convex domain, Tokyo J. Math., **8**, pp. 49-79.

Satake, I. (1972). Linear imbeddings of self-dual homogeneous cones, Nagoya Math. J., **46**, pp. 121-145.

Shima, H. (1975). On locally symmetric homogeneous domains of completely reducible linear Lie groups, Math. Ann., **217**, pp. 93-95.

Shima, H. (1976). On certain locally flat homogeneous manifolds of solvable Lie groups, Osaka J. Math. **13**, pp. 213-229.

Shima, H. (1977a). Homogeneous convex domains of negative sectional curvature,

J. Differential Geometry, **12**, pp. 327-332.

Shima, H. (1977b). Symmetric spaces with invariant locally Hessian structures, J. Math. Soc. Japan, **29**, pp. 581-589.

Shima, H. (1978). Compact locally Hessian manifolds, Osaka J. Math., **15**, pp. 509-513

Shima, H. (1980). Homogeneous Hessian manifolds, Ann. Inst. Fourier, Grenoble, **30**, pp. 91-128.

Shima, H. (1981). Hessian manifolds and convexity, Manifolds and Lie groups, Papers in honor of Y. Matsushima, Progress in Mathematics, Birkhäuser, **14**, pp. 385-392.

Shima, H. (1982). A differential geometric characterization of homogeneous self-dual cones, Tsukuba J. Math., **6**, pp. 79-88.

Shima, H. (1986). Vanishing theorems for compact Hessian manifolds, Ann. Inst. Fourier, Grenoble, **36**-3, pp.183-205.

Shima, H. (1988-1989). Hessian manifolds, Seminaire Gaston Darboux de geometrie et topologie differentiell, Universite Montpellier, pp. 1-48.

Shima, H. (1995a). Harmonicity of gradient mappings of level surfaces in a real affine space, Geometriae Dedicata, **56**, pp. 177-184.

Shima, H. (1995b). Hessian manifolds of constant Hessian sectional curvature, J. Math. Soc. Japan, **47**, pp. 735-753.

Shima, H. (1999). Homogeneous spaces with invariant projectively flat affine connections, Trans. Amer. Math. Soc., **351**, pp. 4713-4726.

Shima, H. and Hao, J. H. (2000). Geometry associated with normal distributions, Osaka J. Math., **37**, pp. 509-517.

Trudinger, N. S. and Wang, X. J. (2000). The Bernstein problem for affine maximal hypersurfaces, Invent. Math. **140**, pp. 399-422.

Urakawa, H. (1999). On invariant projectively flat affine connections, Hokkaido Math. J., **28**, pp. 333-356.

Vey, J. (1968). Une notion d'hyperbolicité sur les variétés localement plates, C. R. Acad. Sc. Paris, **266**, pp. 622-624.

Vinberg, E. B. (1960). Homogeneous cones, Soviet Math. Dokl., pp. 787-790.

Vinberg, E. B. (1961). The Morozov-Borel theorem for real Lie groups, Soviet Math. Dokl., pp. 1416-1419.

Vinberg, E. B. (1963). The Theory of convex homogeneous cones, Trans. Moscow Math. Soc., pp. 340-403.

Vinberg, E. B. (1965). The structure of the group of automorphisms of a homogeneous convex cone, Trans. Moscow Math. Soc., pp. 63-93.

Vinberg, E. B., Gindikin, S. G. and Pyatetskii-Shapiro, I.I. (1965). On the classification and canonical realization of complex homogeneous bounded domains, Proc. Moscow Math. Soc. **12**, pp. 404-437.

Weil, A. (1958). Introduction à l'Étude des Variétés Kählériennes, Hermann, Paris.

Wells, R. O. (1979). Differential Analysis on Complex Manifolds, Graduate Texts in Math., Springer.

Yau, S, T. (1977). Calabi's conjecture and some new results in algebraic geometry, Proc. Nat. Acad. Sci. USA, **74**, pp. 1798-1799.

Yagi, K. (1981). On Hessian structures on affine manifolds, Manifolds and Lie groups, Papers in honor of Y. Matsushima, Progress in Mathematics, Birkhäuser, **14**, pp. 449-459.

Index